OXFORD Revise

Revision & Practice

AQA A LEVEL PHYSICS

⚙ **Knowledge** ⮂ **Retrieval** ✏ **Practice**

Alom Shaha
Helen Reynolds
Catherine Jones
Carol Davenport

3 steps to access your free online copy of this book

1 Visit Education Bookshelf (http://bookshelf.oxfordsecondary.co.uk).

2 Create an account by clicking the *register* button or *sign in* if you already have an account.

3 Click on the *Activate Your Publication* button and enter your unique access code (printed on the inside cover).

OXFORD
UNIVERSITY PRESS

Contents

Shade in each level of the circle as you feel more confident and ready for your exam.

How to use this book

This book uses a three-step approach to revision: **Knowledge**, **Retrieval**, and **Practice**. It is important that you do all three; they work together to make your revision effective.

1 Knowledge

Knowledge comes first. Each chapter starts with a **Knowledge Organiser**. These are clear, easy-to-understand, concise summaries of the content that you need to know for your exam. The information is organised to show how concepts relate to each other so you can understand how the knowledge fits together, rather than learning lots of disconnected facts.

2 Retrieval

The **Retrieval questions** help you learn and quickly recall the information you've acquired. Memorise the short questions and answers about the content in the Knowledge Organiser, then cover the answers with some paper and write as many as you can from memory. Check back to the Knowledge Organiser for any you got wrong, then attempt the questions again until you can answer all the of them correctly.

The **Skills** boxes cover either knowledge and understanding from a **Practical Activity Group** or a key **Maths skill**. Read through the worked example then have a go at the practice questions.

Storing these basic facts in your long-term memory through regular retrieval practice will mean you will find it easier to apply them to complex situations and difficult questions.

3 Practice

Once you think you know the Knowledge Organiser and Retrieval answers really well you can move on to the final stage: **Practice**.

Each chapter has lots of **exam-style questions**, that test your ability to:

- apply your knowledge and understanding, including from the Practical Activity Groups
- analyse and evaluate information
- combine knowledge from different parts of the course (synoptic questions).

Questions with the link icon will have **synoptic links**. This means that they will assess the content from the current chapter along with knowledge from elsewhere in the course.

Questions with the apparatus icon test your **practical skills**. At least 15% of the marks in your exams will be about practical skills.

Questions with the calculator icon will test your **maths skills**

Questions with the pie-chart icon will test your skills in **analysis and evaluation** of data and information.

Exam Tip

Exam tips show you how to interpret the questions, what you need to do in your answers, and advice on how to secure as many marks as possible.

kerboodle

All the **answers** are on Kerboodle and the website.
www.oxfordrevise.com/scienceanswers

⚙ Knowledge

1 Measurements and errors

Base quantities and units

Scientists use a system of units, known as the Système International. The system defines base units for physical quantities.

The units of all other quantities are derived from the base units. They are known as **derived** units.

Different units can be used for the same quantity. It is always possible to convert between them.

The fact that the units must be the same on both sides of an equation can be used to express a derived unit in terms of base units.

For example, to express the unit of force, the newton, in base units:

$$F = ma$$

Units on left-hand side = newtons (N).
Units on right-hand side = $kg \times ms^{-2}$.

$$1\,N = 1\,kg\,m\,s^{-2}$$

Physical quantity	SI unit
mass	kilogram (kg)
length	metre (m)
time	second (s)
electric current	ampere (A)
temperature	kelvin (K)
amount of substance	mole (mol)

Standard form and prefixes

Standard form is a way of writing very big or very small numbers.

Numbers written in standard form must be written as:

$$A \times 10^n$$

where A is a decimal number ≥ 1 or < 10; n is a whole number.

If n is positive, the number represented is greater than one.
If n is negative, the number represented is less than one.

Converting a number bigger than one	Converting a number smaller than one
1 Count the digits *after* the first digit – this is n 2 Put a decimal point after the first digit – this is A (ignore zeroes at the end) 3 Write the number as $A \times 10^n$.	1 Count the digits *after* the decimal point, up to and including the first non-zero digit – this is n 2 Put a decimal point after the first non-zero digit – this is A (ignore zeroes at the start) 3 Write the number as $A \times 10^{-n}$.
Converting a large number from standard form	**Converting a small number from standard form**
move the decimal point n places to the right, adding zeroes if necessary.	move the decimal point n places to the left, adding zeroes if necessary.

A **prefix** in front of a unit tells you what factor of ten the unit is multiplied by. The SI prefixes are given in the table.

Prefix	Symbol	Multiplying factor	Prefix	Symbol	Multiplying factor
tera	T	1×10^{12}	milli	m	1×10^{-3}
giga	G	1×10^{9}	micro	μ	1×10^{-6}
mega	M	1×10^{6}	nano	n	1×10^{-9}
kilo	k	1×10^{3}	pico	p	1×10^{-12}
centi	c	1×10^{-2}	femto	f	1×10^{-15}

Order of magnitude

An **order of magnitude** is a factor of 10. It is often useful to compare the sizes of numbers by comparing their orders of magnitude.

Two numbers are of the same order of magnitude if dividing the bigger number by the smaller number gives an answer less than 10.

If numbers are written in standard form, their orders of magnitude can be compared by dividing the larger power of ten by the smaller power of ten.

Estimation is a way of finding an approximate numerical value for something (usually within the correct order of magnitude) based on limited information or knowledge.

Limitations of physical measurements

Uncertainty

All measurements have an **uncertainty**, expressed as a ± value. This is the possible difference between the measured value and the true value of the quantity. It can be given as a fixed number (**absolute uncertainty**), percentage (**percentage uncertainty**), or fraction (**fractional uncertainty**).

Combining uncertainties

To find the uncertainty when two or more measurements are:

- added or subtracted, add the *absolute* uncertainties
- multiplied or divided, add the *percentage* uncertainties.

To find the uncertainty when two or more measurements are multiplied or divided by each other, add the *percentage* uncertainties.

To find the uncertainty when a measurement is raised to a power, multiply the percentage uncertainty by the power.

Uncertainty in graphs

Error bars represent uncertainties in measurement data on graphs. For example, the error bars on the graph show that the voltage readings have an uncertainty of ±0.03 V and the current readings have an uncertainty of ±0.6 A.

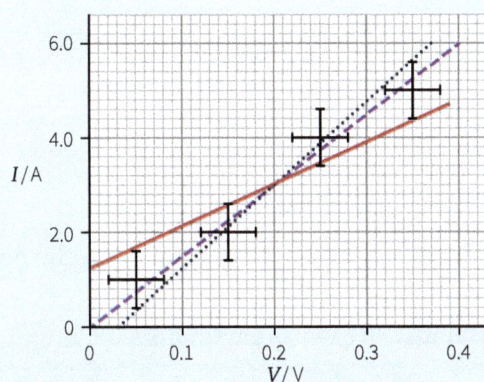

Two **worst-fit lines** can be drawn through the error bars to find the uncertainty in the gradient and intercepts of a straight-line graph: one with the shallowest possible gradient (red, solid line), and one with the steepest possible gradient (blue, dotted line).

The uncertainty in the *gradient* of a straight-line graph can be found by taking the biggest difference between the gradients of the best-fit and worst-fit lines.

The uncertainty in the *intercept* of a straight-line graph can be found by taking the biggest difference between the intercepts of the best-fit and worst-fit lines.

Standard form and prefixes

Errors

Two types of experimental error can affect the uncertainty in measurements.

Random errors cause measurements to be spread about the true value, due to results varying in an unpredictable way from one measurement to the next. Random errors are present when any measurement is made and cannot be corrected.

The effect of random errors can be reduced by making more measurements and calculating a mean.

Systematic errors cause readings to be different from the true value by a consistent amount, such as due to a faulty instrument. A zero error occurs when an instrument does not read zero when it should.

Systematic errors need to be identified before they can be reduced. They can be reduced by repeating an experiment with different apparatus and/or measurement techniques.

Accuracy and precision

Accuracy: Accurate measurements are close to the true (accepted) value.

Precision: Precise measurements have very little spread about the mean; they are not necessarily accurate as they can be affected by random or systematic errors.

Repeatability: If the same measurements are obtained when the same person repeats the experiment using the same equipment and method, the measurements are repeatable.

Reproducibility: If similar measurements are obtained when a different person repeats the experiment using a different method and/or equipment, the measurements are reproducible.

Resolution: The resolution of a measuring instrument is the *smallest* change in the quantity being measured that the instrument can detect (usually the smallest reading the instrument can take).

Retrieval

Learn the answers to the questions below, then cover the answers column with a piece of paper and write as many as you can. Check and repeat.

Questions	Answers
1 Name the six SI base quantities and their units.	mass (kilogram), length (metre), time (second), electric current (ampere), temperature (kelvin), amount of substance (mole)
2 How is a number written in standard form?	$A \times 10^n$, where A is a decimal number between 1 and 10 (not including 10) and n is a whole number
3 What does a prefix in front of a unit mean?	factor of ten by which it is multiplied
4 State the name, symbol, and multiplying factor for each of the SI prefixes from tera to femto.	tera (T) 1×10^{12}; giga (G) 1×10^9; mega (M) 1×10^6; kilo (k) 1×10^3; centi (c) 1×10^{-2}; milli (m) 1×10^{-3}; micro (μ) 1×10^{-6}; nano (n) 1×10^{-9}; pico (p) 1×10^{-12}; femto (f) 1×10^{-15}
5 What is the uncertainty in a measurement of a quantity?	difference between the measured value and the true value of the quantity
6 What is the absolute uncertainty of a measurement?	possible difference from the true value given as a fixed number in the unit of measurement
7 What is the percentage uncertainty of a measurement?	possible difference from the true value given as a percentage of the measurement
8 How are the uncertainties combined when two or more measurements are added or subtracted?	absolute uncertainties are added together
9 How are the uncertainties combined when two or more measurements are multiplied or divided by each other?	percentage uncertainties are added together
10 How is the uncertainty found when a measurement is raised to a power?	percentage uncertainty is multiplied by the power
11 What do error bars on a graph represent?	uncertainties in measurement data
12 What can worst-fit lines on a graph be used for?	finding the uncertainty in the gradient and intercepts of a straight-line graph
13 How can the uncertainty in the gradient of a straight-line graph be found?	from the biggest difference between the gradients of the best-fit and worst-fit lines
14 What is a random error?	error that causes readings to be spread about the true value, due to results varying in an unpredictable way between measurements
15 What is a systematic error?	error that causes readings to be different from the true value by a consistent amount
16 What is a zero error?	an instrument does not read zero when it should
17 What is an accurate measurement?	measurement that is close to the true (accepted) value

(Put paper here)

18 What is a precise measurement?

19 What are repeatable measurements?

20 What are reproducible measurements?

21 What is the resolution of a measuring instrument?

Put paper here measurement where there is very little spread about the mean

Put paper here measurements where the same results are obtained if the original experimenter repeats the measurements using the same equipment and method

Put paper here measurements where similar results are obtained if a different experimenter carries out the measurements using a different method and/or equipment

smallest change in the quantity being measured that the instrument can detect

Maths skills

Practise your maths skills using the worked example and practice questions below.

Uncertainty measurements

To calculate the uncertainty in a final calculated quantity, combine the uncertainties in the initial measurements.

- If adding or subtracting measurements, add the absolute uncertainties together to find the final uncertainty.

- If multiplying or dividing measurements, add the percentage uncertainties together to find the final uncertainty.

- When a measurement is raised to a power, multiply the percentage uncertainty by the power.

Worked example

Question

The current through a bulb is measured as $1.38\,A \pm 0.01\,A$ when the potential difference is $7.50\,V \pm 0.1\,V$. Calculate the resistance of the bulb at this voltage, and give the uncertainty of the resistance value.

Answer

$$R = \frac{V}{I} = \frac{7.50\,V}{1.38\,A} = 5.43\,\Omega$$

To calculate the total uncertainty for R, add the percentage uncertainties for each quantity.

$$\% \text{ uncertainty} = \frac{\text{absolute uncertainty}}{\text{value}} \times 100\%$$

$$V = \frac{0.1}{7.5} \times 100\% = 1.3\%$$

$$I = \frac{0.01}{1.38} \times 100\% = 0.72\%$$

% uncertainty in $R = (1.3 + 0.72)\%$ $= 2.0\%$, so $R = 5.4\,\Omega \pm 2\%$

absolute uncertainty $= 5.4 \times 2\%$ $= 5.4\,\Omega \pm 0.1\,\Omega$

Practice

1 Calculate the percentage uncertainty in these measurements:
   a $6.7\,m \pm 0.1\,m$;
   b $450\,kg \pm 10\,kg$;
   c $366\,000\,J \pm 1000\,J$.

2 Calculate the absolute uncertainty in these measurements:
   a $34.3\,W \pm 6.5\%$;
   b $10\,k\Omega \pm 10\%$;
   c $12\,742\,km \pm 0.3\%$.

3 A student measures the dimensions of a small cube of aluminium as $25.0\,mm \times 25.0\,mm \times 25.0\,mm$, and estimates that the uncertainty in their measurement is $\pm 0.2\,mm$.

Using a top pan balance, they measure the mass of the cube as $42.19\,g \pm 0.01\,g$.

   a Calculate the volume of the cube and the absolute uncertainty.

   b Calculate the density of the cube and the absolute uncertainty.

Exam-style questions

01 In an experiment to determine acceleration due to gravity, a student drops a timing card through a light gate and records the time t of the card through the light gate. They then repeat the experiment at an increased height s and plot a graph of v^2 against s.

Their results are shown in **Table 1**.

Synoptic link

3.4.1.3

Table 1

s / m	t / s	$v / m\,s^{-1}$	$v^2 / m^2\,s^{-1}$
0.1	0.059	1.76	3.11
0.2	0.046	2.26	5.11
0.3	0.041		
0.4	0.036	2.89	8.35
0.5	0.033	3.15	9.93

01.1 Data are missing for $s = 0.3\,m$.
Complete the table. The timing card is 0.104 m long. **[2 marks]**

01.2 **Figure 1** is a graph of v^2 against s.
Draw a line of best fit on **Figure 1** by first plotting the missing data point from **01.1**. **[2 marks]**

Figure 1

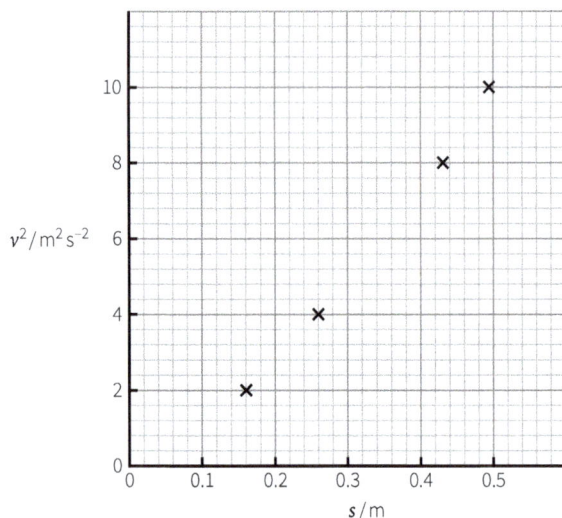

01.3 Describe the error suggested by these results and give **one** possible cause. **[2 marks]**

_____

_____

01.4 Determine the gradient of the line of best fit in **01.2**. **[2 marks]**

_____

_____

Exam tip

You would expect to see a straight line through the origin to show v^2 directly proportional to s. Give a reason why the line might not go through the origin.

01.5 Calculate the acceleration of free fall g using the gradient from **01.4**.

[3 marks]

acceleration of free fall = _____ m s^2

01.6 Calculate the percentage difference between your calculated value of g and the actual value. [1 mark]

percentage difference = _____ %

02 This question is about finding the refractive index of glass using a semi-circular glass block.

02.1 Describe a method you could use to achieve accurate values for the angle of incidence and the angle of refraction.
You may draw a labelled diagram to help you explain the method.

[3 marks]

_____

_____

_____

Synoptic link

3.3.2.3

Exam tip

Well-drawn, labelled diagrams help explain methods.

02.2 **Figure 2** is a graph of sin i against sin r, where i is the angle of incidence and r is the angle of refraction.
Calculate the refractive index of the glass used in the experiment.

[2 marks]

Figure 2

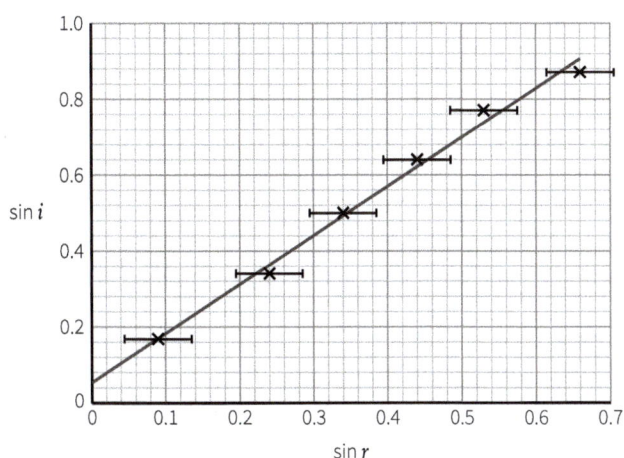

refractive index = _____

02.3 Calculate the absolute uncertainty in your value for the refractive index.

[3 marks]

① Exam tip

Figure 2 has error bars. How can you use those to determine the absolute uncertainty?

absolute uncertainty = _____

02.4 The actual value of the refractive index is 1.5.

Determine whether this lies within the uncertainty of the experiment.

[1 mark]

_____

_____

_____

_____

03 This question is about investigating the relationship between volume and temperature at a constant pressure.

03.1 Describe how you would use the apparatus in **Figure 3** to obtain accurate readings to investigate the relationship between the volume of trapped air and the temperature.

[4 marks]

⊗ Synoptic link

3.6.2.2

Figure 3

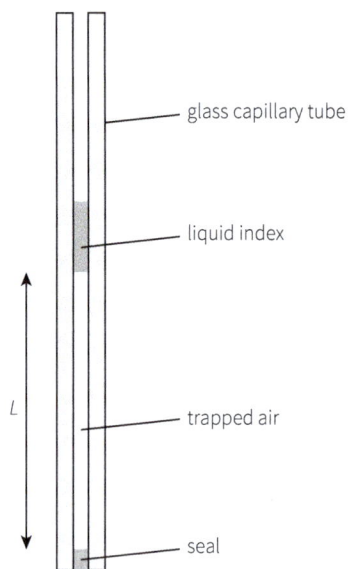

glass capillary tube

liquid index

L

trapped air

seal

03.2 The relationship between volume and temperature can be used to determine absolute zero.

Explain what is meant by absolute zero. [2 marks]

03.3 **Figure 4** shows the results from an experiment investigating the relationship between volume and temperature.

Calculate absolute zero in °C using the results from **Figure 4**.

[4 marks]

Figure 4

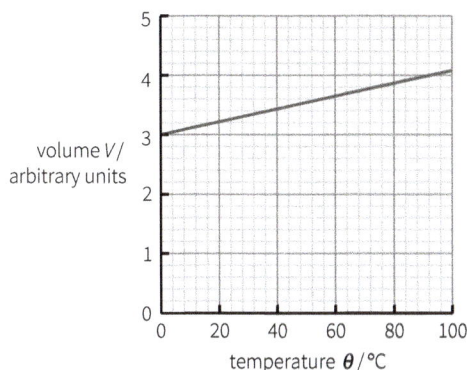

03.4 The experiment is repeated but with the water bath cooling quickly. These results give a value for absolute zero of −340 °C.

Compare this value to the value calculated in **03.3**, and explain why the two values might differ. [3 marks]

04 A student takes some measurements to determine the resistivity of a piece of wire: length = 0.500 ± 0.001 m, diameter = 0.22 ± 0.01 mm, resistance = $7.0 \pm 0.4 \, \Omega$.

04.1 Describe how the student measured the diameter of the wire.

[2 marks]

04.2 Calculate the resistivity of the wire. Give the units. [3 marks]

04.3 Calculate the percentage uncertainty of the resistivity. [2 marks]

04.4 A second student collected a series of readings for the resistance of different lengths of the same wire.

Explain how you could determine the resistivity from these results, and why this is a more accurate method than that used in **04.2**.

[3 marks]

05 A student investigates the relationship between the depth and speed of a wave in a ripple tank.

05.1 Describe how the student could accurately measure the depth and speed of a wave. [4 marks]

05.2 The student finds an equation relating depth h of the water to velocity v of the wave, where g is the acceleration due to gravity:

$$v = \sqrt{gh}$$

Describe how they should process the results of this experiment, using the equation above, to confirm the relationship between depth and velocity and the accuracy of their results. [3 marks]

05.3 Describe how the student could check the reproducibility of their results. [1 mark]

06 28 students are sitting in a classroom.

06.1 Calculate the mass of air in a typical classroom in kg.
You will need to make a sensible guess at the dimensions of the room.

density of air = 1.2 kg m^{-3} **[2 marks]**

06.2 Each student in the classroom is emitting 80 W.
Calculate the thermal energy transferred to the room in J s^{-1}.

[1 mark]

06.3 Calculate the rise in temperature of the room in 20 minutes if the windows and doors are kept closed.

specific heat capacity of air = 960 J kg^{-1} K^{-1} **[2 marks]**

06.4 Suggest whether you could use the warming effect of people as a method of heating schools in winter. **[2 marks]**

Synoptic links

3.6.2.1   3.4.1.7

Exam tip

'Suggest' questions don't necessarily have a correct answer. You simply need to back your answer with an explanation.

07 A student is investigating the relationship between the length of a pendulum and the time period.

07.1 Describe, with a labelled diagram, how an experiment could be conducted safely to obtain accurate results for small amplitude oscillations. **[4 marks]**

07.2 The results for the student's investigation are shown in **Table 2**.

Complete **Table 2** by calculating the values for T^2. Correct the **two** mistakes in the table. **[3 marks]**

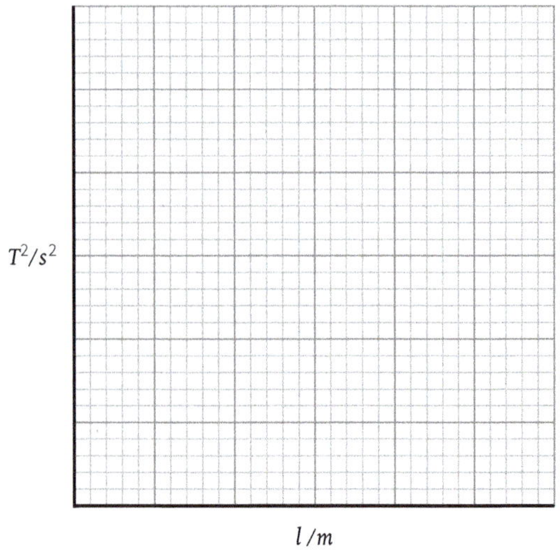

Synoptic link

3.6.1.3

Exam tip

Make sure your answer to **07.1** refers to safety and how you can ensure small amplitude oscillations, as these are both mentioned in the question.

Table 2

Length / m	Time for one oscillation / s	T^2 / s^2
2	2.85	
1.91	2.77	
1.82	2.72	
1.73	2.67	
1.64	2.57	

07.3 Plot a graph of T^2 against length on the axes in **Figure 5**. Determine the gradient. **[3 marks]**

Figure 5

T^2/s^2

l / m

07.4 Determine the acceleration due to gravity g in m s^{-2} using the gradient of the graph in **Figure 5**. [2 marks]

07.5 Comment on the accuracy and precision of the results obtained. [2 marks]

08 A student investigates the e.m.f. and internal resistance of a cell.

08.1 The unit of e.m.f. is the volt.

Show that the volt may be written in base units as kg m^2 A^{-1} s^{-3}. [3 marks]

Synoptic link

3.5.1.6

08.2 Draw a suitable diagram to show how to conduct this investigation. [2 marks]

08.3 **Figure 6** shows the results of the student's investigation.

Exam tip

Start by writing out an equation including volts, for example:
$W = VQ$

Figure 6

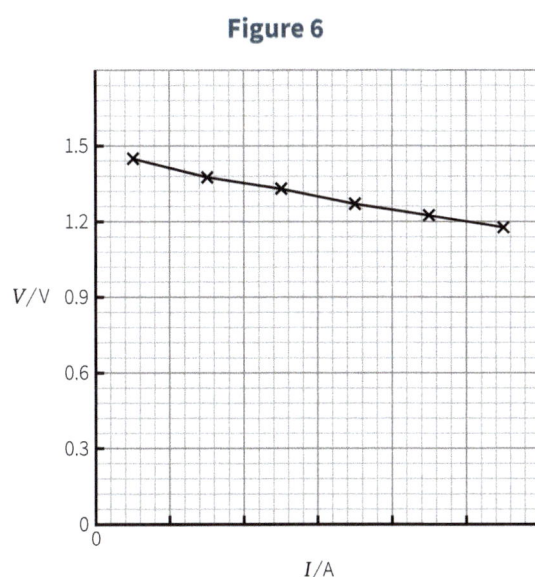

Calculate the e.m.f. in volts and the internal resistance in ohms of the cell using the results from **Figure 6**. [2 marks]

08.4 The voltmeter is found to have zero error and is reading 0.1 V too high. Explain the effect this will have on your answers to **08.3**. [3 marks]

Knowledge

2 Particles and radiation

Constituents of the atom

According to the nuclear model of the atom, an atom consists of a positively charged **nucleus** orbited by **electrons**.

The nucleus contains **protons** and **neutrons**, also known as **nucleons**. The table shows the charge and mass of the proton, neutron, and electron.

The **specific charge** of a particle is the ratio of its charge to mass:

$$\text{specific charge} = \frac{\text{charge}}{\text{mass}}$$

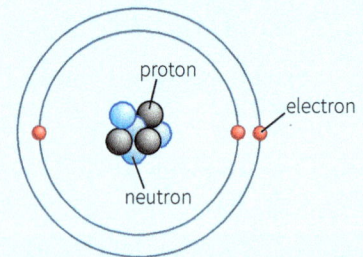

Particle	Charge / C	Charge relative to proton	Mass / kg	Mass relative to proton
proton	$+1.60\times10^{-19}$	+1	1.67×10^{-27}	1
neutron	0	0	1.67×10^{-27}	1
electron	-1.60×10^{-19}	−1	9.11×10^{-31}	0.0005

Protons, nucleons and isotopes

The **proton number** Z is the number of protons in a nucleus.

All atoms of the same element have the same number of protons in the nucleus, so they have the same proton number.

The **nucleon number** A is the number of protons plus neutrons in a nucleus.

Atoms of the same element can have different numbers of neutrons, so they can have different nucleon numbers.

Atoms can be represented like this:

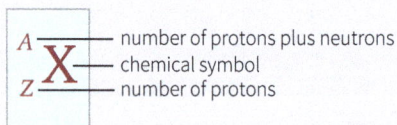

$$^{A}_{Z}X$$

- A — number of protons plus neutrons
- X — chemical symbol
- Z — number of protons

for example, the uranium isotope wih 92 protons and 146 neutrons can be represented like this:

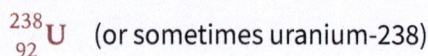

$$^{238}_{92}U \quad \text{(or sometimes uranium-238)}$$

Isotopes are atoms of the same element with the same number of protons but a different number of neutrons.

Isotopic data about a material provide the ratios of different isotopes occurring in it.

Isotopic data about the amount of carbon-14 in organic matter can be used to calculate the ages of archaeological finds.

Stable nuclei and the strong nuclear force

A **stable nucleus** is one where the protons and neutrons are held together strongly so that it does not disintegrate.

The **strong nuclear force** balances the electrostatic repulsion between protons and holds the nucleus together.

It has short-range *attraction* (up to ~3 fm), and very-short range *repulsion* (closer than ~0.5 fm).

It has the same effect between two protons as is does between two neutrons, or between a proton and a neutron.

The graph shows how the strong nuclear force varies with separation between two nucleons.

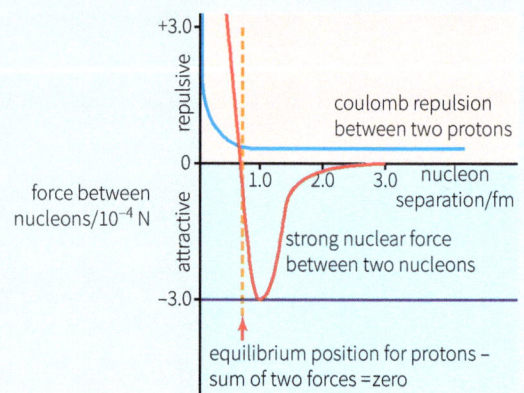

Unstable nuclei and radioactive decay

Nuclear radiation is emitted by **unstable** atomic nuclei to become more stable.

This process is called **radioactive decay**. It is a **random** process – it is not possible to predict when a nucleus will decay.

An unstable nucleus can decay by emitting an alpha particle (α), a beta particle (β), or a gamma ray (γ).

Alpha radiation tends to be emitted from large nuclei, which the strong nuclear force cannot keep stable.

Beta minus decay occurs when neutrons are unstable; a neutron is transformed into a proton, and an electron is released. The presence of protons reduces the chance of neutron decay, so isotopes with too many neutrons compared to protons undergo beta decay.

	Alpha decay $^4_2\alpha$	Beta minus decay $^0_{-1}\beta$	Gamma radiation γ
Radiation			
Emitted from nucleus	a particle – two protons and two neutrons	fast-moving electron and an antineutrino	short λ, high f electromagnetic radiation
Change in nucleus	A decreases by four; Z decreases by two	a neutron changes into a proton, so A stays the same; Z increases by one	loses energy
Ionising power	highest	high	low
Range in air	~5 cm	~1 m	a few km
Stopped by	sheet of paper	a few millimetres of aluminium	several centimetres of thick lead
Equation	$^A_ZX \rightarrow ^{A-4}_{Z-2}Y + ^4_2\alpha$	$^A_ZX \rightarrow ^A_{Z+1}Y + ^0_{-1}\beta + \bar{v}$	

The neutrino

The existence of the neutrino v was hypothesised by Wolfgang Pauli to account for conservation of energy in beta decay.

A certain amount of energy should be transferred by beta decay, but beta particles had a range of energies instead of one value. Pauli suggested that another particle (later named the neutrino) might be emitted with the beta particle to carry away the excess energy. He suggested this particle had:

- no charge, to conserve charge
- very little mass, making it hard to detect.

The neutrino's antiparticle, the **antineutrino** $\bar{v}$, is emitted in beta minus decay.

Mass, energy, and photons

Since $E = mc^2$, mass can be converted into energy and vice versa.

The **rest energy** of a particle is equal to its **rest mass** (mass when it is stationary). Rest energy is usually measured in MeV, where:

$$1 \text{ MeV} = 1 \times 10^6 \text{ eV}$$

$$1 \text{ eV} = 1.6 \times 10^{-19} \text{ J}$$

In certain situations, electromagnetic radiation behaves like particles called **photons**, which have no mass but do have energy and momentum.

The energy E of a photon depends on the frequency of the electromagnetic radiation:

photon energy $E = hf = h\dfrac{c}{\lambda}$

where h is the planck constant, 6.63×10^{-34} J s.

Particles and antiparticles

Every particle has a corresponding **antiparticle** that has:

- the *same* rest mass as the particle
- the *opposite* charge as the particle.

Particle	Antiparticle
electron e; relative charge −1	positron ē; relative charge +1
proton p; relative charge +1	antiproton p̄; relative charge −1
neutron n; relative charge 0	antineutron n̄; relative charge 0
neutrino v; relative charge 0	antineutrino v̄; relative charge 0

Annihilation and pair production

Annihilation occurs when a particle and its antiparticle meet, and all the mass of the particle and antiparticle is converted into photons of electromagnetic radiation.

There must be at least two photons released by annihilation, because one photon would not allow momentum to be conserved.

The minimum energy of each photon produced by annihilation that produces two photons is given by:

$$hf_{min} = E_0$$

where E_0 is the rest energy of particle/antiparticle.

PET scanners
Positron emission tomography (PET) scanners offer a practical application of annihilation. PET scanners use a positron-emitting isotope, and detect the radiation produced by electron–positron annihilation.

Pair production is the opposite process to annihilation. Electromagnetic radiation in the form of a gamma photon is converted to matter in the form of a particle–antiparticle pair.

The minimum energy needed for a photon to undergo pair production is the total rest energy of the particle–antiparticle pair:

$$hf_{min} = 2E_0$$

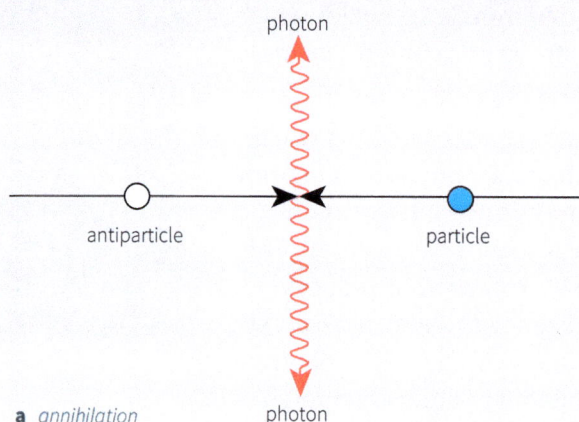

a *annihilation*

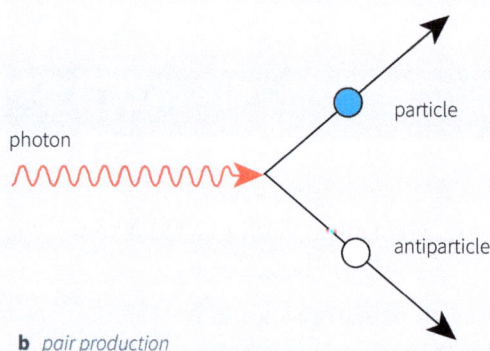

b *pair production*

Exchange particles

Exchange particles, or **gauge bosons**, are particles that travel between two particles when they exert a force on (interact with) each other.

Exchange particles are all **virtual particles** – they only exist for a short time, during which the interaction takes place.

Particle interactions

The table shows the four fundamental forces (interactions).

Force	Properties	Exchange particle
Gravity	attractive force between any two objects due to their mass	**graviton**
Electromagnetic	acts between objects with charge – repulsive when charge is the same; attractive when charges are opposite	virtual photon γ
Strong nuclear	only acts on a set of particles called hadrons; holds the protons and neutrons together in a stable nucleus	**gluons** for quark interactions; **pions** for all other hadrons
Weak nuclear	extremely short range (<0.001 fm) force that acts on hadrons and leptons, and is involved in β^- and β^+ decay, electron capture, and electron–proton collisions	W^+ and W^- **bosons**

Interaction diagrams

Interactions between particles can be represented by simplified **Feynman diagrams**, in which:

- wavy lines represent exchange particles
- straight lines represent all other particles
- arrows on straight lines represent the direction of their travel in time
- the sequence of time is up the page
- the angles between the lines do not represent the paths of the particles.

Key examples:

Electron capture/ electron-proton collision

$\beta-$ decay
$p \rightarrow n + e^+ + v_e$

$\beta+$ decay
$n \rightarrow p + e^- + \bar{v}_e$

Electromagnetic repulsion between two protons

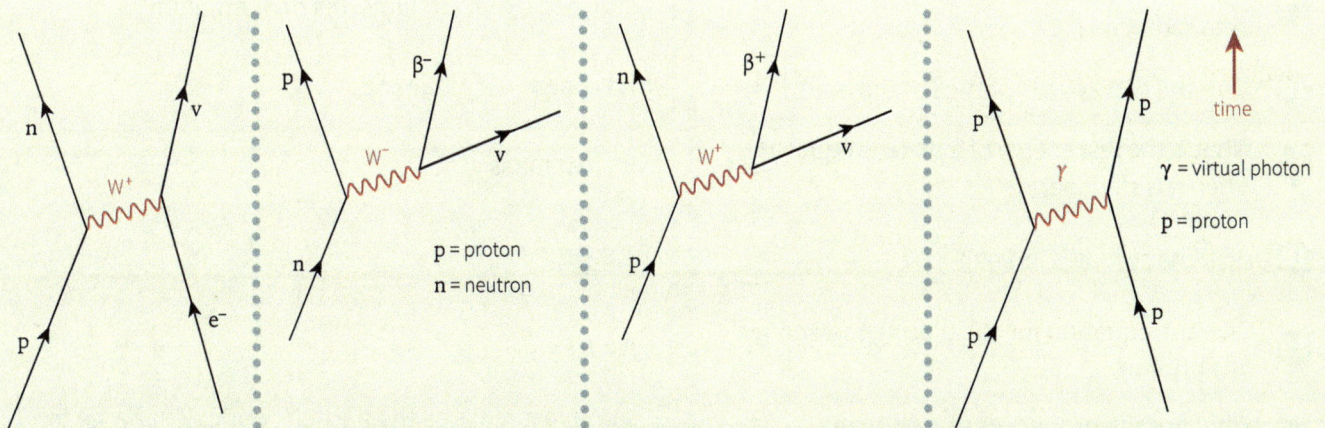

p = proton
n = neutron

γ = virtual photon
p = proton

Retrieval

Learn the answers to the questions below, then cover the answers column with a piece of paper and write as many as you can. Check and repeat.

	Questions	Answers
1	Compare the masses of a proton, neutron, and electron.	protons and neutrons have a similar mass; mass of an electron is about 0.0005 the mass of a proton
2	What is the specific charge of a particle?	ratio of its charge to mass
3	What is the proton number of a nucleus?	number of protons in the nucleus, Z
4	What are isotopes?	atoms of the same element with the same number of protons, but different numbers of neutrons
5	What can isotopic data for the amount of carbon-14 in organic matter be used for?	calculate the age of archaeological finds
6	What does the strong nuclear force do in the nucleus?	balances the electrostatic repulsion between protons and holds the nucleus together
7	What kind of nuclei emit nuclear radiation?	unstable atomic nuclei, to become more stable
8	Name the three types of nuclear radiation, their symbols, and their constituent parts.	alpha α – two protons and two neutrons beta β – fast-moving electron and an antineutrino gamma γ – short λ, high f electromagnetic radiation
9	State the change in a nucleus when beta radiation is emitted.	neutron changes into a proton, so nucleon number A stays the same but proton number Z increases by one
10	What change occurs in a nucleus when gamma radiation is emitted?	nucleus loses energy
11	What is the penetrating power of beta radiation?	stopped by few millimetres of aluminium
12	Give the two key properties of the neutrino.	no charge, very little mass
13	What is the rest energy of a particle equal to when it is stationary?	its rest mass
14	How many joules in 1 eV?	1.6×10^{-19}
15	Give the equation for calculating the energy of a photon.	$E = hf = \dfrac{hc}{\lambda}$
16	What are the properties of a particle's antiparticle?	same rest mass as the particle, opposite charge to the particle
17	What happens when a particle and an antiparticle meet?	annihilation – all the mass of the particle and antiparticle is converted into photons of electromagnetic radiation

Put paper here

18	How many photons are released in annihilation, and why?	at least two photons, because one photon would not allow momentum to be conserved
19	Give a practical application of annihilation.	positron emission tomography (PET) scanners
20	What is pair production?	opposite process to annihilation, where electromagnetic radiation in the form of a gamma photon is converted into matter in the form of a particle–antiparticle pair
21	Name the four fundamental interactions.	gravity, electromagnetic, weak nuclear, strong nuclear
22	What are exchange particles?	particles that travel between two particles when they exert a force on each other

Put paper here · *Put paper here*

Maths skills

Practise your maths skills using the worked example and practice questions below.

Significant figures (s.f.)	Worked example	Practice
You should give the final answer for a calculation to the same number of s.f. as the least accurate measured quantity. When answering a written question, you should use the same number of s.f. (or one fewer) than for the values given in the question. Standard form can clarify how many s.f. a number has.	**Question** Protons have a rest mass of 1.67×10^{-27} kg and a charge of 1.60×10^{-19} C. Calculate the specific charge of a proton. **Answer** $$\text{specific charge} = \frac{\text{charge}}{\text{mass}}$$ $$\text{specific charge} = \frac{1.60 \times 10^{-19}\,\text{C}}{1.67 \times 10^{-27}\,\text{kg}}$$ Both values in the question are given to 3 s.f., so the answer should be given to 3 s.f. specific charge $= 9.58 \times 10^7$ C kg^{-1}	**1** State the number of s.f. given in the following measurements: **a** 17 µm **b** 0.126 nm **c** 1.27×10^5 m. **2** Answer each question to a suitable number of s.f. **a** Electrons have a mass of 9.11×10^{-31} kg and a charge of -1.6×10^{-19} C. Calculate the specific charge of an electron. **b** A photon of ultraviolet light has a frequency of 750 THz. Calculate the energy of the photon.

01 The atomic number of caesium-137 is 55.

01.1 Determine the number of protons, neutrons, and electrons in an atom of caesium-137. **[3 marks]**

_____

_____

_____

_____

01.2 A small proportion of caesium-137 decays to the stable isotope barium-137 by β^- decay.

Complete the equation for this decay by filling in the boxes. **[3 marks]**

$$^{137}_{55}Cs \rightarrow \, ^{137}_{\square}Ba + {}^{\square}_{\square}\square + \square$$

> **!** **Exam tip**
>
> Remember that numbers in equations obey conservation laws because they represent the conservation of mass and charge.

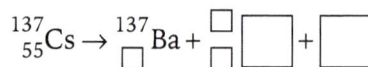

01.3 State which fundamental interaction is responsible for the decay in **01.2**. **[1 mark]**

_____

_____

01.4 An isotope of barium with an atomic mass of 133 is unstable and decays by electron capture.

Compare the nucleus of the unstable barium isotope with the nucleus of the stable barium isotope. **[1 mark]**

_____

_____

_____

01.5 Suggest and explain the atomic mass of the isotope produced by the electron capture in **01.4**. **[2 marks]**

_____

_____

_____

01.6 Compare the changes to the nucleus from β^- decay to the changes to the nucleus from electron capture. **[2 marks]**

_____

_____

_____

02 Humans continuously take in radioactive material in the food that they eat. Some of the carbon in food is radioactive. One atom in every 10^{12} carbon atoms is carbon-14.

02.1 Describe the difference between a carbon-12 atom and a carbon-14 atom in terms of particles in the nucleus. **[1 mark]**

_____

_____

02.2 Calculate the specific charge of a carbon-12 nucleus. Show your working. **[3 marks]**

specific charge = _____ $C\,kg^{-1}$

02.3 The strong nuclear force prevents the nucleus of a stable isotope, such as carbon-12, from decaying.

Sketch the graph of the strong nuclear force between two nucleons on the axes in **Figure 1**.

Add a scale to the x-axis.

Use the graph to explain why the nucleus does not implode. **[4 marks]**

> (!) **Exam tip**
>
> There are four marks available for **02.3**, so carefully work through what the question asks you to do.

Figure 1

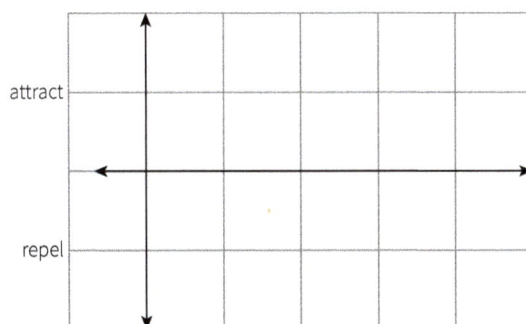

_____

_____

_____

_____

02.4 Describe the constituent parts of an antimatter atom of carbon-12. **[1 mark]**

_____

_____

02.5 Suggest and explain any differences between the graph of the strong nuclear force in **02.3** for nucleons of an atom of matter and an atom of antimatter. [3 marks]

_____

_____

_____

_____

_____

03 In the discovery of alpha and beta decay, Ernest Rutherford used the penetrative power of the radiation to show that they were different.

03.1 State **one** similarity and **one** difference in the changes to the mass number and atomic number of an unstable nucleus decaying by alpha and beta decay. [2 marks]

_____

_____

_____

03.2 Rutherford used radium as a source of alpha particles. In the decay of radium-226 to lead-206, five α particles and a number of β^- particles are emitted. The atomic number of radium is 88. The atomic number of lead is 82.

Calculate the number of β^- particles emitted.
Show your working. [3 marks]

_____

03.3 Explain why the neutrino was hypothesised to account for the conservation of energy in beta decay but not in alpha decay. [3 marks]

_____

_____

_____

_____

(!) **Exam tip**

'Suggest' means that there might be a range of possible answers to this question.

03.4 It is possible to detect alpha particles but not neutrinos with a cloud chamber.
Suggest why. [2 marks]

_____

_____

_____

_____

04 In a collision between an electron and a positron, the particles annihilate to produce two gamma ray photons.

04.1 Compare electrons and positrons. **[1 mark]**

_____

_____

04.2 Calculate the total energy in joules produced in the annihilation using the rest masses of the electron and positron.

Calculate the frequency in Hz of each gamma ray produced use the total energy you found above.

$1\,\text{MeV} = 1.6 \times 10^{-13}\,\text{J}$ **[3 marks]**

> **! Exam tip**
> The formula sheet contains data about particles. Make sure you are familiar with what is and what is not on the sheet.

_____

04.3 Compare the frequency of the photons produced in the annihilation of a proton–antiproton pair with that of the photon calculated in **04.2**.

Explain your answer. **[2 marks]**

_____

_____

_____

_____

04.4 Calculate the wavelength in metres of a photon that could produce a proton–antiproton pair. **[3 marks]**

_____

05 Two of the fundamental interactions are the electromagnetic interaction and gravity.

05.1 A pair of students model the electromagnetic interaction by standing on skateboards and throwing a ball between them. One student catches it, moves back, then throws it. Describe how this model can be used to explain the electromagnetic interaction.

Suggest **one** weakness of the model. **[3 marks]**

05.2 State the other **two** fundamental interactions. **[2 marks]**

> **! Exam tip**
> All models have strengths and weaknesses. When you come across a model, make sure that you can provide one strength and one weakness.

05.3 A student draws the diagram in **Figure 2** to show an interaction.

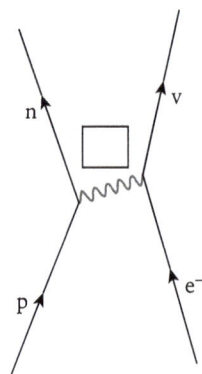

Figure 2

Complete the diagram in **Figure 2** by writing the missing symbol in the box. **[1 mark]**

05.4 Write the equation that is represented by **Figure 2**.

Explain why the particle symbol that you wrote in the box in **Figure 2** does not appear in the equation. **[3 marks]**

05.5 Compare the range of the exchange particle in **05.3** to that of the particle modelled in **05.1**. **[1 mark]**

06 Sodium has a number of isotopes.

06.1 One atom of a particular sodium isotope has an atomic mass of 24. An atom of this isotope contains 11 protons. Calculate the number of neutrons in the nucleus. **[1 mark]**

06.2 Calculate the specific charge in $C\,kg^{-1}$ of the nucleus. **[3 marks]**

06.3 Another nucleus decays by alpha decay to form sodium-24.

Calculate the number of protons and neutrons in the nucleus that decays.

Write the atom in $^{A}_{Z}X$ notation, using the symbol **X** to represent the element. **[2 marks]**

06.4 The protons in the nucleus repel each other because of electrostatic repulsion, but the strong nuclear force holds the nucleus together.

Compare the forces on two protons with the forces on two neutrons, in terms of their direction and the distances over which they act. **[4 marks]**

06.5 Sodium-24 decays by beta decay to $^{24}_{12}Mg$. Write the equation for this decay. **[2 marks]**

06.6 $^{24}_{12}Mg$ is in an excited state with 2.76 MeV of excess energy.

Calculate the wavelength in m of the gamma ray emitted by this nucleus. 1 MeV = 1.6×10^{-13} J **[2 marks]**

06.7 Sodium street lamps glow with a characteristic yellow light. Compare the emission of gamma rays and visible light from an atom of sodium. **[4 marks]**

> **Synoptic links**
>
> 3.2.2.3   3.2.2.4

> **!** **Exam tip**
>
> Remember that you can, and should, use data from the formula and equations sheet.

07 Radium used to be added to paint for clock and watch faces as it makes the paint luminescent.

07.1 The most common isotope of radium is $^{266}_{88}Ra$.

Calculate the number of protons and neutrons in a nucleus of $^{266}_{88}Ra$. **[1 mark]**

Synoptic links

3.2.2.2   3.2.2.3

07.2 $^{266}_{88}Ra$ decays by alpha decay to radon, Rn.

Complete the equation for this decay. **[2 marks]**

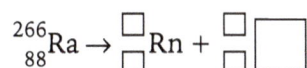

$$^{266}_{88}Ra \rightarrow {}^{\square}_{\square}Rn + {}^{\square}_{\square}\square$$

07.3 Luminescent paint made with radium glows with a green colour. The typical wavelength of light perceived to be green is 532 nm.

Calculate the energy of a photon of green light in eV.
$1\,eV = 1.6 \times 10^{-19}\,J$ **[2 marks]**

07.4 Suggest the mechanism by which the decay of radium results in the emission of green light. **[2 marks]**

07.5 Describe the evidence for the mechanism that you have described in **07.4**. **[1 mark]**

> **Exam tip**
>
> Atoms emit or absorb photons depending on the movement of electrons. You should be clear about the link between the change in energy and the frequency of the photon emitted or absorbed.

08 When matter and antimatter interact, gamma rays are produced.

08.1 Calculate the frequency in Hz and wavelength in m of each gamma ray photon produced when an electron and positron annihilate.
$1\,MeV = 1.6 \times 10^{-13}\,J$ **[4 marks]**

Synoptic links

3.2.1.5   3.2.1.6   3.2.1.7

08.2 Positron emission tomography is used in hospitals to locate structures in the brain. A positron-emitting material is taken into the body of a patient and travels to the brain.

When a positron annihilates with an electron at a specific location, two gamma rays are produced.

Explain why the gamma rays are detected on opposite sides of the patient's head. **[2 marks]**

> **Exam tip**
>
> Remember that two particles annihilate, and two photons are produced.

08.3 Write down the class of particles to which positrons and electrons belong. **[1 mark]**

08.4 Kaons and antikaons are mesons.

Compare mesons with the class of particles to which positrons and electrons belong. **[1 mark]**

08.5 A K^+ with a quark structure of $u\bar{s}$ decays into three pions.

Determine the charge on the third pion if two of the pions have the quark structure $u\bar{d}$. **[2 marks]**

08.6 Write an equation for the decay of the anti-strange quark suggested by the decay in **08.5**.

Explain your answer. **[2 marks]**

3 Properties of particles

Classification of particles

All particles and antiparticles can be classified into groups based on properties such as their mass.

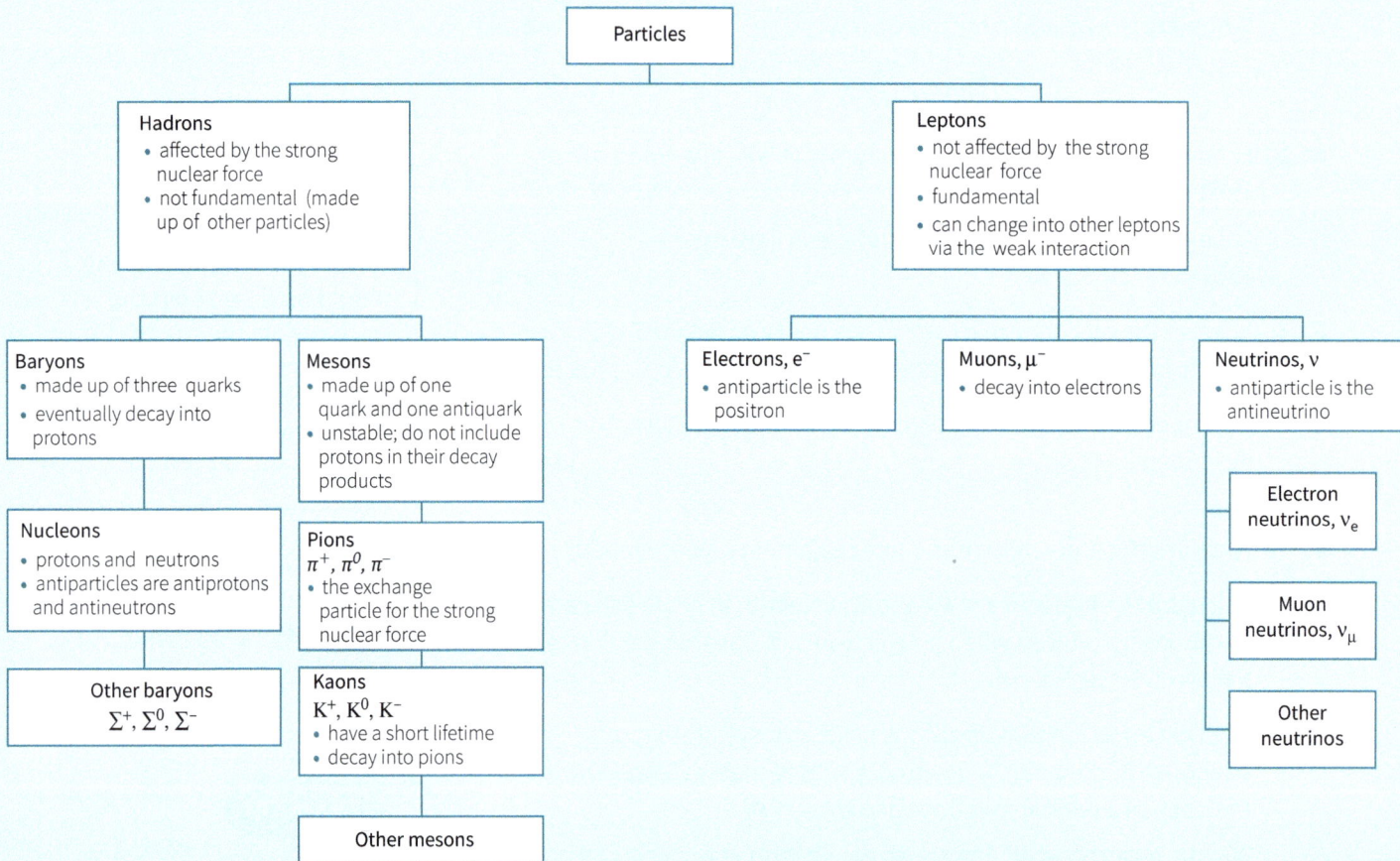

```
                              ┌─────────────┐
                              │  Particles  │
                              └─────────────┘
```

Hadrons
- affected by the strong nuclear force
- not fundamental (made up of other particles)

Leptons
- not affected by the strong nuclear force
- fundamental
- can change into other leptons via the weak interaction

Baryons
- made up of three quarks
- eventually decay into protons

Mesons
- made up of one quark and one antiquark
- unstable; do not include protons in their decay products

Electrons, e^-
- antiparticle is the positron

Muons, μ^-
- decay into electrons

Neutrinos, ν
- antiparticle is the antineutrino

Nucleons
- protons and neutrons
- antiparticles are antiprotons and antineutrons

Pions
π^+, π^0, π^-
- the exchange particle for the strong nuclear force

Electron neutrinos, ν_e

Muon neutrinos, ν_μ

Other baryons
$\Sigma^+, \Sigma^0, \Sigma^-$

Kaons
K^+, K^0, K^-
- have a short lifetime
- decay into pions

Other neutrinos

Other mesons

Strangeness

Strangeness is a property of hadrons only. Strangeness number s is a value from −3 to +3. Leptons have a strangeness of 0.

Strangeness is always conserved in the strong interaction but can change by 0, +1, or −1 in weak interactions. Strange particles:

- are always produced through the strong interaction, so are always created in pairs to conserve strangeness
- always decay through the weak interaction.

Conservation laws and quantum numbers

Conservation laws state that energy, momentum, and charge must be conserved in all particle interactions.

Quantum numbers are used to describe other quantities that are conserved in particle interaction.

Baryon number B	+1 for any baryon	−1 for any antibaryon	0 for mesons or leptons
Lepton number L	+1 for any lepton	−1 for any antilepton	0 for any non-lepton

There are two types of lepton number:
- electrons and electron neutrinos, L_e
- muons and muon neutrinos, L_μ.

Quarks, antiquarks and quark combinations

Hadrons are made of fundamental particles called **quarks**, and their antiparticles **antiquarks**.

You need to know the names and properties of up, down, and strange quarks (and their antiquarks). Quarks always have fractional values for charge and baryon number, but particles composed of quarks always have integer values.

Properties	Quarks			Antiquarks		
type	Up u	Down d	Strange s	Up $\bar{u}$	Down $\bar{d}$	Strange $\bar{s}$
charge Q	$+\frac{2}{3}$	$-\frac{1}{3}$	$-\frac{1}{3}$	$-\frac{2}{3}$	$+\frac{1}{3}$	$+\frac{1}{3}$
strangeness s	0	0	−1	0	0	+1
Baryon number B	$+\frac{1}{3}$	$+\frac{1}{3}$	$+\frac{1}{3}$	$-\frac{1}{3}$	$-\frac{1}{3}$	$-\frac{1}{3}$

A hadron's properties can be explained in terms of the quarks it is made up of.

Its total charge, baryon number, and strangeness equal the sum of these values for the three quarks it is made up of.

Baryons

Baryons are made up of three quarks, and antibaryons are made up of three antiquarks.

Baryon	Quark combination
proton	uud
neutron	udd
antiproton	$\bar{u}\bar{u}\bar{d}$
antineutron	$\bar{u}\bar{d}\bar{d}$
Σ (sigma particle)	any combination of three quarks; must contain a strange quark

Mesons

Mesons and antimesons are made up of a quark and an antiquark. The figure shows the possible combinations and corresponding mesons and antimesons.

- π^0 can be any combination of a quark and its corresponding antiquark
- π^+ is the antiparticle of π^-, and K^+ is the antiparticle of K^-
- there are two uncharged kaons, K^0 and $\bar{K}^0$.

Quarks and beta decay

In β^- decay, a neutron (udd) changes into a proton, releasing an electron and an electron antineutrino. A down quark changes into an up quark via the weak interaction.

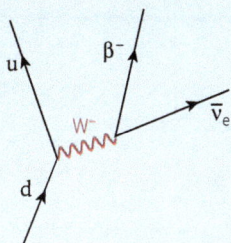

In β^+ decay, a proton (uud) changes into a neutron, releasing a positron and an electron neutrino. An up quark changes into a down quark via the weak interaction.

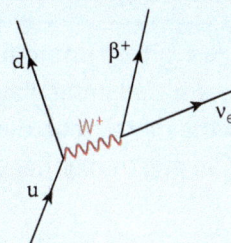

Electromagnetic radiation and quantum phenomena

The **photoelectric effect** is the emission of electrons from the surface of a metal when electromagnetic radiation above a certain frequency is incident on it.

Trying to explain the photoelectric effect led Albert Einstein to propose the photon model that treats electromagnetic radiation as packets of energy:

$$E = hf = \frac{hc}{\lambda}$$

where:

h = Planck constant, 6.63×10^{-34} J s

f = frequency of electromagnetic radiation

c = speed of electromagnetic radiation

λ = wavelength of electromagnetic radiation.

The electrons emitted due to the photoelectric effect are called **photoelectrons**.

The minimum energy an electron needs to break bonds and escape from the metal surface is called the **work function** ϕ of the metal. The work function is different for different metals.

The minimum frequency for photoelectric emission of electrons to take place is called the **threshold frequency**. The threshold frequency is different for different metals.

$$\text{threshold frequency } f_{\min} = \frac{\phi}{h}$$

The photoelectric effect equation is:

$$\begin{array}{ccccc} \text{energy} & & \text{work} & & \text{max. kinetic} \\ \text{transferred} & = & \text{function} & + & \text{energy of} \\ \text{to electron} & & & & \text{photoelectron} \\ hf & = & \phi & + & E_{k(\max)} \end{array}$$

Emission can take place from a metal surface if:

$$E_{k(\max)} > 0 \text{ or } hf > \phi$$

The photon model

Some aspects of the photoelectric effect cannot be explained by a wave model of light, but can be explained by the **photon model**.

Below threshold frequency, no electrons are emitted from the metal
When light is incident on a metal surface, lots of photons hit the metal but a single electron in the metal will only absorb energy from a single photon, gaining energy hf.

If the energy gained by the electron is greater than the metal's work function ($hf > \phi$), the electron is emitted from the surface.

Photoelectrons are emitted with a range of kinetic energies
The intensity of radiation is the number of photons per second per unit area.

Increasing intensity results in more photoelectrons because there are more individual photons that can be absorbed by individual electrons, but each photon has the same energy and can only interact with one electron.

If frequency is increased beyond threshold frequency, excess energy gained by the photoelectron goes to the electron's kinetic energy store, increasing $E_{K(\max)}$.

A range of kinetic energies occur because electrons lose different amounts of energy leaving the metal, depending on how deep below the surface they are initially.

Photoelectrons are emitted immediately
Photoelectrons are emitted immediately once the appropriate frequency light is incident on the metal, regardless of intensity. One electron interacts with one photon – an electron either gains enough energy to be emitted or it doesn't

Stopping potential

A **photocell** uses the photoelectric effect to produce the current in a circuit.

Light of an appropriate frequency is shone upon the cathode, which emits electrons that travel across a vacuum to the anode (or collecting electrode).

If the anode of the photocell is made negative by applying an external potential difference, the photoelectrons can be slowed down and stopped.

The stopping potential V_s is the minimum potential needed to stop photoelectric emission.

At the stopping potential, the work done to stop the fastest moving photoelectrons is $e\,V_s$ so:

$$e\,V_s = E_{k(max)}$$

A graph of stopping potential V_s against frequency f is a straight line with:

- gradient $= \dfrac{h}{e}$

- y-intercept $= -\dfrac{\phi}{h}$

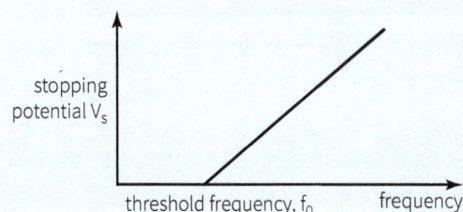

Energy levels and photon emission

Electrons in an atom can only exist in discrete energy levels. The lowest energy an electron can have in an atom is known as the **ground state** and is labelled energy level $n = 1$.

Electrons cannot exist between energy levels.

- Electrons move up energy levels if they absorb energy in the form of a photon or from a collision. Electrons moving up energy levels is known as **excitation**.

- Electrons move down energy levels if they emit energy in the form of a photon. Electrons moving down energy levels is known as **de-excitation**.

The figure shows the energy levels of a hydrogen atom.

Electrons can only absorb or emit photons that have an energy equal to the energy difference between energy levels.

When an electron moves from energy level E_1 to E_2, the energy of the emitted or absorbed photon is equal to the difference in energy between the levels:

$$\text{energy of the emitted or absorbed photon} = hf = E_1 - E_2$$

where:

h = Planck constant, 6.63×10^{-34} J s

f = frequency of the photon in Hz

E_1 and E_2 = energies in J.

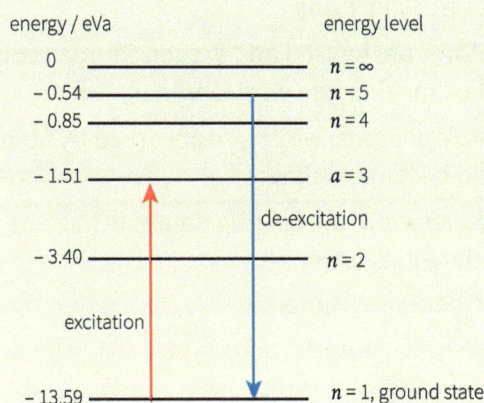

Line spectra

Line spectra provide evidence for transitions between discrete energy levels in atoms.

- Each coloured line in a line emission spectrum corresponds to a wavelength (and frequency) of light emitted by the source.
- The photons that produce each line all have the same energy unique to the line.
- Each photon is emitted when the atom is de-excited.
- An atom only emits particular wavelengths of light because the electrons in it can only emit photons with energies equal to the difference between two of its energy levels.

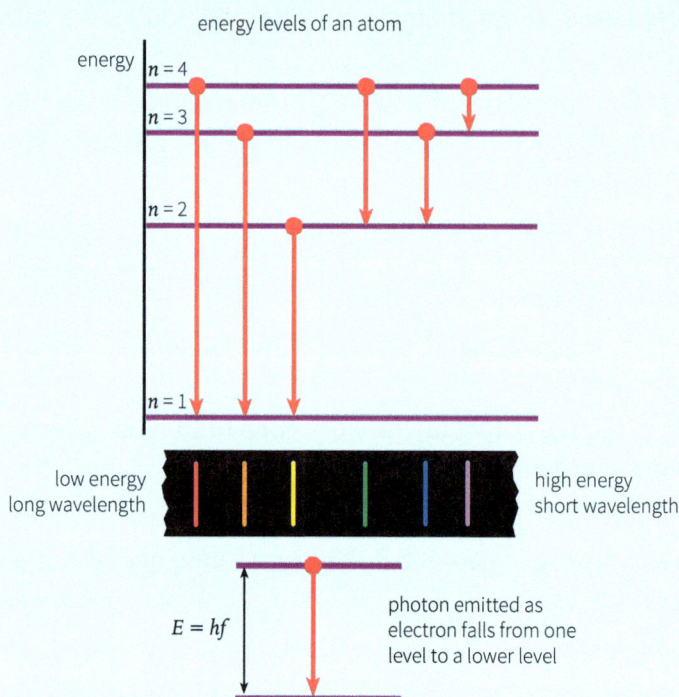

No two elements produce the same pattern of wavelengths because the arrangement of energy levels in atoms are unique to each element.

Similarly, a line absorption spectrum has black lines that arise from only particular wavelengths of light being absorbed by the source, because the electrons in an atom can only absorb photons with energies equal to the difference between two energy levels.

energy levels of an atom

energy
$n = 4$
$n = 3$
$n = 2$
$n = 1$

low energy
long wavelength

high energy
short wavelength

$E = hf$

photon emitted as electron falls from one level to a lower level

Fluorescence

When fluorescent substances absorb UV radiation, their electrons absorb UV photons and move to higher energy levels.

They emit photons of visible light (they glow) as they drop back to lower energy levels.

A **fluorescent tube** is a glass tube with a fluorescent coating on its inner surface. It is filled with mercury vapour.

When in use, a high voltage is applied across the mercury vapour, which accelerates free electrons.

Mercury atoms are ionised and excited through collisions with each other and with fast moving free electrons in the tube.

They emit UV photons, which are absorbed by atoms of the fluorescent coating and become excited.

The excited atoms in the coating de-excite in steps and emit visible photons (the tube glows).

Ionisation

Ionisation occurs when an electron gains enough energy to leave the atom.

Electrons are defined as having *zero* energy when they are at the ionisation level, so the energy values of all other levels in an atom are negative because an electron needs to gain energy in order to leave the atom.

The **ionisation energy** is the amount of energy needed to remove an electron from the ground state of an atom.

Values for energy levels may be quoted in joules (J) or electron volts (eV).

1 eV is the energy gained by an electron accelerated through a potential difference of 1 V:

$$1\,eV = 1.6 \times 10^{-19}\,J$$

Diffraction of electrons

The observation that electrons can be diffracted and produce **interference patterns** provides evidence of their wave properties.

Electron diffraction can be observed by firing a beam of electrons at a thin sheet of graphite in a vacuum.

The pattern formed on the screen is a result of electrons being diffracted as they pass through rows of atoms. They then interfere with each other to produce maxima and minima in the |same way light diffracted through a slit does.

The separation of the maxima in the diffraction pattern depends on the wavelength of the electrons – the bigger the wavelength, the bigger the spread of the lines.

$$\text{wavelength } \lambda = \frac{h}{mv}$$

So, the wavelength of a particle is inversely proportional to its momentum (and inversely proportional to its mass and velocity).

The momentum and, therefore, wavelength of electrons can be changed by changing their velocity. Velocity can be altered by changing the potential difference through which they are accelerated.

Wave–particle duality

Light (electromagnetic radiation) has a dual nature – it can behave like a wave or a particle depending on the circumstances.

- Diffraction and interference can only be explained using a wave model of light.
- The photoelectric effect can only be explained using a particle model of light.

In 1924, Louis de Broglie hypothesised that particles of matter, like electrons, should have wave properties.

de Broglie suggested that all matter particles have a wavelength associated with their wave properties and related to their momentum p by the formula:

$$\text{de Broglie wavelength } \lambda = \frac{h}{p} = \frac{h}{mv}$$

de Broglie's work was evaluated by other scientists through the peer review process, and his hypothesis was confirmed by experiments that showed electrons and other types of particles can be diffracted.

de Broglie's work is an example of how scientists' knowledge and understanding of the nature of matter has changed over time.

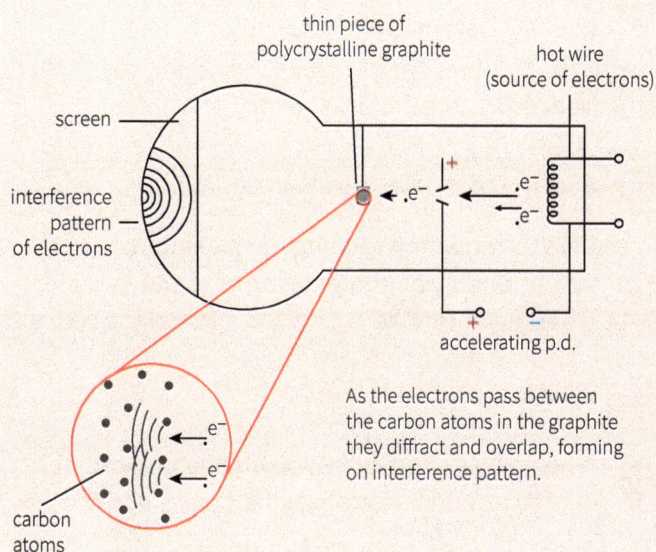

As the electrons pass between the carbon atoms in the graphite they diffract and overlap, forming on interference pattern.

Retrieval

Learn the answers to the questions below, then cover the answers column with a piece of paper and write as many as you can. Check and repeat.

Questions	Answers
1 Which of the two main groups of particles contains only fundamental particles?	leptons
2 Which particles can be classified as leptons?	electrons, muons, electron neutrinos, muon neutrinos
3 Which quantities are always conserved in particle interactions?	energy E, momentum p, charge Q, baryon number B, lepton numbers L_e and L_μ
4 Which quantity is conserved in strong interactions but not in weak interactions?	strangeness s
5 What values does strangeness take?	from −3 to +3 for hadrons, 0 for leptons
6 What are antihadrons made up of?	antiquarks
7 What must the total charge, baryon number, and strangeness of a hadron be equal to?	sum of these values for the three quarks of which it is composed (always integers)
8 Give the quark composition of a proton.	uud
9 What is the quark composition of a sigma Σ particle?	any combination of three quarks, must contain a strange quark
10 What is the photoelectric effect?	emission of electrons from the surface of a metal when electromagnetic radiation above a certain frequency is incident on it
11 Give the equation relating energy of a photon of em radiation to its frequency.	$E = h f = \dfrac{hc}{\lambda}$
12 Define the threshold frequency of a metal.	minimum frequency of photons needed for photoelectric emission of electrons from the metal
13 Give the photoelectric effect equation.	$h f = \phi + E_{k(max)}$
14 Give the equation relating the maximum kinetic energy of photoelectrons to the stopping potential.	$e V_s = E_{k(max)}$
15 What is excitation of an atom?	an electron moves up one or more of the energy levels in an atom
16 When an electron moves down one or more energy levels in an atom, what does it emit?	energy in the form of a photon
17 What do each of the coloured lines in a line emission spectrum correspond with?	a particular wavelength (and therefore frequency) of light emitted by the source
18 When does ionisation occur?	when an electron gains enough energy to leave the atom

Put paper here

19	Define an electron volt.	energy gained by an electron accelerated through a potential difference of 1 V 1 eV = 1.6×10^{-19} J
20	Identify the phenomena that can only be explained by the wave model of light.	diffraction and interference
21	What did de Broglie propose about the nature of particles?	all matter particles have wave properties
22	Give the equation for calculating the de Broglie wavelength of a particle.	$\lambda = \dfrac{h}{p} = \dfrac{h}{mv}$

Put paper here · *Put paper here*

Maths skills

Practise your maths skills using the worked example and practice questions below.

Rearranging equations

To rearrange an equation:

- decide which variable to make the subject

- carry out the same operations on both sides of the equation until that variable is isolated.

For example to find the frequency f using:

$$E_{k(max)} = hf - \phi$$

you need to isolate f, so add ϕ to each side, then divide each side by h.

$$hf = E_{k(max)} + \phi$$

$$f = \frac{(E_{k(max)} + \phi)}{h}$$

Worked example

Question

The work function ϕ of potassium is 2.29 eV. A photon is incident on the surface of a potassium sheet, and an electron is ejected with a kinetic energy of 0.21 eV.

Calculate the frequency of the electromagnetic radiation.

Answer

$$hf = E_{k(max)} + \phi$$

Step 1: convert energy values from eV to J

1 eV = 1.6×10^{-19} J, therefore:

$$\phi = 2.29 \times 10^{-19} \text{ J}$$
$$= 3.66 \times 10^{-19} \text{ J}$$

$$E_{k(max)} = 0.21 \times 1.6 \times 10^{-19} \text{ J}$$
$$= 3.4 \times 10^{-20} \text{ J}$$

Step 2: rearrange equation

$$f = \frac{(E_{k(max)} + \phi)}{h}$$

$$= \frac{3.4 \times 10^{-20} \text{ J} + 3.66 \times 10^{-19}}{6.63 \times 10^{-34} \text{ J s}}$$

$$= 6.0 \times 10^{14} \text{ Hz}$$

Practice

1 A photon of electromagnetic radiation is incident on an aluminium surface ($\phi = 4.08$ eV). The emitted photon has a kinetic energy of 1.10 eV.

Calculate the frequency of the incident electromagnetic radiation.

2 The work function of calcium is 2.87 eV.

Calculate the minimum frequency of light that will cause a photoelectron to be emitted (the threshold frequency).

3 UV radiation with a photon energy of 3.2×10^{-19} J is incident on the surface of a metal. Electrons are emitted from the metal with a maximum kinetic energy of 1.3 eV.

Calculate the work function of the metal. Give your answer in eV.

Exam-style questions

01 Neutrons that are not bound in the nucleus are unstable. They decay by the emission of a beta particle and one other particle.

When scientists first observed this decay, they could not detect the other particle.

01.1 Describe how the scientists knew that this particle was emitted.

[1 mark]

_____

_____

01.2 Write the equation for neutron decay. [1 mark]

_____

_____

> **!** **Exam tip**
>
> In all reactions, consider the conservation of charge, mass, and quantum numbers.

01.3 State the class of particle to which the beta particle belongs.

[1 mark]

_____

01.4 A student considers a different decay for a neutron, and writes the following equations.

$$n \rightarrow e^+ + e^-$$
$$n \rightarrow p^+ + e^-$$

Suggest **one** reason why the student might think these decays are possible, and explain why they are not observed. [3 marks]

> **!** **Exam tip**
>
> Check the number of marks and instructions carefully; here there should be an explanation for each equation.

_____

_____

_____

_____

_____

_____

01.5 A neutron is a baryon but is not stable.

State the only stable baryon. [1 mark]

_____

01.6 Hadrons are subject to the strong interaction.

Discuss the role of the strong interaction in neutron decay. [1 mark]

_____

_____

_____

_____

02 In nuclear medicine, the isotope iodine-124 can be used to study cancer and treat tumours. It decays by positron emission to an isotope of tellurium, as shown in the equation below.

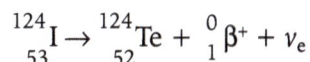

$$^{124}_{53}\text{I} \rightarrow \,^{124}_{52}\text{Te} + \,^{0}_{1}\beta^+ + \nu_e$$

02.1 Write an equation to show the change in quark content of the nucleus as a result of the decay in the equation above. **[1 mark]**

_____

_____

02.2 Show how the following are conserved in this decay. **[3 marks]**

lepton number: _____

_____

charge: _____

_____

baryon number: _____

_____

> **! Exam tip**
>
> You must include all the particles when you evaluate a reaction in terms of quantum numbers, and write down each number even if it is zero.

02.3 State **one** other quantity that is conserved in this decay. **[1 mark]**

_____

02.4 An antimuon, μ^+, also decays into a positron.

Write the equation for this decay. **[1 mark]**

_____

_____

02.5 Compare the products of the reaction in **02.1** with the products in **02.4**. **[2 marks]**

_____

_____

_____

_____

03 A kaon is a meson.

03.1 Compare a meson to a baryon. **[2 marks]**

03.2 Describe the difference between a kaon and a proton in terms of interactions. **[2 marks]**

03.3 The K^+ meson has a strangeness of +1.

State the quark composition of a K^+ meson. **[1 mark]**

03.4 A K^+ meson decays into two particles.

$$K^+ \rightarrow \mu^+ + \nu_\mu$$

Suggest the interaction responsible for this decay.
Give a reason for your answer. **[2 marks]**

03.5 Kaons were discovered by examining tracks in a cloud chamber.

Suggest why a single scientist would not have been able to discover the kaon. **[1 mark]**

Exam tip

Reduce equations to quarks and leptons to help you work out the interactions responsible for decays.

04 Antimatter particles can be made in particle accelerators.

04.1 State the quark structure of a proton and an antiproton. **[2 marks]**

04.2 Explain in terms of quarks why the proton and antiproton have opposite charges. **[2 marks]**

04.3 When a proton and antiproton collide, two mesons are produced.

$$p + \bar{p} \rightarrow X + Y$$

Suggest the identities of **X** and **Y**.

Explain your answer in terms of quarks. **[3 marks]**

Exam tip

Learning some of the properties of particles that are given on the formula sheet will save you time in the exam.

04.4 Baryon number is conserved in the collision in **04.3** even though there are different numbers of quarks present before and after the collision.

Explain why baryon number is conserved. **[2 marks]**

05 A pion interacts with a proton.

$$p + \pi^- \rightarrow \pi^+ + \pi^- + n$$

05.1 State the reason why this reaction involves the strong interaction only. **[1 mark]**

05.2 In another interaction between a proton and a pion, a K^+ and a sigma particle are produced.

Determine the charge on the sigma particle using conservation of charge.

Give a reason for your answer. **[2 marks]**

Exam tip

Conservation is important throughout physics. Make a list of quantities that are always conserved, and those that are not.

05.3 Determine the quark content of the sigma particle.
Show that it has the charge you deduced in **05.2**. **[2 marks]**

05.4 The kaon from the interaction in **05.2** then decays.

$$K^+ \rightarrow \mu^+ + \nu_\mu$$

Compare the interaction that produced the kaon with the interaction by which it decays. **[2 marks]**

06 Two particle classifications are hadrons and leptons.

06.1 State **one** example of a particle in each of these particle classifications. **[2 marks]**

06.2 Compare the two particles classifications in **06.1** in terms of their interactions. Consider all four fundamental interactions. **[2 marks]**

06.3 Another distinction in particle physics is between baryons and mesons. State why a pion is a meson and not a baryon. **[1 mark]**

06.4 A negative pion interacts with a proton to produce a neutral pion and another particle, **X**.

Determine the identity of particle **X**.

Show how the interaction obeys two conservation laws apart from energy and momentum. **[3 marks]**

06.5 Particle **X** in part **06.4** can be used to determine the structure of a material by scattering the particles from a sample of the material. The particles are diffracted.

Explain how particles can be diffracted by atoms. **[2 marks]**

06.6 Calculate the momentum in $kg\,m\,s^{-1}$ of a particle that can be diffracted by an atom. Assume a typical size of an atom. **[2 marks]**

07 Protons are accelerated in particle accelerators.

07.1 Explain why an electric field can be used to accelerate a proton but not a neutron. **[2 marks]**

07.2 Early experiments in particle physics involved analysing cloud chamber tracks of particle interactions of cosmic rays.
Suggest why these early experiments used cloud chambers and not particle accelerators. **[1 mark]**

07.3 A proton interacts with a negative kaon.
State the quark content of a proton. **[1 mark]**

07.4 A proton and a kaon are both hadrons, but they are not both mesons. Explain why this is true. **[2 marks]**

07.5 In an interaction, two kaons and another particle are produced.

$$p + K^- \rightarrow K^+ + K^0 + X$$

Deduce the charge on **X**. **[1 mark]**

07.6 State, with a reason, whether particle **X** is a lepton, baryon, or meson. **[2 marks]**

07.7 The quark content of K^+ and K^0 are shown in **Table 1**.

Table 1

K^+	$u\bar{s}$
K^0	$d\bar{s}$
K^-	$s\bar{u}$

If strangeness is conserved then the quark content of **X** in **07.5** could be sss.
Explain why. **[2 marks]**

Synoptic link

3.2.2.4

! Exam tip

You need to know all the definitions of the categories of particles. You should learn them along with an example of each.

Synoptic link

3.7.3.2

! Exam tip

When explaining the conservation of quantum numbers, you should always write an equation showing the numbers on both sides.

08   An iron nucleus in a cosmic ray collides with a nitrogen nucleus. The velocity of the iron nucleus before the collision is $7.1 \times 10^7 \, m \, s^{-1}$ in the x-direction, and it moves at an angle of 25° to the x-direction after the collision, with a speed of $4.8 \times 10^7 \, m \, s^{-1}$. The nitrogen nucleus moves at an angle of 36° to the x-direction.

Synoptic links

3.2.1.1   3.2.1.3
3.4.1.4   3.4.1.6

Figure 1

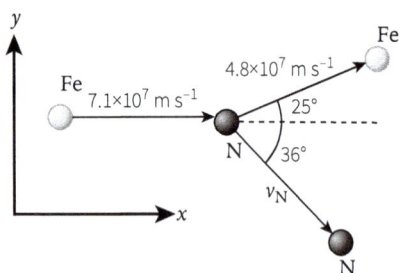

08.1 Calculate the mass of the iron nucleus, $^{56}_{26}$ Fe, in kg.   **[3 marks]**

08.2 Calculate the final velocity in $m \, s^{-1}$ of the nitrogen nucleus. The mass of the nitrogen nucleus is $2.34 \times 10^{-26} \, kg$.

    Give your answer to an appropriate number of significant figures.

    [4 marks]

08.3 As a result of this interaction, some gamma rays are produced. Explain why this means that the collision is not elastic.   **[2 marks]**

08.4 One of the gamma rays has sufficient energy to produce a pion/anti-pion pair.

    Calculate the wavelength in m of the gamma ray that produces this pair.   **[2 marks]**

> **! Exam tip**
>
> Remember that the types of collisions are linked to the conservation of energy.

09   When photons interact with electrons they can cause them to be ejected from the surface of a metal. The equation that describes the photoelectric effect is:

$$hf = \phi + KE$$

09.1 Describe what each part of the equation represents.   **[3 marks]**

09.2 The work function of nickel is 5.01 eV.

    Show that radiation with a minimum frequency of $1.2 \times 10^{15} \, Hz$ is needed to remove electrons from the surface of nickel.   **[2 marks]**

> **! Exam tip**
>
> Ensure you learn the meaning of all the symbols in key equations.

09.3 Calculate the speed of an electron in $m \, s^{-1}$ emitted from the surface of nickel when it absorbs a photon with a frequency of $4.2 \times 10^{15} \, Hz$.

    [3 marks]

09.4 Electrons exhibit wave-like behaviour.

    Calculate the wavelength in m of an electron with the speed calculated in **09.3**.   **[2 marks]**

09.5 Discuss whether electrons with the wavelength in **09.4** would be diffracted by a thin piece of graphite.   **[1 mark]**

09.6 In an electron diffraction experiment, electrons are accelerated to speeds that produce diffraction by a piece of graphite. In an electron diffraction tube rings are produced on a screen.

Describe and explain the effect of decreasing the potential difference through which the electrons are accelerated on the rings produced.

[4 marks]

10 Light of wavelength 188 nm ejects electrons from a photocathode that are stopped by a potential difference of 4.5 V. **Table 2** shows some metals and their work functions.

Table 2

Metal	Work function / eV
caesium	2.1
aluminium	4.08
platinum	6.35

10.1 Determine the metal from which the cathode is constructed. Explain your answer. **[3 marks]**

10.2 Describe and explain the effect on the stopping potential of using one of the other metals in **Table 2**. **[2 marks]**

10.3 Photocathodes can be used in photomultiplier tubes which, together with scintillators, are used to detect radiation. In the earliest scintillator, alpha and beta radiation impacted a zinc sulfide screen and produced a flash of light that was visible through a microscope in a dark room.

Explain how the light was produced. **[2 marks]**

> **! Exam tip**
>
> You need to be able to convert between energy in eV and energy in J.

10.4 Compare the process by which light is produced in the scintillator described in **10.3** with the process that produces light when a current passes through mercury vapour in a fluorescent tube. **[3 marks]**

10.5 Some of the energy levels for the electrons in a mercury atom are shown in **Figure 2**.

Figure 2

> **! Exam tip**
>
> Be precise about the direction of electron transition between levels.

$E / 10^{-19}$ J

———————— A

-7.94 ———————— B

-8.86 ———————— C

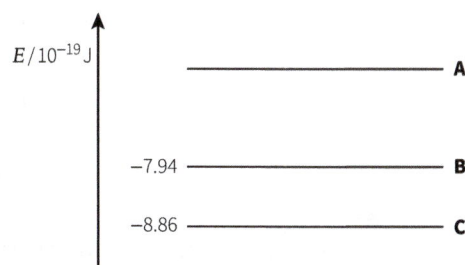

Compare the transition of an electron from **B** to **C** with the transition of an electron from **C** to **B**. **[3 marks]**

10.6 Calculate the energy of **A** in J if the transition between **A** and **C** produces a photon with an energy of 1.8 eV. **[4 marks]**

10.7 It is dangerous to look at mercury lamps because they emit ultraviolet radiation.

Suggest, using a calculation, whether the transition described in **10.3** is responsible for the emission of ultraviolet photons. **[3 marks]**

Knowledge

Progressive waves

Waves are caused by something oscillating the particles in the medium (or oscillating the field) through which the wave travels.

A progressive wave transfers energy from one place to another without transferring matter. Wave motion is described by a number of properties.

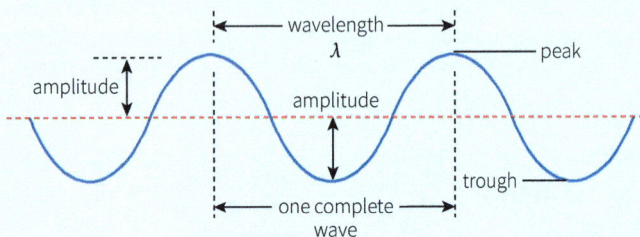

Frequency and period are related by the equation:

$$f = \frac{1}{T}$$

The **speed**, frequency, and **wavelength** of a wave are related by the equation:

$$\text{wave speed } c = f\lambda$$

Property	Description	Unit
displacement	distance and direction an oscillating particle moves from its rest position	metre (m)
amplitude A	maximum displacement an oscillating particle in the medium moves from its rest position	metre (m)
wavelength λ	distance between any point on a wave to an adjacent one that is exactly in phase with it	metre (m)
period T	time taken for one complete wave to pass a fixed point	second (s)
frequency f	number of complete waves passing a fixed point per second	hertz (Hz)

Phase difference

The **phase** of a particle on a wave is the fraction of a cycle it has completed since the start of the cycle. The phase difference between two particles on a wave, or two waves, is the fraction of a cycle by which their oscillations are separated.

Phase difference can be expressed as a fraction of a wavelength λ, or in degrees or radians, where 1 cycle = 360° = 2π radians. If two points on a wave are separated by a horizontal distance d:

$$\text{phase difference in radians} = \frac{2\pi d}{\lambda}$$

A **wavefront** is a line or a surface on which the vibrations of the medium have the same phase at all points.

Types of wave

Mechanical waves require a physical substance to move through.

Electromagnetic waves can travel through a vacuum – they are oscillations in electric and magnetic fields that permeate space. All electromagnetic waves travel at the same speed in a vacuum: $3.00 \times 10^8 \text{ m s}^{-1}$.

Longitudinal waves – oscillations of the medium are *parallel* to the direction of energy transfer. For example, sound waves.

Transverse waves – oscillations of the medium are *perpendicular* to the direction of energy transfer. For example, electromagnetic waves and waves on a string.

Superposition

The principle of superposition states that when two waves meet, the total displacement at a point is equal to the sum of the individual displacements at that point.

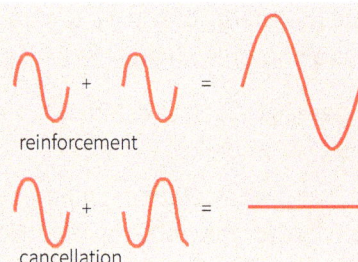

reinforcement

cancellation

Polarisation

If the oscillations of a transverse wave are restricted into a single plane, they are polarised. Longitudinal waves cannot be polarised.

Transverse waves passing through a slit in a board become polarised because only the oscillations parallel to the slit pass through it. If a wave can be polarised, that means it is a transverse wave.

Light reflected from some surfaces, like water or glass, is partially polarised. This reflected light causes **glare**, which can be reduced by filtering out the polarised light using polarising filters.

Electromagnetic waves used for television and radio are polarised, so the aerials for detecting them need to be aligned in the appropriate direction to get a strong signal.

unpolarised waves on rope — slit in board — polarised waves — slit in board — no waves get through second slit

unpolarised light — polaroid — polarised light — polaroid at right angles to first polaroid — no light gets through second polaroid

Stationary waves

A stationary wave is formed when two progressive waves with the same frequency and amplitude, moving in opposite directions, superpose.

Stationary waves are often the result of the reflection of a progressive wave superposing with the original wave.

A **node** is a point where there is no oscillation – the two waves always meet with a phase difference of 180° (π radians). An **antinode** is a point where oscillation is at maximum amplitude – the two waves always meet in phase at that point.

$$\text{distance between two nodes} = \frac{1}{2}\lambda$$

All the points between two nodes oscillate in phase, meaning they reach their maximum displacements at the same time.

Stationary waves that oscillate freely do not transfer energy to their surroundings. When stationary waves are formed:

- by sound waves, there is silence at the nodes and maximum volume at the antinodes
- in a microwave oven, there is no heating at the nodes, and maximum heating at the antinodes.

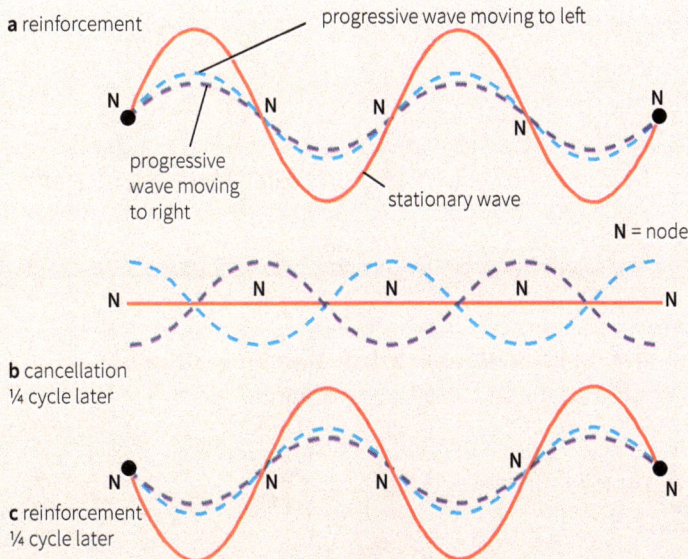

a reinforcement

progressive wave moving to left

progressive wave moving to right

stationary wave

N = node

b cancellation ¼ cycle later

c reinforcement ¼ cycle later

Stationary waves on a vibrating string

The first harmonic

As the frequency of oscillation is increased from zero, the first harmonic is seen at the lowest frequency that gives a pattern. The frequency f of the first harmonic is related to the length l, tension T, and mass per unit length μ of the string by the equation:

$$f = \frac{1}{2l}\sqrt{\frac{T}{\mu}}$$

The second and third harmonic

The second harmonic occurs at twice the frequency of the first harmonic; the third harmonic occurs at three times the frequency of the first harmonic; etc.

N = node A = antinode
(dotted line shows string half a cycle earlier)

a first harmonic

b second harmonic

c third harmonic

Retrieval

Learn the answers to the questions below, then cover the answers column with a piece of paper and write as many as you can. Check and repeat.

Questions	Answers
1 What is a progressive wave?	wave that transfers energy from one place to another without transferring matter
2 Define the displacement of a wave and state its unit.	distance and direction an oscillating particle moves from its rest position, in metres
3 Define the amplitude of a wave and state its unit.	maximum displacement an oscillating particle in the medium moves from its rest position, in metres
4 Define the period of a wave and state its units.	time taken for one complete wave to pass a fixed point, in seconds
5 Define the frequency of a wave and state its units.	number of complete waves passing a fixed point per second, in hertz
6 State the relationship between the frequency and period of a wave.	$f = \dfrac{1}{T}$
7 State the relationship between the speed, frequency, and wavelength of a wave.	$c = f\lambda$
8 What is the phase of a particle on a wave?	fraction of a cycle it has completed since the start of the cycle
9 What is the phase difference between two particles on a wave?	fraction of a cycle by which their oscillations are separated
10 What units is phase difference expressed in?	fraction of a wavelength λ, or degrees or radians, where 1 cycle = 360° = 2π radians
11 State the phase difference in radians for two points on a wave separated by a horizontal distance d.	phase difference in radians = $\dfrac{2\pi d}{\lambda}$
12 Define a wavefront.	line or a surface on which the vibrations of the medium have the same phase at all points
13 What are mechanical waves?	waves requiring a physical substance to move through
14 Why are electromagnetic waves not mechanical waves?	they can travel through a vacuum
15 What is a longitudinal wave?	wave where the oscillations of the medium are parallel to the direction of energy transfer
16 What is a transverse wave?	wave where the oscillations of the medium are perpendicular to the direction of energy transfer
17 Define polarisation.	restriction of the oscillations of a transverse wave into a single plane
18 Describe a node on a stationary wave.	point where there is no oscillation

Put paper here

19	State the principle of superposition.	when two waves meet, the total displacement at a point is equal to the sum of the individual displacements at that point
20	How is a stationary wave formed?	when two progressive waves with the same frequency and amplitude, moving in opposite directions, superpose
21	Describe an antinode on a stationary wave.	point that oscillates with maximum amplitude
22	When stationary waves are formed on a string, what is the first harmonic?	stationary wave that occurs at the lowest frequency

Put paper here *Put paper here*

Practical skills

Practise your practical skills using the worked example and practice questions below.

Stationary waves on a string

To investigate the variation of the frequency f of stationary waves on a string with length l, tension T, and mass per unit length μ, use the apparatus:

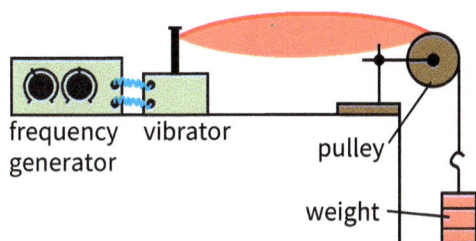

frequency generator   vibrator   pulley   weight

To find the frequency of the first harmonic:

1 set the frequency generator to zero
2 increase frequency until first harmonic appears on string
3 adjust frequency to get antinode with maximum possible amplitude; record this frequency
4 increase or decrease frequency so the first harmonic disappears; adjust frequency back to maximum amplitude; record this frequency
5 repeat, and average the readings.

The l, T, or μ can then be systematically increased or decreased and these steps repeated to investigate how the frequency varies.

Worked example

Question

A student wants to investigate the relationship between frequency f and tension T of a string. Describe the method the student should use, including the data they should collect and the graph they should plot.

Answer

1 Set up the apparatus as shown; ensure mass per unit length and length of string constant.
2 Set known T by adding masses to the mass holder; record this T.
3 Adjust f until first harmonic appears, and antinode has the maximum possible amplitude; record the f.
4 Change T of string by adding or removing masses in set intervals; repeat step 4.
5 Repeat with different T values, then repeat entire experiment again and find mean f for each T.

$$f = \frac{1}{2\,l}\sqrt{\frac{T}{\mu}}$$

6 Plot a graph of f against $\sqrt{T}$. This produces a straight line through the origin, confirming the relationship between f and T.

Practice

1 Describe two safety precautions that should be taken when carrying out this practical.

2 Describe how you would investigate the relationship between f and l of a string, including what data you would collect and the graph you would plot.

3 A string of known length and unknown mass per unit length is used to investigate how f varies with T. Show how the mass per unit length of the string can be found from a graph of f^2 against T.

Exam-style questions

01 **Figure 1** shows a progressive wave moving from left to right on a rope.

Figure 1

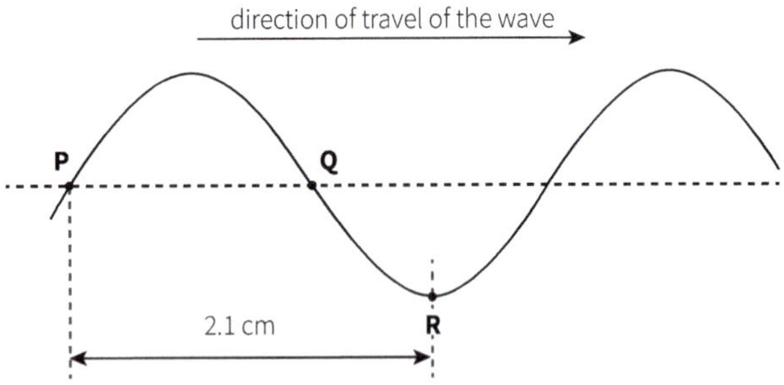

direction of travel of the wave

P Q

2.1 cm R

01.1 Calculate the wavelength of the wave. **[1 mark]**

wavelength = _____ cm

01.2 The speed of the wave is $8.4 \, cm \, s^{-1}$.

Calculate the frequency of the wave. **[1 mark]**

frequency = _____ Hz

01.3 State the phase difference between points **Q** and **R** on the rope.

State an appropriate unit for your answer. **[2 marks]**

phase difference = _____

unit = _____

01.4 Describe how the displacement of point **P** varies over the next half period. **[2 marks]**

_____

_____

_____

> ⚠ **Exam tip**
>
> Make sure you use scientific terminology when answering 'describe' questions.

02 **Figure 2** shows a progressive wave on a spring.

Figure 2

02.1 Annotate **Figure 2** to show a complete wavelength. **[1 mark]**

02.2 Write the definition for the amplitude of a wave. **[1 mark]**

_____

_____

02.3 State the type of progressive wave shown in **Figure 2**.
Give a reason for your answer. **[2 marks]**

_____

_____

_____

02.4 Several points on the spring appear to stop oscillating.
Explain what conditions would cause this to happen. **[3 marks]**

_____

_____

_____

_____

03 A website advises people who fish to wear polarising sunglasses on sunny days to enable them to see the fish underwater more clearly.

03.1 Compare polarised and unpolarised light. **[2 marks]**

_____

_____

_____

03.2 Explain why polarising sunglasses will help a person fishing to see the fish more clearly. **[2 marks]**

_____

_____

_____

03.3 Describe how a polarising filter can be used to test whether a pair of sunglasses is polarising. **[3 marks]**

_____

_____

_____

_____

03.4 Explain why sound waves cannot be polarised. **[2 marks]**

_____

_____

_____

03.5 When looking at the fish at an angle in the water, the fish appear to be higher in the water than they are.
Sketch a diagram to show the rays of light when a ray is reflected from a fish, and explain why the fish appears higher in the water.

[3 marks]

_____

_____

04 A microwave transmitter is pointed towards a metal plate. As a microwave detector is moved between the transmitter and the metal plate, a series of maxima and minima are detected. The distance between five minima is 54 mm.

Figure 3

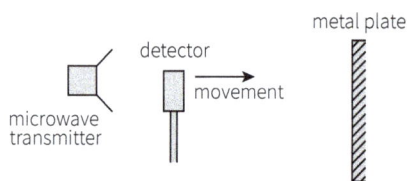

04.1 Explain how the maxima and minima are formed. **[3 marks]**

04.2 Calculate the wavelength in mm of the transmitted microwaves.
[2 marks]

04.3 Calculate the frequency of the microwaves.
State your answer in GHz. **[2 marks]**

04.4 The minima remain the same distance apart as the detector is moved, but they vary in amplitude.
Annotate **Figure 3** with the label **P** to show the position of the lowest minimum. **⌊1 mark⌋**

04.5 Explain the position of **P** in **04.4**. **[2 marks]**

04.6 A student attempted to repeat the experiment, but no microwaves were detected. They checked the transmitter and receiver were both working.
Suggest what the student may have done wrong. **[1 mark]**

05 A layer of mini marshmallows is spread out evenly in a microwave oven. The microwave oven is switched on for 10 seconds. Only some of the marshmallows melt.

05.1 Explain why only some of the marshmallows melt. **[3 marks]**

05.2 The microwave oven is labelled 'frequency 2450 MHz'.
Calculate the expected distance in m between two melted marshmallows. **[3 marks]**

> **! Exam tip**
>
> Use key words carefully. When you have finished **04.1**, mark it yourself to see what the examiners are looking for when they ask you to describe how stationary waves form.

> **! Exam tip**
>
> Think about the path difference for the transmitted and reflected wave. What will happen to the amplitude of the wave as it travels through the air?

> **Synoptic link**
>
> 3.1.2

05.3 In an actual demonstration, the distance between two melted marshmallows is 7.1 cm.

Calculate the speed in m s^{-1} of the microwaves.

Determine the percentage error in this reading. **[3 marks]**

05.4 Describe **one** change that would have to be made to a microwave oven for this demonstration to work. **[1 mark]**

05.5 Compare **two** properties of progressive and stationary waves. **[4 marks]**

> **! Exam tip**
>
> Make sure you write something about both types of wave for **05.5**. It is not enough to only say something about one of the waves.

06 **Figure 4** shows the side view of a guitar string vibrating at the first harmonic.

Figure 4

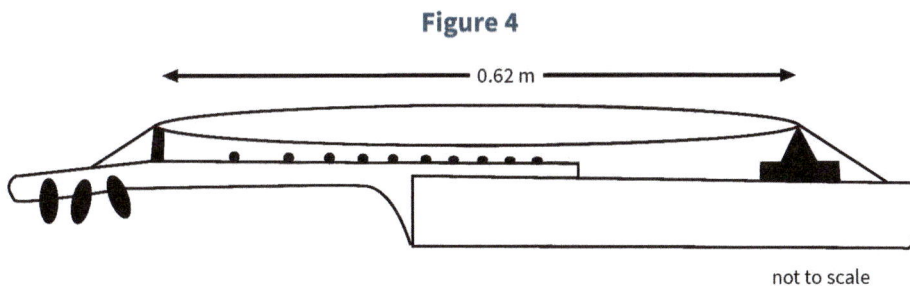

0.62 m

not to scale

06.1 Explain how a stationary wave is formed on the guitar string when it is plucked. **[3 marks]**

06.2 Calculate the wave speed in m s^{-1} when the string vibrates at a frequency of 82.41 Hz. **[2 marks]**

06.3 Placing a finger on the 12th fret of a guitar halves the length of the vibrating string.
Determine the new frequency in Hz, assuming the wave speed is the same. **[1 mark]**

06.4 The string described in **06.2** is the lowest frequency string and has a mass per unit length of 6.44×10^{-3} kg m^{-1}.

Calculate the mass per unit length in kg m^{-1} of the string with frequency 196 Hz.

Assume the same tension and length of string. **[2 marks]**

> **! Exam tip**
>
> Use relationships to solve problems like these. What variables are changing? How can you use that to predict the new mass per unit length?

07 **Figure 5** shows a stretched string driven by an oscillator at end **A** and fixed to a wall at end **B**. A stationary wave forms between **A** and **B**.

Figure 5

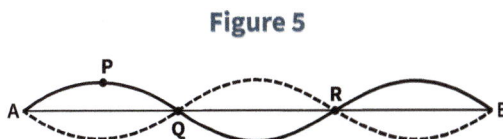

07.1 State the harmonic represented in **Figure 5**. **[1 mark]**

07.2 State the names given to points **P** and **R** on the wave and describe how each were formed. **[2 marks]**

> **! Exam tip**
>
> Make sure you give two answers for each point in **07.2** – the name and the description.

07.3 Compare the phase and amplitude of points **P** and **Q**. [2 marks]

07.4 The length of string between points **A** and **B** is 1.20 m. The speed of the wave is 13.6 m s⁻¹.

Calculate the frequency in Hz of the oscillation. [2 marks]

07.5 Sketch the wave formed if the frequency were doubled. [1 mark]

08 A student investigates the relationship between the tension of the string and fundamental frequency of the stationary wave formed using the apparatus shown in **Figure 6.**

They gather the data shown in **Table 1**.

Synoptic links

| 3.1.2 | MS 3.2 |
| MS 3.3 | MS 3.4 |

Figure 6

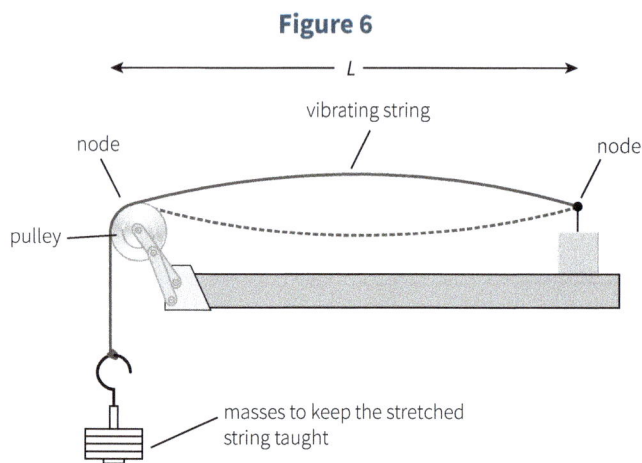

Table 1

Mass / g	Time period / ms	Frequency / Hz	Frequency² / Hz²
40	148	6.8	46.2
80	100	10.0	100.0
120			
160	72	13.9	193.0
200	64	15.6	243.0

The student checked the frequency of the oscillator using an oscilloscope. The trace for the 120 g reading is shown in **Figure 7**.

Figure 7

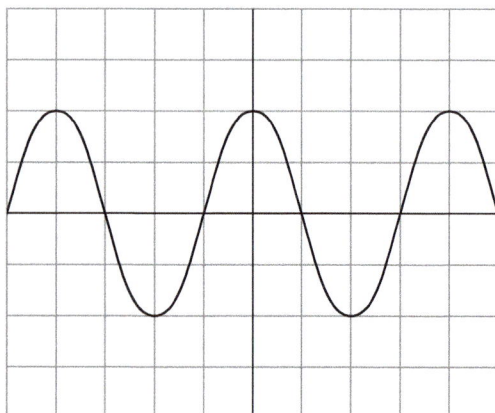

08.1 The oscilloscope is set to 20 ms per division.

Complete **Table 1** by using **Figure 7** to determine the time period for a mass of 120 g. **[2 marks]**

08.2 Show that the equation for the first harmonic predicts that $f^2 \propto m$. **[3 marks]**

08.3 **Figure 8** shows the graph of f^2 against m.

Plot the missing data point and draw a line of best fit.

Calculate the gradient of the graph. **[4 marks]**

> **!** **Exam tip**
>
> Always draw clear large triangles on the graph and show your working when calculating a gradient.

Figure 8

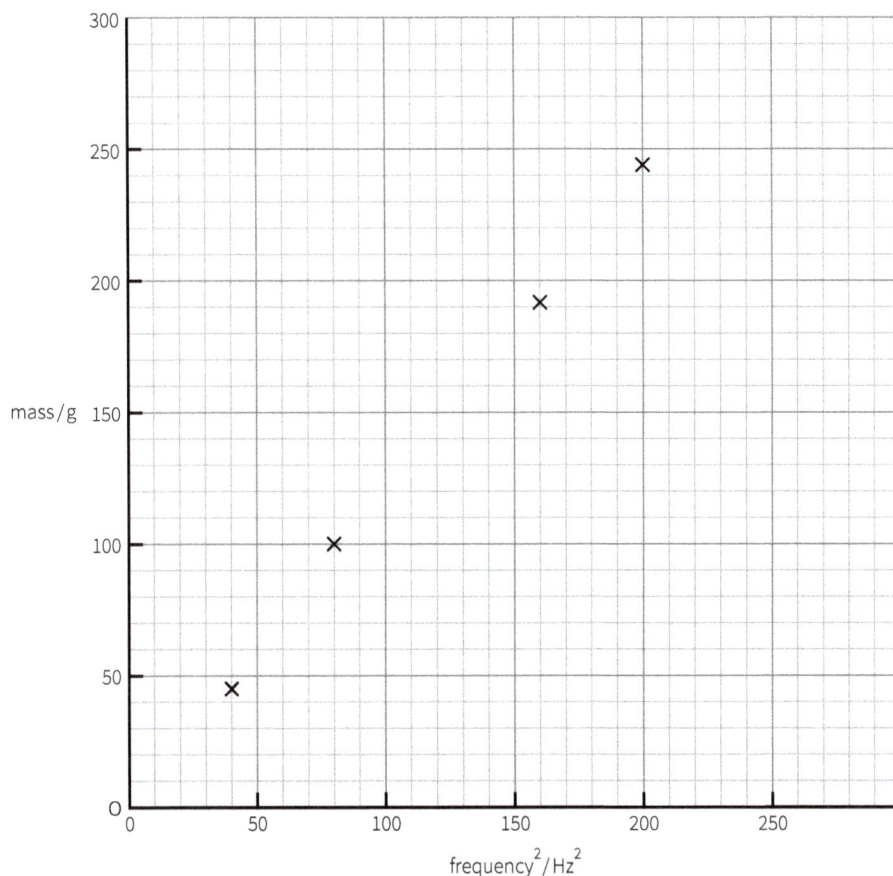

08.4 Calculate μ in $g\,m^{-1}$, using the value for the gradient from **08.3**.
The length of the string is 1.00 m. **[2 marks]**

08.5 The student measures the mass of the 1.00 m length of string.
The mass is 1.7 g, giving a value of $\mu = 1.7\,g\,m^{-1}$.

Calculate the absolute uncertainty in the value of μ. **[3 marks]**

08.6 Calculate the percentage difference between the actual and experimental value of μ. **[1 mark]**

Knowledge

5 Refraction, diffraction, and interference

Refraction of light

Refraction is the change of direction that occurs when light passes at an angle across a boundary between two transparent substances.

Visible light and other waves in the electromagnetic spectrum travel at a speed of $3.00 \times 10^8 \text{ m s}^{-1}$ in a vacuum. Light travels slower in all other substances.

The **refractive index** n of a substance is the ratio of the speed of light in a vacuum c and the speed of light in that substance c_s:

$$n = \frac{\text{speed of light in a vacuum}}{\text{speed of light in substance}} = \frac{c}{c_s}$$

The refractive index of air is approximately 1.

The greater the refractive index of a substance, the more slowly light travels through it, and the higher its **optical density**.

When light travels from one substance into another, the **angle of incidence**, **angle of refraction**, and refractive index of each substance are related by **Snell's law**:

$$n_1 \sin \theta_1 = n_2 \sin \theta_2$$

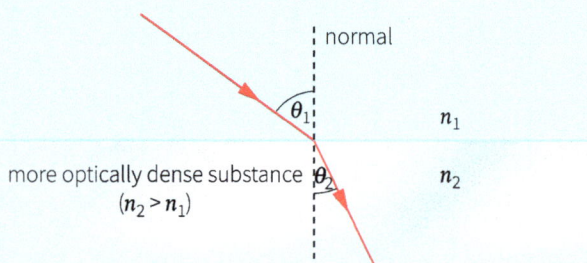

Rays of light travelling from:

- a less optically-dense substance into a more optically-dense one, refract towards the normal
- a more optically-dense substance into a less optically-dense one, refract away from the normal.

Total internal reflection and optical fibres

Total internal reflection occurs when light is travelling from an optically-dense material to a less optically-dense material, and the angle of incidence is greater than the critical angle.

The **critical angle** θ_c is the angle at which the angle of refraction is 90° for light going from an optically-dense material to a less optically-dense material:

$$\sin \theta_c = \frac{n_2}{n_1}$$

where $n_2 < n_1$

Optical fibres make use of total internal reflection to transmit information encoded in light over long distances and a round corners.

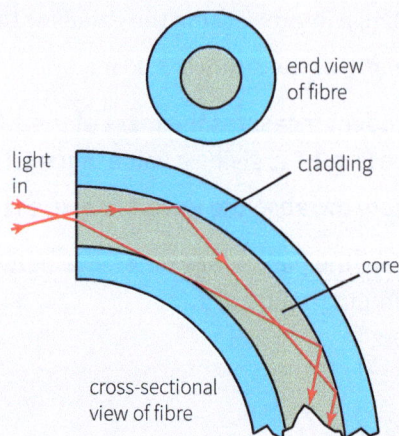

The core of an optical fibre is made of glass with a higher refractive index than the cladding. The cladding protects the core from being damaged and prevents light from leaking out.

Signal degradation in optical fibres

Signal degradation in an optical fibre can cause information to be lost. It can occur due to absorption or pulse broadening.

Absorption

Some of the light is absorbed by the optical fibre every time it reflects, resulting in a decrease in the amplitude of the signal.

Pulse broadening

The signal pulse gets broader as it travels along the fibre, reducing the quality of the signal and leading to loss of information if the pulses overlap.

It can be caused by modal or material dispersion.

Modal dispersion

If the core is wide enough, rays of light entering it at different angles will take different paths through it and travel different distances. Modal dispersion can be reduced by using an optical fibre with a narrower core, known as a single-mode fibre.

Material dispersion

If white light is used for the signal, the different wavelengths in it travel at slightly different speeds through the fibre, causing them to separate as they travel along the fibre. Material dispersion can be eliminated by using monochromatic light.

Interference

Waves are coherent when they have the same frequency and a constant phase difference (phase difference is always the same).

Constructive interference occurs when two waves *in phase* superpose to give a wave with a larger amplitude.

Destructive interference occurs when two waves *out of phase* superpose to cancel each other out.

An **interference pattern** occurs when waves with a constant phase difference superpose, producing cancellation and reinforcement at fixed positions.

The **path difference** between two waves is the difference in the distance travelled by the waves to the point at which they superpose.

- If the path difference is a whole number of wavelengths $n\lambda$, the waves always arrive in phase at that point and constructive interference occurs, producing a maximum.

- If the path difference is $\frac{1}{2}$, $1 + \frac{1}{2}$, $2 + \frac{1}{2}$ ..., or $(n + \frac{1}{2})\lambda$ then destructive interference occurs, producing a minimum.

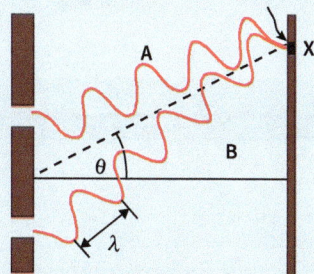

distance travelled by wave **A** to **X** = 4.5λ
distance travelled by wave **B** to **X** = 5.5λ
path difference = 5.5 − 4.5 = 1.0λ
∴ constructive interference occurs at **X**

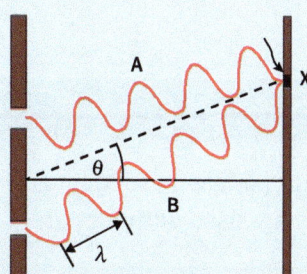

distance travelled by wave **A** to **X** = 4.5λ
distance travelled by wave **B** to **X** = 5.5λ
path difference = 5.0 − 4.5 = 0.5λ
∴ distructive interference occurs at **X**

Two-source interference occurs when coherent waves from two sources superpose to produce an interference pattern.

Young's double slit experiment

Young's double slit experiment provides evidence for the wave nature of light because the resulting fringe pattern is due to diffraction and interference, which cannot happen with particles.

A single beam of light is split into two coherent beams as it passes through the closely spaced double slits.

This produces a pattern of dark and bright fringes on a screen.

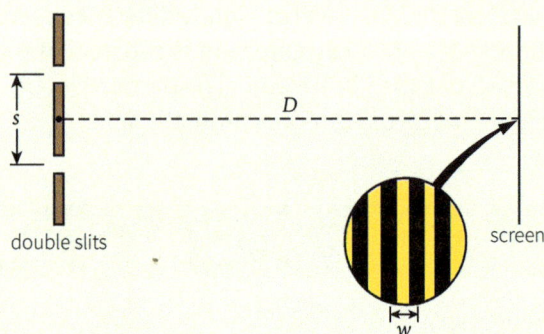

- Bright fringes form where the path difference between light from each slit is $n\lambda$, so the waves arrive in phase and superpose constructively.
- Dark fringes form where the path difference is $(n+\frac{1}{2})\lambda$, so the waves arrive out of phase and superpose destructively.

Light sources

Light sources include a light bulb shone through a narrow single slit to illuminate the double slits, or a laser beam shone directly through the double slits.

Fringe separation

Fringe separation w is the distance between the centres of two adjacent maxima or minima (bright or dark fringes). It is related to the wavelength of the light λ, the distance from the screen D, and the slit separation s by the equation:

$$w = \frac{\lambda D}{s}$$

Fringe separation also depends on the colour of light used. It is greater for red light than blue light because red light has a longer wavelength.

The image below shows the fringe pattern produced by white light. Each component colour of white light produces its own fringe pattern with blue light on the inner edge and red light on the outer edge. The central fringe is white because every colour contributes at the centre of the pattern.

Lasers

Lasers are a source of monochromatic light (single frequency) and coherent light waves. Lasers can cause harm if they are used without care. To use a laser safely in the laboratory:

- turn it off when not in use
- never shine it at a person
- avoid shining it at reflective surfaces
- display a warning sign on lab door
- wear laser safety goggles.

Diffraction

Diffraction is the spreading of waves when they pass through a gap or round an obstacle.

Light passing through a narrow **single slit** is diffracted, producing an interference pattern on a screen.

For monochromatic light:

- the central fringe is much brighter and is twice as wide as the outer fringes
- the outer fringes are the same width and their brightness decreases with distance from the centre.

For white light, the different wavelengths are diffracted by different amounts, resulting in:

- a bright white central maximum
- less bright fringes that are spectra, with blue light on the inner edge and red light on the outer edge.

Diffraction grating 🧪

Monochromatic light passing through a diffraction grating will produce an interference pattern where the maxima are much sharper than for a single or double slit. The central maximum is the brightest and is called the **zero order maximum**.

As light from the source is diffracted through each slit of a grating, the diffracted waves constructively superpose in certain directions only, and destructively superpose in all other directions.

When white light passes through a diffraction grating, the central maximum is white; every other order of interference is a spectrum.

Diffraction gratings are used in **spectrometers** to study the spectrum of light from any source. Spectrometers are used in chemical analysis and to study the composition and movement of stars.

Deriving the diffraction grating formula 🧪

Each maximum is formed where the light from each slit arrives exactly in phase. The path difference for:

- the central maximum $= 0\lambda$
- the first order $= 1\lambda$
- the second order $= 2\lambda$

and so on, where λ is the wavelength of the light waves.

The figure shows the direction of the wavefronts from adjacent slits, P and Q, which contribute to the nth order maximum.

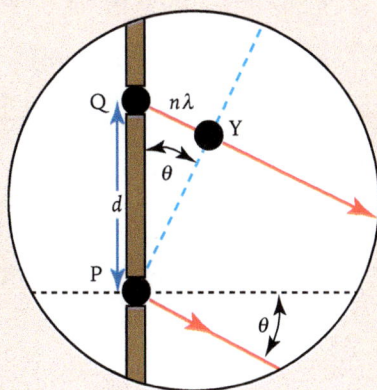

- The path difference QY between the wavefronts from P and Q is a whole number of wavelengths, so $QY = n\lambda$.
- The beams of light travel away at an angle θ from the normal to the grating, so angle QPY is also θ.
- The slit separation $PQ = d$.
- For the triangle PQY, $\sin\theta = \dfrac{n\lambda}{d}$

Rearranging this results in the diffraction grating formula:

$$d\sin\theta = n\lambda$$

The maximum value of n occurs when $\sin\theta = 1$. So, for any wavelength of light using a grating with slit separation, the maximum number of orders produced is given by d/λ is rounded down to the nearest whole number.

Retrieval

Learn the answers to the questions below, then cover the answers column with a piece of paper and write as many as you can. Check and repeat.

	Questions	Answers
1	What is the refractive index of a substance?	$n = \dfrac{\text{speed of light in a vacuum}}{\text{speed of light in substance}} = \dfrac{c}{c_s}$
2	Give the refractive index of air.	approximately 1
3	What is an optically-dense substance?	substance with a high refractive index, in which light travels slowly
4	State Snell's law.	$n_1 \sin\theta_1 = n_2 \sin\theta_2$
5	Which direction do rays of light bend when they travel from a less optically-dense medium to a more optically-dense one?	towards the normal
6	What is the critical angle θ_c in total internal reflection?	angle at which the angle of refraction is 90°
7	What does the cladding of an optical fibre do?	allows total internal reflection to take place; protects the core from being damaged; prevents light from leaking out of it
8	Give two ways in which signal degradation in an optical fibre occurs.	absorption, pulse broadening
9	What is the consequence of pulse broadening in an optical fibre?	reduces the quality of signal and can lead to loss of information if pulses overlap
10	Define coherent waves.	waves with the same frequency and a constant phase difference
11	What is an interference pattern?	when waves with a constant phase difference superpose to produce cancellation and reinforcement at fixed positions
12	What is the path difference between two waves if they constructively interfere at a point?	a whole number of wavelengths $n\lambda$
13	When does two-source interference occur?	when coherent waves from two sources superpose to produce an interference pattern
14	In the double slit experiment, what is the path difference for light from each slit where a bright fringe is formed?	$n\lambda$
15	What is the fringe separation in a fringe pattern?	distance between the centres of two adjacent maxima or minima
16	Why are the fringes produced by red light farther apart than those produced by blue light in the double slit experiment?	red light has a longer wavelength
17	State the diffraction grating formula.	$d\sin\theta = n\lambda$

Put paper here

🔬 Practical skills

Practise your practical skills using the worked example and practice questions below.

Diffraction grating interference

The angle at which the nth order maximum occurs for a diffraction grating can be calculated using:

$$d \sin \theta = n\lambda$$

where:

- d = slit separation (m)
- θ = angle between normal to grating and nth order beam (°)
- n = number of orders of maximum
- λ = wavelength (m)

If the number of slits per metre on a grating is N,

d can be found using $d = \dfrac{1}{N}$

If N is the number of slits per millimetre, d can be calculated using $d = 10^{-3} \times \dfrac{1}{N}$

For any λ and grating with slit separation d, the maximum number of orders produced $= \dfrac{d}{\lambda}$ rounded down.

Worked example

Question

A laser beam is shone through a diffraction grating with 300 lines per millimetre. When a screen is placed 1.5 m from the grating, the distance between the two first-order maxima is 0.6 m.

Calculate the wavelength of the laser light.

Answer

Step 1: write equation: $d \sin \theta = n\lambda$

Step 2: sketch diagram

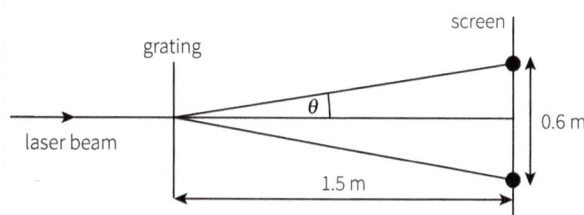

Step 3: match info in question to equation

$$d = 10^{-3} \times \frac{1}{N} = \frac{1}{300} \times 10^{-3}$$

$$\tan \theta = \frac{0.3}{1.5}, \text{ so } \theta = \tan^{-1} \frac{0.3}{1.5} = 11.3°$$

$$n = 1; \lambda = ?$$

Step 4: make λ subject $\quad \lambda = \dfrac{d \sin \theta}{n}$

$$\lambda = \frac{\left(\dfrac{10^{-3}}{300}\right) \times \sin 11.3}{1}$$

$$\lambda = 6.5 \times 10^{-7} \text{ m (650 nm)}$$

Practice

1 A laser beam of wavelength 520 nm is shone through a diffraction grating with 600 lines per millimetre.

Calculate:

a the angle of diffraction for the first-order maxima

b the angle of diffraction for the second-order maxima

c the number of diffracted orders produced.

2 A laser beam of wavelength 650 nm is shone through a diffraction grating. The first-order maximum occurs at an angle of 40.5° to the zero-order beam.

Calculate the number of lines per millimetre on the grating.

01 A student uses Young's double-slits and a screen to determine the wavelength of a laser. The measurements are shown in **Figure 1**.

Figure 1

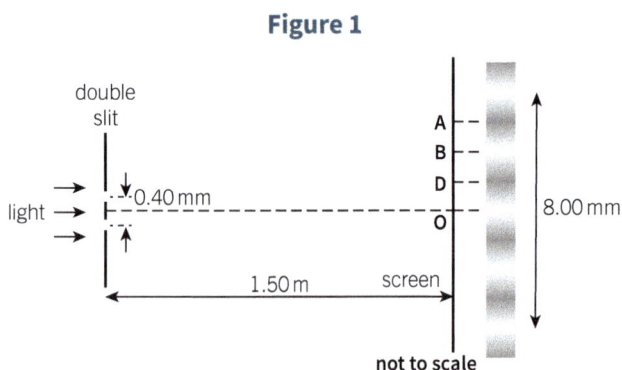

not to scale

Synoptic link

3.1.2

A laser is a source of coherent and monochromatic light.

01.1 Describe what is meant by a coherent light source. **[1 mark]**

_____

_____

01.2 Describe what is meant by a monochromatic light source. **[1 mark]**

_____

_____

01.3 Suggest **one** safety precaution you should take when using a laser.

[1 mark]

_____

_____

> **Exam tip**
>
> Learn definitions of key words carefully; there are very few simple recall questions. Make sure you get the marks.

01.4 Determine the wavelength of the laser by taking readings from **Figure 1**. **[3 marks]**

wavelength = _____ m

01.5 Calculate the percentage uncertainty in your answer for **01.4**.

[3 marks]

> **Exam tip**
>
> Look at the number of decimal places to determine the resolution of the measuring instrument. For example, 0.40 mm means the uncertainty is ± 0.01 mm.

percentage uncertainty = _____ %

01.6 The laser is replaced with a laser of longer wavelength.
State and explain how the spacing of the maxima would change.

[2 marks]

_____

_____

_____

02 A ray of green light is incident on side **JK** of a glass block with
refractive index of 1.5, as shown in **Figure 2**.

Figure 2

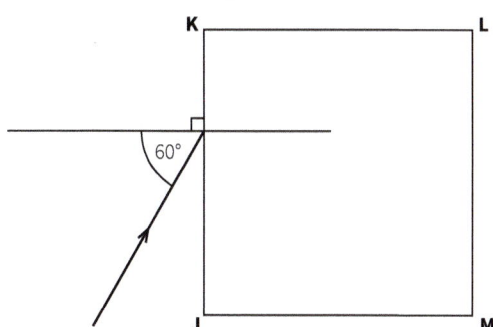

02.1 Calculate the angle of refraction as the ray of light enters the glass.

[2 marks]

angle of refraction = _____ °

02.2 Calculate the critical angle for the glass block. [1 mark]

critical angle = _____ °

02.3 The refracted ray then strikes surface **KL**.

Determine whether the ray will be internally reflected or
refracted at this surface. [2 marks]

_____

_____

_____

> **! Exam tip**
>
> It may help to sketch the ray
> on the diagram to determine
> the angle.

02.4 A thin film of liquid with a refractive index of 1.4 is coated on side **KL**. Calculate the new critical angle. **[1 mark]**

critical angle = _____ °

03 Two speakers are set up 10 m apart on a stage at a festival concert. The speakers produce a test sound of frequency 1500 Hz.

03.1 A sound engineer walks along a row of seats parallel to the stage and 20 m in front of the speakers. As they walk, they hear changes in the intensity of the sound.

Explain the sound engineer's observation. **[3 marks]**

_____

_____

_____

Synoptic link

3.3.1.1

Exam tip

Make sure you include key words such as superposition and path difference in your explanation.

03.2 Calculate the wavelength in m of the sound produced by the speakers.
Speed of sound in air = 340 m s^{-1} **[1 mark]**

wavelength = _____ m

03.3 Calculate the distance in m between the maxima. **[1 mark]**

distance = _____ m

03.4 Describe what the sound engineer would hear if the frequency doubled. **[2 marks]**

_____

_____

04 A diffraction grating can be used to determine the wavelength of a laser.

04.1 Describe how you could use a diffraction grating to measure the wavelength of a laser.

Include in your description the measuring instruments needed and how you would ensure the results were as accurate as possible. **[4 marks]**

Exam tip

A clearly labelled diagram may help you here.

_____

_____

_____

_____

04.2 A diffraction grating is labelled 330 lines per mm.

Calculate the grating spacing in m. **[1 mark]**

grating spacing = _____ m

04.3 Using the 330 lines per mm diffraction grating, a student calculates θ_1 to be 12.5°.

Determine the wavelength of the laser they used in nm. **[2 marks]**

wavelength = _____ m

04.4 Describe and explain how the diffraction pattern would change if the student swapped the laser for a white light source. **[3 marks]**

_____

_____

_____

05 **Figure 3** shows the path of a laser beam through a glass block.

The refractive index of the glass is 1.6.

Figure 3

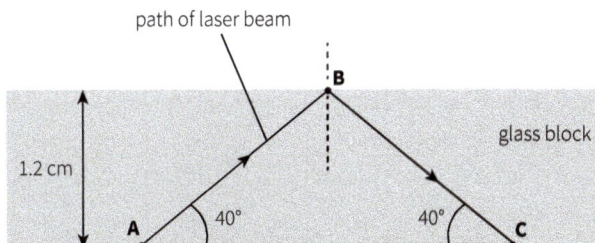

05.1 Determine the critical angle in degrees for the glass block.

Explain why the light is totally internally reflected at point **B**.

[3 marks]

05.2 Calculate the velocity in m s^{-1} of the laser beam inside the glass block. **[1 mark]**

05.3 Calculate the time taken in s for light to travel from **A** to **B** to **C**. **[2 marks]**

05.4 Cladding is added to the glass block. The cladding has a refractive index of 1.4.

Calculate the new critical angle in degrees at the glass–cladding boundary. **[1 mark]**

> **! Exam tip**
>
> What will be the speed of light in the glass?

05.5 Explain why optical fibre needs to have cladding. **[2 marks]**

05.6 **Figure 4** shows a pulse of monochromatic light input into a long length of optical fibre. The shape of the pulse as it leaves the fibre is also shown.

Explain why the amplitude of the pulse has decreased, and why the pulse is broader. **[2 marks]**

Figure 4

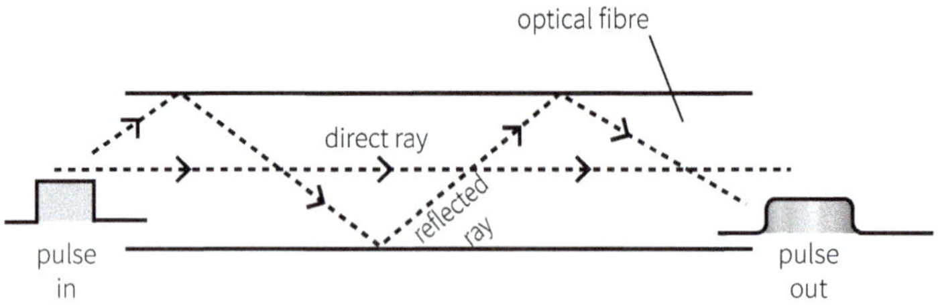

06 A single-slit diffraction pattern is produced on a screen using a monochromatic light source.

Figure 5 shows the intensity of the central maximum.

Figure 5

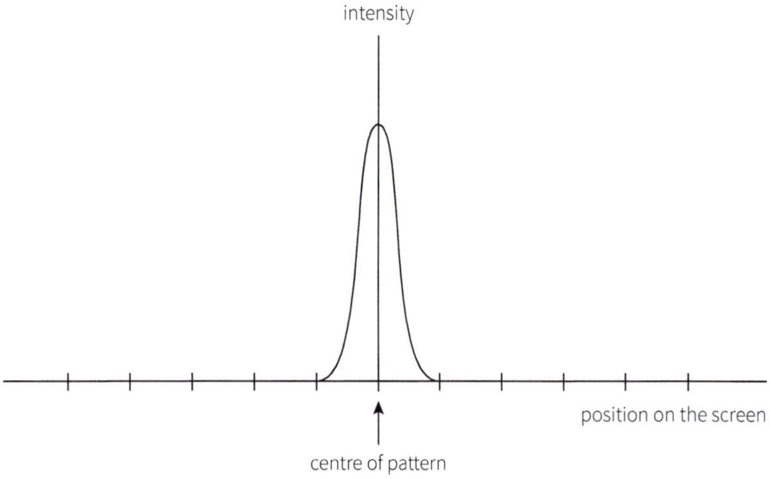

06.1 Sketch on **Figure 5** to show how the intensity varies to the left of the central maximum. **[2 marks]**

06.2 Describe how the appearance of the central maximum would change if a shorter wavelength was used. **[1 mark]**

06.3 Describe how the appearance of the central maximum would change if a narrower slit was used. **[1 mark]**

06.4 The experiment is repeated using a white light source.

Describe how the pattern would change using this white light source. **[3 marks]**

> **! Exam tip**
>
> Sketch carefully. Can you think of two changes to the appearance of the intensity as you move to the left?

07 **Figure 6** shows a layer of oil floating on water. A ray of monochromatic light incident upon the surface is both reflected and refracted.

Figure 6

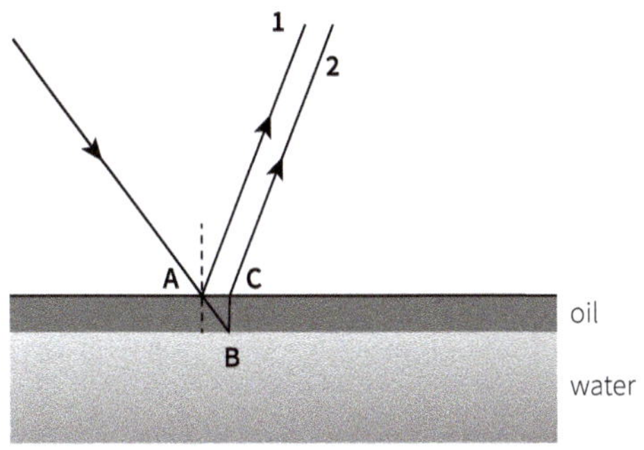

oil

water

07.1 Describe what is meant by monochromatic light. **[1 mark]**

07.2 Determine the critical angle in degrees at the oil–water boundary. refractive index of oil = 1.45; refractive index of water = 1.30 **[2 marks]**

07.3 When the oil layer is observed from different angles, dark and light bands appear.

Explain how this could occur by considering the difference in the paths of the ray when it splits at **A**. **[3 marks]**

07.4 Bright colours often appear when light is reflected from a thin layer of oil. The colours that appear brightest depend on the viewing angle.

Suggest a reason for this phenomenon. **[2 marks]**

Exam tip

In an unfamiliar situation, think about other situations in waves when bright and dark fringes appear, such as superposition. Could that explain this?

08 **Figure 7** shows a fine beam of electrons fired at a thin slice of graphite. This results in a diffraction pattern on the screen.

Figure 7

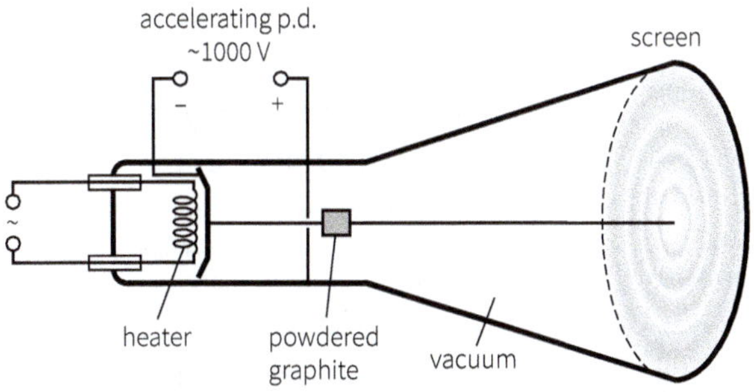

accelerating p.d.
~1000 V

screen

heater   powdered graphite   vacuum

Synoptic links

3.2.2.4   3.4.1.8   3.5.1.1

08.1 Suggest why the demonstration in **Figure 7** is important. **[1 mark]**

08.2 Explain why the bright circles appear on the screen. **[3 marks]**

08.3 Calculate the kinetic energy in J of an electron accelerated by 1000 V. **[1 mark]**

08.4 Show that the speed of the electron is about $1.9 \times 10^7 \, \text{m s}^{-1}$. **[2 marks]**

08.5 Calculate the de Broglie wavelength in m of the electron. **[1 mark]**

08.6 Explain how increasing the voltage would change the appearance of the diffraction pattern. **[4 marks]**

Exam tip

Quote your answer to one more significant figure than you must prove.

6 Force, energy, and momentum 1

Scalars and vectors

Scalar quantities only have magnitude, and include energy, temperature, mass, time, distance, and speed.

Vector quantities have direction *and* magnitude, and include force, weight, displacement, velocity, acceleration, and momentum.

Vector addition

By calculation

To add two vectors, V_1 and V_2, at right angles to each other:

1 sketch a right-angled triangle formed by adding V_1 and V_2 (the resultant is the hypotenuse of the triangle)

2 calculate magnitude of resultant using Pythagoras' theorem:

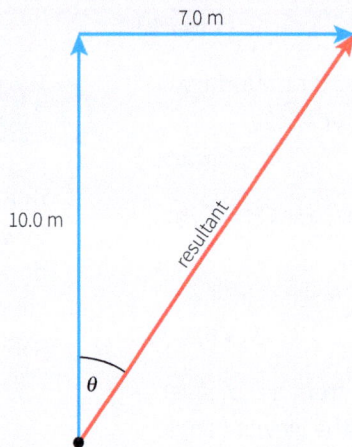

$$V_R = \sqrt{(V_1^2 + V_2^2)}$$

3 calculate angle θ between V_1 and resultant using:

$$\tan\theta = \frac{V_1}{V_2}$$

By scale drawing

To add two or more vectors:

1 choose an appropriate scale

2 use a protractor and a ruler to draw the first vector to scale and in the correct direction

3 draw the next vector starting at the end of the first vector

4 repeat with any other vectors

5 draw a line from the start of the first vector to the end of the last vector (= resultant vector)

6 measure angle of resultant

7 measure length of the final line and use scale to determine magnitude of resultant vector

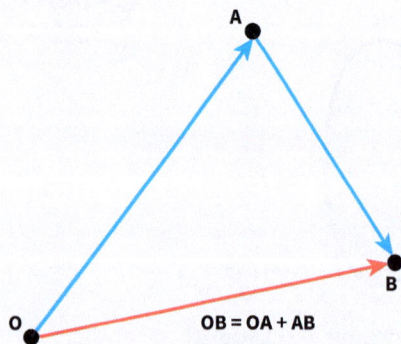

Resolving vectors

All vectors can be resolved into two components at right angles to each other.

Resolving a vector V into horizontal (V_x) and vertical components (V_y), gives:

$$V_x = V\cos\theta$$

and

$$V_y = V\sin\theta$$

To resolve a vector:

1 sketch right-angled triangle where the vector is the hypotenuse and the perpendicular components form the other two sides

2 apply trigonometry to obtain the value for each component.

(a) by vector mathematics
example: block being pushed on rough surface

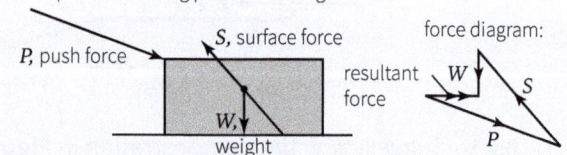

(b) by components
example: block sliding down a smooth slope

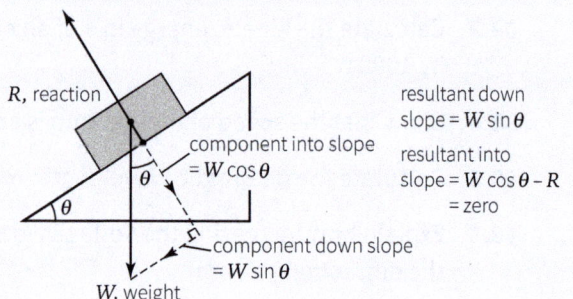

resultant down slope $= W\sin\theta$

resultant into slope $= W\cos\theta - R$ $=$ zero

Free-body force diagrams

A free-body force diagram shows the forces acting on a single object.

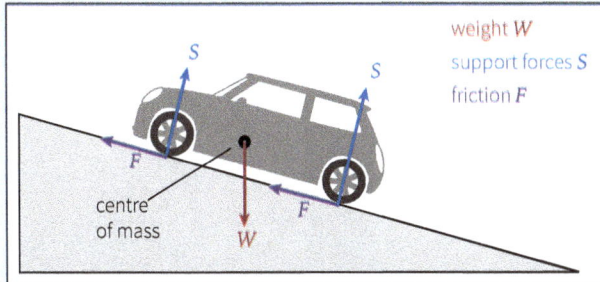

weight W
support forces S
friction F

centre of mass

The arrows start at the point where the forces act, and represent the direction and size of the forces. All forces act in the same plane.

Moments

The moment of a force about a point is defined as *force × perpendicular distance from the point to the line of action of the force*. It is measured in newton metres (Nm).

A couple is a pair of equal and opposite coplanar forces acting on a body, but not along the same line, so they produce a turning effect.

moment of a couple = $F \times d$

where: F = magnitude of one of the forces
d = perpendicular distance between the lines of action of the forces

The principle of moments states that, for an object in equilibrium, the sum of the clockwise moments about any point = the sum of the anticlockwise moments about that point.

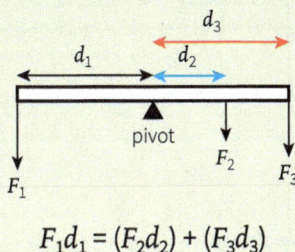

$$F_1d_1 = (F_2d_2) + (F_3d_3)$$

Conditions for equilibrium

An object in **equilibrium** is either at rest or moving with a constant velocity.

For an object to be in equilibrium:

- the resultant force on it must be zero
- the total clockwise moment about any point on the object must equal the total anticlockwise moment about that point.

If an object in equilibrium has:

- two forces acting on it, the force vectors will be equal in size and act in opposite directions
- three forces acting on it, adding the vectors will form a closed triangle when placed nose to tail

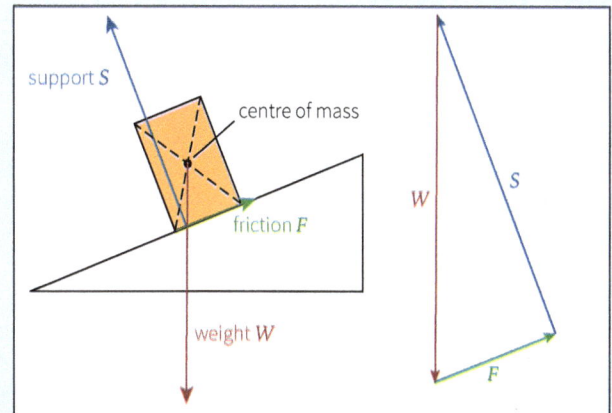

support S
centre of mass
friction F
weight W

- n forces acting on it, adding the vectors will result in a closed polygon of forces with n sides.

Centre of mass

The centre of mass of an object is the point through which a single force has no turning effect when applied – it is the point at which the weight of a body acts.

The centre of mass of a regular solid of uniform density is at its geometric centre. An object will topple over if the line of action of its weight is outside of its base.

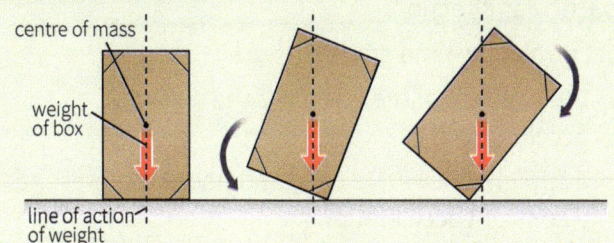

centre of mass
weight of box
line of action of weight

Displacement–time graphs

A **distance–time** graph shows how the total distance travelled by an object changes with time.

A **displacement–time** graph shows the same, but from a particular point and in a fixed direction. A negative displacement indicates a distance travelled in the opposite direction.

Distance–time and displacement–time graphs of an object thrown in the air

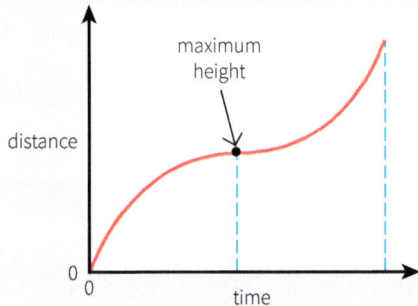

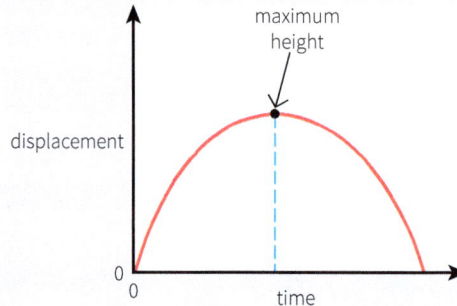

On a displacement–time graph for an object:

- the gradient of the line gives its velocity
- a straight line with a constant gradient represents steady speed
- a curved line represents acceleration.

For a curved displacement–time graph, the instantaneous velocity at a particular time can be found by drawing a tangent to the curve and calculating its gradient.

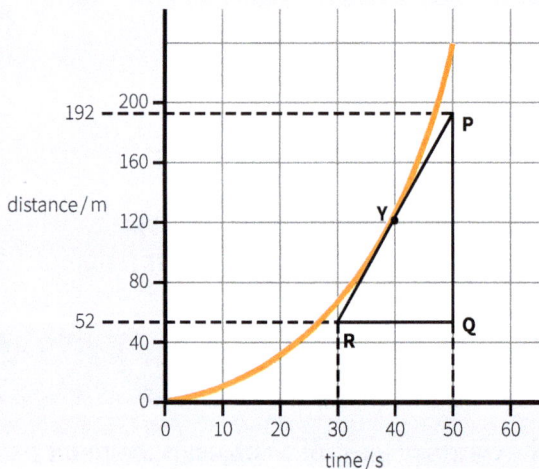

$$\text{speed at } \mathbf{Y} = \frac{\mathbf{PQ}}{\mathbf{QR}} = \frac{192-52}{50-30}$$

$$= 7 \, \text{ms}^{-1}$$

Velocity–time graphs

A **speed–time** graph shows how the speed of an object changes with time.

A **velocity–time** graph shows the same, but in a particular direction.

On a velocity–time graph:

- the gradient of the line gives the object's acceleration
- a straight line with a constant gradient represents a constant acceleration
- a horizontal line represents a steady speed.

Velocity time graph of a bouncing ball

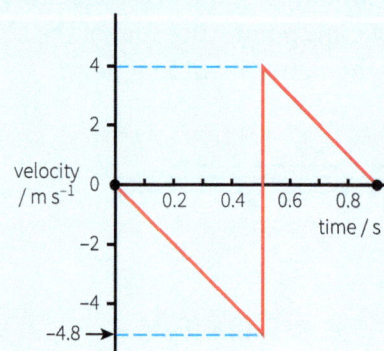

Interpreting velocity–time graphs

A curved line on a velocity–time graph represents changing acceleration. The instantaneous acceleration at a particular time can be found by drawing a tangent to the curve and calculating its gradient.

The area under a velocity–time graph represents the displacement of the object.

To estimate the area of curved sections:

1 calculate distance a small square on the graph represents
2 count squares under curved part of line
3 number of squares × area per square

each square = distance of $2\,m\,s^{-1} \times 10\,s = 20\,m$

area under curve ≈ 29.5 squares

distance ≈ $29.5 \times 20 = 590\,m$

Acceleration–time graphs

An **acceleration–time** graph shows how the acceleration of an object changes with time.

On an acceleration–time graph for an object:

- the area under the line represents the change in velocity of the object
- a horizontal line represents uniform acceleration
- a value of zero indicates the object is moving with constant velocity.

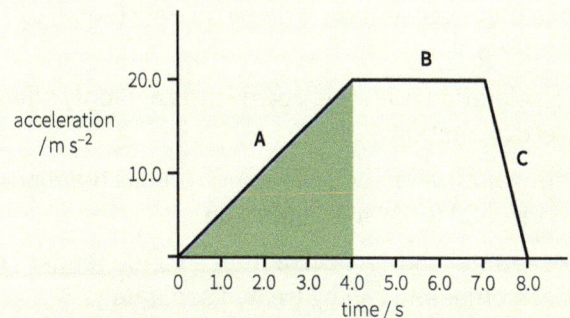

A: acceleration is constantly increasing, rate of change of velocity is increasing

B: object at constant acceleration of $20\,m\,s^{-2}$ velocity still changing all the time

C: rate of acceleration is decreasing, but velocity continues to increase

Shaded area: change in velocity $= \frac{1}{2} \times 4 \times 20 = 40\,m\,s^{-1}$

Motion along a straight line

velocity v = change in displacement per unit time $= \dfrac{\Delta s}{\Delta t}$

acceleration a = change in velocity per unit time $= \dfrac{\Delta v}{\Delta t}$

The *suvat* equations describe motion on a straight line for **uniform (constant) acceleration**:

$v = u + at$ $\qquad s = ut + \frac{1}{2}at^2$

$s = \left(\dfrac{u+v}{2}\right)t$ $\qquad v^2 = u^2 + 2as$

s = displacement (m)
u = initial velocity (m s⁻¹)
v = final velocity (m s⁻¹)
a = acceleration in initial velocity (m s⁻²)
t = time (s)

For motion along a straight line, a positive sign (+) shows forwards or upwards movement, and a negative sign (−) shows backwards or downwards movement.

Acceleration due to gravity

An object in **free fall** only has its weight (the force of gravity) acting on it. All objects in free fall accelerate at the same rate, regardless of their mass. The acceleration due to gravity g is $9.81\,\mathrm{m\,s^{-2}}$ on and near the surface of the Earth.

Projectile motion

For any projectile:

- the only acceleration is due to gravity (equal to $9.81\,\mathrm{m\,s^{-2}}$)
- acceleration is always downwards and only affects vertical motion
- horizontal velocity is constant because there is no horizontal force or acceleration
- horizontal and vertical motion are independent of each other and can be treated separately.

Problems involving a projectile can be solved using the *suvat* equations and using $9.81\,\mathrm{m\,s^{-2}}$ g for acceleration a.

For an object that is only moving vertically, the upwards direction is positive (+), so the acceleration is $-9.81\,\mathrm{m\,s^{-2}}$ because gravity acts downwards.

For an object dropped vertically, the initial velocity u is always zero.

For an object launched horizontally with an initial velocity of u:

- initial vertical velocity u_y is zero
- the object accelerates downwards at $9.81\,\mathrm{m\,s^{-2}}$
- vertical distance travelled s_y in time t is given by

$$s_y = \frac{1}{2}gt^2$$

- horizontal distance s_x travelled in time t is given by

$$s_x = ut$$

object projected horizontally at initial velocity u

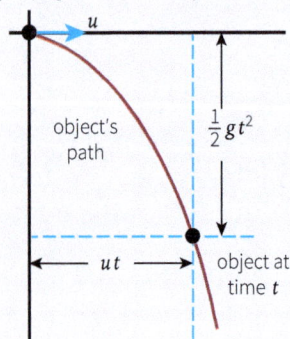

To solve problems involving objects launched with an initial velocity u at an angle θ to the horizontal:

1. resolve initial velocity of projectile into horizontal and vertical components
2. apply equations of motion for uniform acceleration separately to horizontal and vertical components of the object's motion.

For an object with an initial velocity of u at an angle θ to the horizontal:

- vertical velocity $= u\sin\theta$
- horizontal velocity $= u\cos\theta$

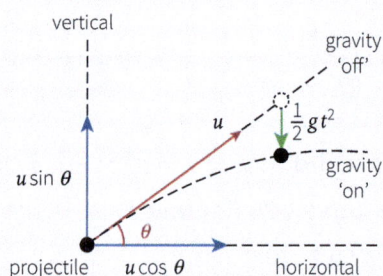

Any problem involving an object experiencing uniform acceleration in a different direction to its velocity can be treated like a projectile motion problem. For example:

- a ball rolling down an inclined plane
- an electron beam on a parabolic path.

Friction

Friction is the force that acts to oppose the relative motion between two surfaces in contact with each other, or an object moving through a fluid.

Drag is the general name for the frictional forces acting on an object when it moves through a fluid. Drag:

- acts in the opposite direction to an object's movement
- increases with increasing viscosity of a fluid
- increases with increasing speed of an object
- depends on the shape and size of an object
- causes a moving object to slow down and stop if there is no force propelling it.

Streamlining an object reduces the drag acting on it.

Air resistance is the drag acting on an object when moving through air. If air resistance is acting on a projectile:

- its horizontal speed, range, and maximum height will be reduced
- its downward trajectory will be steeper than its upward trajectory.

Lift

A lift force can arise when an object moves through a fluid. It depends on the shape of the object and the angle at which the object moves through the fluid.

Lift acts at right angles to the direction of fluid flow over the object. It is a result of **Newton's third law** – the object exerts a force on the fluid in one direction, so the fluid exerts a force on the object in the opposite direction.

Terminal speed

If an object is travelling at a steady speed in a fluid, the driving force and drag will be equal in size but act in opposite directions.

If an object in a fluid has a constant driving force F, the resultant force on it at any instant = F – drag.

For an object falling through a fluid, two forces are acting – its weight due to gravity and the drag due to the fluid:

- weight remains constant
- drag force starts small but increases as it speeds up
- resultant force will decrease as drag force increases
- acceleration will decrease as it falls
- if it falls for a long enough time, the object will reach **terminal velocity**.

Terminal velocity

Terminal velocity is the steady speed a falling object reaches when the drag force D on it is equal to its weight, so the resultant force is zero.

The figure shows the velocity–time graph for a shuttlecock falling through air.

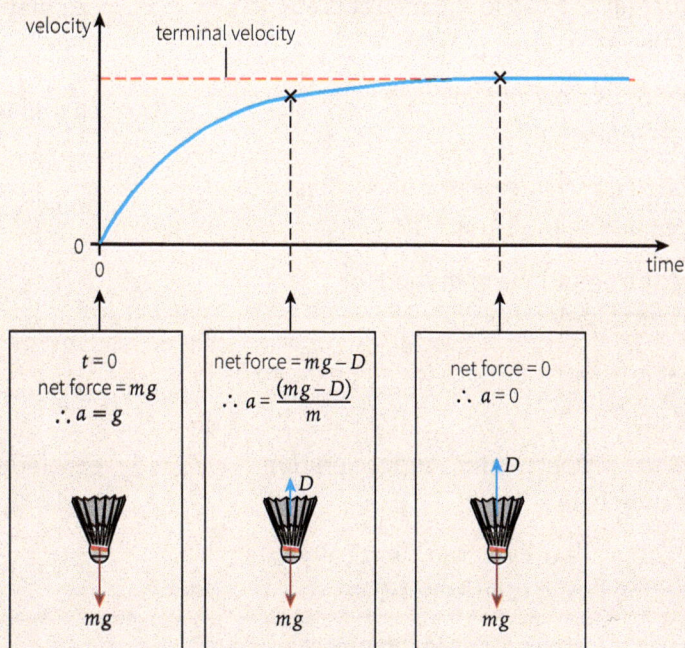

At $t=0$, net force = mg ∴ $a = g$

net force = $mg - D$ ∴ $a = \dfrac{(mg-D)}{m}$

net force = 0 ∴ $a = 0$

The maximum speed of a road vehicle is reached when the maximum driving force provided by its engine is equal to the sum of all the resistive forces acting on it – the resultant force is zero.

For answers and more practice questions visit www.oxfordrevise.com/scienceanswers
Even more practice and interactive revision quizzes are available on **kerboodle**
6 Knowledge 65

Retrieval

Learn the answers to the questions below, then cover the answers column with a piece of paper and write as many as you can. Check and repeat.

Questions | Answers

1 Define scalar and vector quantities.

scalar quantities have magnitude only; vector quantities have magnitude and direction

2 How is a vector resolved?

split it into two components at right angles to each other

3 What do the arrows on a free-body diagram represent?

direction and size of the forces acting on an object

4 What is an object at equilibrium?

object at rest or moving with a constant velocity

5 Define the moment of a force about a point.

moment = force × perpendicular distance from the point to the line of action of the force

6 What does the principle of moment state?

sum of clockwise moments about any point = sum of anticlockwise moments about that point

7 What does a displacement–time graph show?

how distance travelled by an object from a fixed point and direction changes with time

8 How is steady speed represented on a displacement–time graph?

straight line with a constant gradient

9 How can instantaneous velocity at a particular time be found on a curved displacement–time graph?

drawing a tangent to the curve at that time and calculating its gradient

10 What does a curved line on a velocity–time graph represent?

changing acceleration

11 What does a value of zero on an object's acceleration–time graph indicate?

object is moving with constant velocity

12 Give the equations of motion for uniform acceleration.

$v = u + a\,t; s = \left(\dfrac{u+v}{2}\right)t; s = u\,t + \frac{1}{2}a\,t^2; v^2 = u^2 + 2\,a\,s$

13 What are the conditions for an object to be in free fall?

only has its weight (force of gravity) acting on it

14 Give the properties for the acceleration of a projectile.

acceleration is due to gravity, is always downwards, and only affects the vertical motion of the object

15 If an object is only moving vertically, what direction is regarded as positive?

upwards

16 What is the acceleration for an object launched horizontally with an initial velocity U?

accelerates downwards at $9.81\,\mathrm{m\,s^{-2}}$

17 What happens to the trajectory of a projectile if air resistance acts on it?

downward trajectory will be steeper than upward trajectory

Put paper here

Practical skills

Practise your practical skills using the worked example and practice questions below.

Determination of g by free-fall

To determine the acceleration due to gravity g, measure the time taken t for an object to fall different heights, then plot these to give a straight-line graph. If you plot h on the y-axis and t^2 on the x-axis, the gradient of the graph will be equal to $\frac{1}{2}g$.

Worked example

Question

A student measured the time taken for a ball bearing to fall through a height h. They repeated the measurement three times for each h and calculated mean t values.

h/m	t/s	t^2/s^2
1.18	0.482	0.232
1.00	0.447	0.200
0.84	0.409	0.167
0.74	0.386	0.149
0.62	0.353	0.125
0.54	0.325	0.106

Plot a graph of h against t^2 and use it to determine g.

Answer

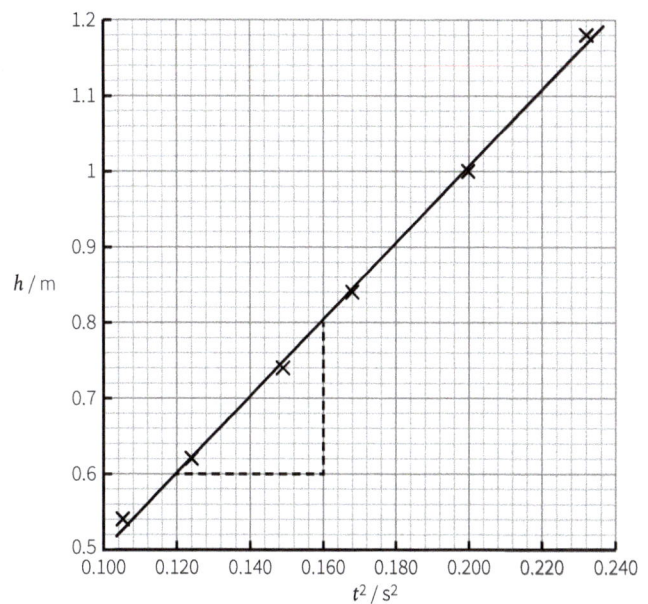

$$\text{gradient} = \frac{(0.8 - 0.6)\,\text{m}}{(0.160 - 0.120)\,\text{s}^2} = 5.10\,\text{m s}^{-2}$$

$$\text{gradient} = \frac{1}{2}g, \text{ so } g = 10\,\text{m s}^{-2}$$

Practice

1 A student measured the time taken for a tennis ball to fall different heights.

h/m	t_1/s	t_2/s	t_3/s
2.0	0.63	0.64	0.64
1.6	0.57	0.55	0.58
1.2	0.49	0.51	0.48
1.0	0.46	0.46	0.47
0.8	0.38	0.42	0.41
0.6	0.38	0.39	0.36

a Calculate the mean t value for each h.

b Plot a graph to calculate g.

2 The student repeats the experiment, but uses A4 sheets of paper instead of a tennis ball. They obtain a smaller value for g. Suggest why the values for g are different.

3 The Moon has a mass much smaller than that of the Earth. Predict and explain what effect this would have on the values of g and t if a similar experiment was carried out on the lunar surface.

Exam-style questions

01 On a NASA Apollo mission, a hammer and a feather were dropped from the same height on the Moon. They both hit the surface of the Moon at the same time.

01.1 Use equations to explain why the hammer and feather hit the surface of the Moon at the same time. **[2 marks]**

_____

_____

_____

Synoptic link

3.4.1.5

01.2 By examining a video of the hammer and feather drop, it is possible to time how long they took to fall. The feather and hammer took 36 frames to fall and the frame speed is 29.97 frames per second.

Show that the time taken for the hammer and feather to fall was 1.20 s. **[1 mark]**

01.3 Show that the acceleration due to gravity on the Moon is $1.7 \, \text{m s}^{-2}$ if the drop height was 1.2 m. **[2 marks]**

01.4 In 1971, an astronaut played golf on the Moon. When they hit their second shot, the astronaut observed that the golf ball flew for 'miles and miles and miles'.

Calculate the initial vertical and horizontal velocities of the golf ball.

The golf ball left the club at a speed of $50 \, \text{m s}^{-1}$ at an angle of 35° to the horizontal. **[2 marks]**

vertical velocity = _____ m s^{-1}

horizontal velocity = _____ m s^{-1}

01.5 Calculate the total time of flight of the golf ball. **[2 marks]**

time of flight = _____ s

> **Exam tip**
>
> To find the time of flight, consider the vertical motion of the golf ball. Fill in each of the *suvat* values you have been given and then choose an appropriate equation.

01.6 Calculate the total horizontal distance travelled by the golf ball in metres before landing.

Compare this value to the statement that the golf ball flew for 'miles and miles and miles'.

1 mile = 1.6 km   **[2 marks]**

_____

_____

_____

02 A student performs an experiment to measure the acceleration of free fall using the apparatus shown in **Figure 1**.

Figure 1

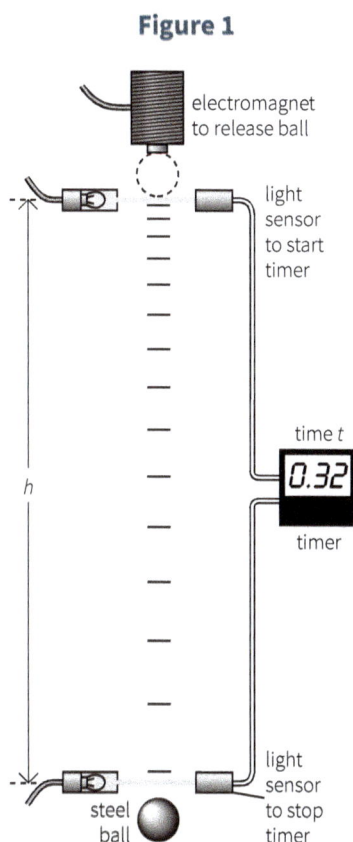

electromagnet to release ball

light sensor to start timer

time *t*

0.32

timer

light sensor to stop timer

steel ball

h

02.1 The height the ball is dropped in **Figure 1** is 0.45 m.

Calculate the acceleration of free fall using these measurements.   **[2 marks]**

Synoptic links

MS 3.2

acceleration of free fall = _____ m s^{-2}

02.2 A teacher suggests the student improves the experiment by recording the time taken for the ball to fall different heights. Suggest **one** reason why this would improve the experiment.   **[1 mark]**

_____

_____

02.3 **Figure 2** shows the results of their experiment.

Figure 2

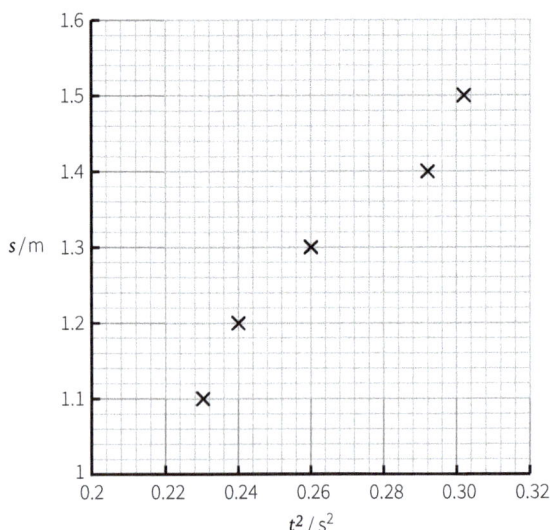

Draw a line of best fit on the graph and calculate the gradient.

[3 marks]

Exam tip

Make sure the points plotted on the graph are equally distributed above and below your line of best fit.

gradient = _____

02.4 Determine a value for g using your gradient from **02.3**. **[3 marks]**

$g =$ _____ $m\,s^{-2}$

Exam tip

Reaction time is not a source of error in this experiment, but how else might the timing go wrong? What other measurements are needed in the experiment that could introduce error?

02.5 The results suggest possible random errors in the experiment. Suggest **one** source of random error. **[1 mark]**

_____

_____

03 **Table 1** shows data for the Apollo 11 rocket launch.

03.1 Draw a velocity–time graph for the rocket launch on **Figure 3** by plotting the data in **Table 1**. **[2 marks]**

Synoptic links

3.6   3.7   3.8

Table 1

Time / s	Velocity / m s^{-1}
0	0
5	11
10	22
15	36
20	55
25	73
30	94

Figure 3

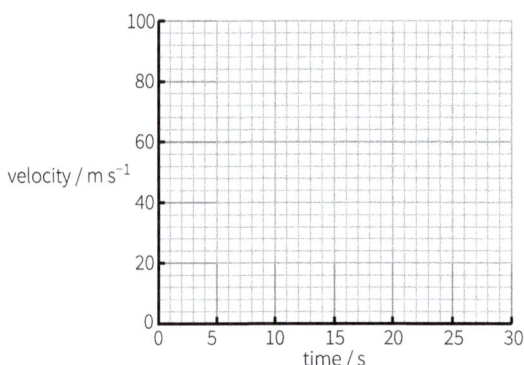

Exam tip

This is not uniform motion. You need the initial acceleration not the average acceleration. How can that be achieved?

03.2 Calculate the initial acceleration of the rocket in m s^{-2}. **[2 marks]**

03.3 Calculate the initial thrust of the rocket in N.
The initial mass of the rocket is 140 000 kg. **[2 marks]**

03.4 Determine the height of the rocket in m in the first 10 s. **[2 marks]**

03.5 Describe how the acceleration of the rocket changes in the first 30 s.
Suggest a possible explanation for your answer. **[2 marks]**

! Exam tip

'Suggest' questions are just that. You have not met the idea before and are not expected to know the true answer. You need to apply your knowledge to give a reasonable guess.

04 **Figure 4** shows a person's arm holding a weight out horizontally.
The arm is stationary.

Figure 4

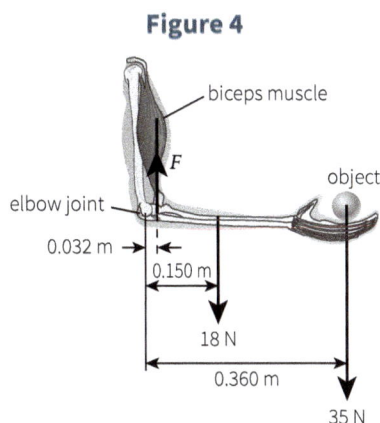

biceps muscle

object

F

elbow joint

0.032 m

0.150 m

18 N

0.360 m

35 N

Synoptic link

3.4.2.2

04.1 State the principle of moments. **[2 marks]**

04.2 Calculate the force F in N in the biceps muscle shown in **Figure 4**. **[2 marks]**

04.3 The person starts to lower their forearm.
Explain how the force F in the biceps muscle varies as the arm moves from a horizontal to a vertical position. **[2 marks]**

04.4 The tendon attaching the biceps muscle to the bone has an ultimate tensile strength of 32.5 MPa. The cross-sectional area of the tendon is 22.7 mm^2.

Calculate the maximum force in N the biceps could withstand before damage to the tendon would occur. **[2 marks]**

! Exam tip

The definition for moments is force × perpendicular distance. Why might the direction of the distance be important here?

05 **Figure 5** shows a child in a swing seat being pulled back by a horizontal force F_1. The child is stationary and the chain supporting the swing seat is at an angle θ to the vertical.

Figure 5

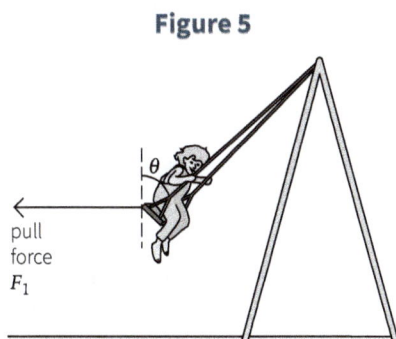

θ

pull force
F_1

Synoptic links

3.6.1.2   3.6.1.3

05.1 Draw a free-body diagram to show the forces acting on the child.

[3 marks]

05.2 The child has a mass of 20 kg and $\theta = 30°$.
Calculate the force F_1 in N that is pulling the child back. [3 marks]

05.3 The parent pulling the swing seat back lets go. The swing moves freely backwards and forwards with simple harmonic motion.
State the meaning of simple harmonic motion. [2 marks]

05.4 The chain length is 1.2 m.
Calculate the period of the oscillations in s. [1 mark]

05.5 The maximum amplitude of the oscillations is 1.0 m.
Sketch a displacement–time graph of the oscillations over the first 6 s. [3 marks]

Exam tip

When sketching graphs, first find some fixed points. Where will the curve start, and at what time will the curve go through the x-axis? Will the amplitude remain the same? When you have some points draw a smooth curve between them.

06 A flying disc-shaped toy generates a lift force as it flies through the air. The toy is launched horizontally from a known height and the time in the air is recorded.

height of launch = 1.50 m; time of flight = 4.20 s; mass = 0.180 kg

06.1 Explain the path of a projectile, ignoring air resistance. [3 marks]

06.2 Assuming projectile motion, show that the vertical acceleration of the toy is $0.2 \, \text{m s}^{-2}$. [2 marks]

06.3 Calculate the resultant vertical force in N. [1 mark]

06.4 Calculate the lift force in N acting on the toy. [2 marks]

Exam tip

In $F = m\,a$, the F is the resultant force. Sketch a free-body diagram of the toy to help you calculate the lift.

07 A hinged shelf is fixed to a wall and held horizontal by a string, as shown in **Figure 6**.

Figure 6

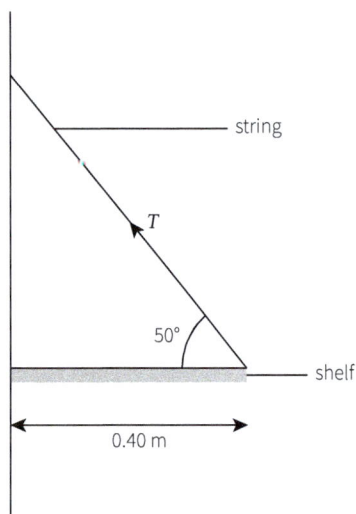

07.1 The shelf is uniform in shape and has a mass of 2.5 kg.
Calculate the weight of the shelf in N.
Draw an arrow on **Figure 6** to represent this weight. [2 marks]

07.2 The forces acting on the shelf are in equilibrium.
Draw an arrow on **Figure 6** to represent the resultant force from the wall acting on the shelf. Label the arrow **F**. [2 marks]

Exam tip

Think about the conditions for equilibrium.

07.3 Calculate the tension T in N in the string. **[2 marks]**

07.4 In a different design of shelf, the string is attached to the middle of the shelf 0.20 m from the hinge.
Explain how this would change the tension in the string. **[2 marks]**

08 A train sets off along a straight horizontal track in a southerly direction. An idealised version of the displacement–time graph is shown in **Figure 7**.

Figure 7

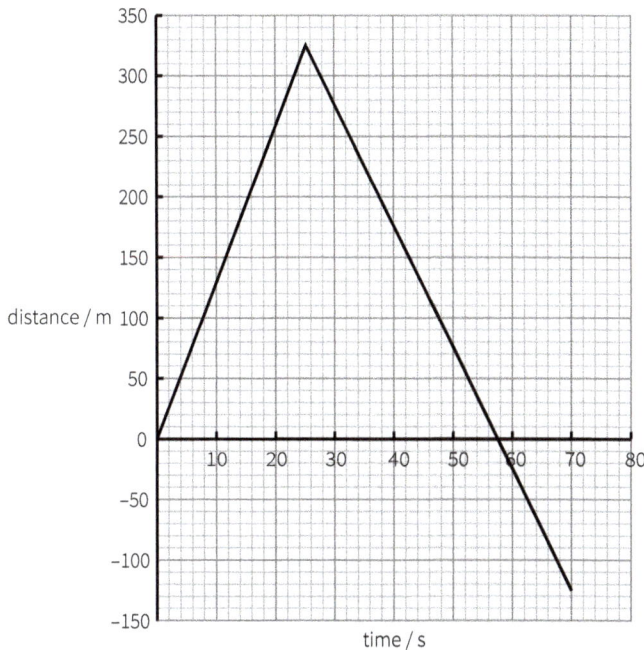

08.1 Compare distance and displacement. **[2 marks]**

08.2 Describe the position of the train at 70 s of the journey relative to the starting position. **[2 marks]**

08.3 Sketch the velocity–time graph for this journey on **Figure 8**. Include values on the velocity axis. **[3 marks]**

Figure 8

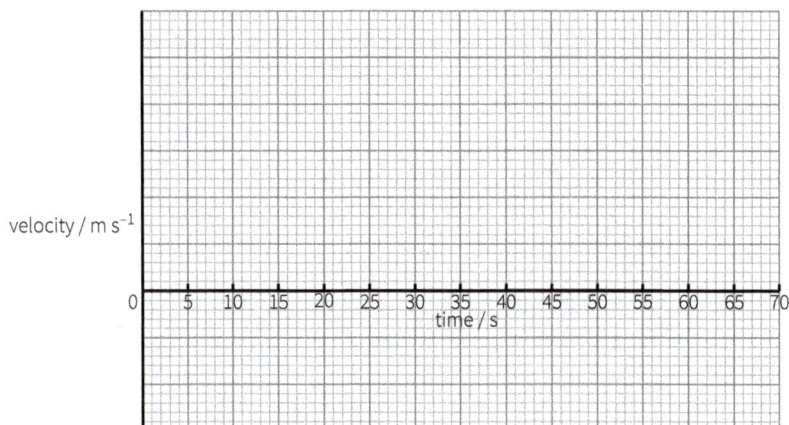

> **! Exam tip**
>
> **08.3** is worth 3 marks. This means you need to add details to you sketch, including calculating the velocity of the train and adding a scale to the axis.

08.4 Explain why the velocity–time graph you have sketched is idealised. **[1 mark]**

⚙ Knowledge

7 Force, energy, and momentum 2

Newton's first law

Inertia of an object is its tendency to remain in a steady state at rest or moving in a straight line at a constant speed.

Newton's first law states that the velocity, speed, and/or direction of an object only changes if a resultant force acts on it. If the resultant force is:

- zero on a stationary object, it will remain stationary
- zero on a moving object, it will continue moving at the same velocity
- *not* zero on an object, its velocity will change.

A change in velocity can mean:

- starting to move
- stopping moving
- speeding up
- slowing down
- changing direction.

If an object is doing any of these things, the forces on it will be unbalanced and a resulting force will be acting on it.

Newton's second law

Newton's second law states that the acceleration of an object is:

- proportional to the resultant force on the object, $a \propto F$
- inversely proportional to the mass of the object, $a \propto \frac{1}{m}$

In situations where the mass of an object is *not* changing, the resultant force, mass, and acceleration are linked by the equation:

resultant force = mass $\times$ acceleration
$$F = ma$$

The acceleration is always in the same direction as the resultant force.

$F = ma$ can be used to show why all objects fall at the same rate if there is no air resistance:

- resultant force F acting on a falling object is its weight W, so $W = F$
- since weight $W = mg$ and resultant force $F = ma$, $mg = ma$
- m cancels out because gravitational mass and inertial mass are the same, leaving $g = a$

Therefore, acceleration of a free-falling object is independent of mass.

Newton's third law

Newton's third law states that whenever two objects interact, they exert equal and opposite forces on each other. This means forces always occur in pairs.

A pair of forces will *always*:

- act on separate objects
- be the same size
- be in opposite directions
- act along the same line
- be of the same type.

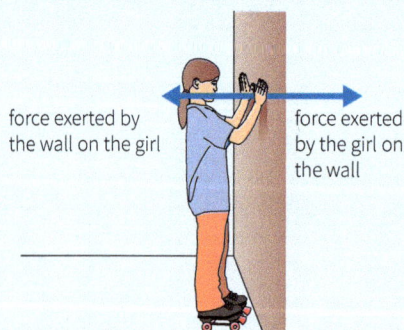

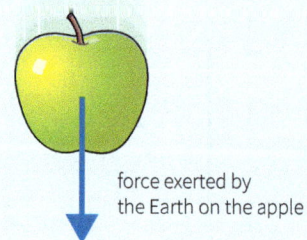

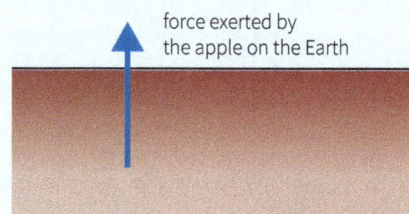

force exerted by the wall on the girl

force exerted by the girl on the wall

force exerted by the Earth on the apple

force exerted by the apple on the Earth

Elastic and inelastic collisions

In an **elastic** collision, kinetic energy is conserved – the objects involved have the same kinetic energy after the collision as before.

In an **inelastic** collision, kinetic energy is not conserved – the objects involved have less kinetic energy after the collision than before.

Momentum

Momentum is a property of all moving objects. It is a vector quantity with the unit $kg\,m\,s^{-1}$. Momentum depends on the **mass** and **velocity** of an object and is defined by the equation:

$$\text{momentum} = \text{mass} \times \text{velocity}$$

$$\rho = m v$$

Conservation of momentum

The principle of **conservation of momentum** states that, for a system of interacting objects, the total momentum remains constant if no external force acts on the system.

This means the total momentum before two objects collide is the same as the total momentum after they collide.

If two moving objects collide, this can be written as:

$$m_1 u_1 + m_2 u_2 = m_1 v_1 + m_2 v_2$$

where:

m_1 = mass of object 1

m_2 = mass of object 2

u_1 = initial velocity of object 1

u_2 = initial velocity of object 2

v_1 = final velocity of object 1

v_2 = final velocity of object 2

Examples

If an object explodes, momentum is conserved because:

- total momentum before is zero

- total momentum afterwards is also zero because different parts of the object travel in different directions, so their momenta cancel out.

If two moving objects recoil from each other, they start with a total momentum of zero and end moving away from each other with velocities v_1 and v_2.

This can be written as:

$$m_1 v_1 + m_2 v_2 = 0$$

If an object is moving, or is able to move, an unbalanced force acting on it will change its momentum.

Momentum and impulse

Since $F = ma$ and $a = \dfrac{\Delta v}{t}$ we can write:

$$F = \frac{\Delta(mv)}{t}$$

The force acting on an object is equal to the rate of change of momentum of the object; this is another way of stating Newton's second law.

The **impulse** of a force is the force multiplied by the time for which the force acts:

impulse = $F\Delta t = \Delta(mv)$

impulse = change of momentum $(mv - mu)$

For a ball rebounding from a wall:

change in momentum = $(-mv) - (mu)$

velocity = +u

wall

Before impact

velocity = −v

wall

After impact

The area under a **force–time** graph represents the impulse of the force.

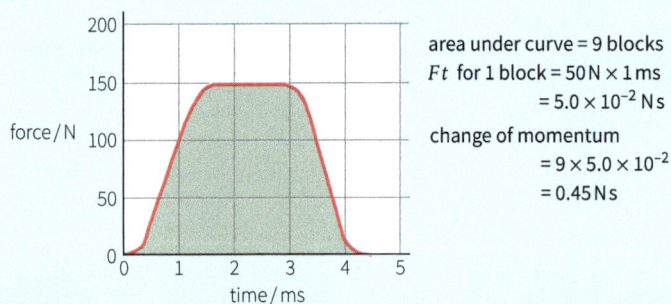

force / N

time / ms

area under curve = 9 blocks
Ft for 1 block = $50\,N \times 1\,ms$
$= 5.0 \times 10^{-2}\,Ns$

change of momentum
$= 9 \times 5.0 \times 10^{-2}$
$= 0.45\,Ns$

The greater the time taken for the change in momentum of an object:

- the smaller its change of momentum
- the smaller the force it experiences.

Ethical transport design

Ethical transport design requires that vehicles are made so that people are protected in a crash. Many everyday safety features work by increasing the time taken for the change in momentum in order to reduce the force of an impact, including:

- air bags, seat belts, and crumple zones in cars
- cycle helmets
- crash mats used for gymnastics
- cushioned surfaces in children's playgrounds
- packaging for fragile items.

Work and energy

Work done and **energy transferred** mean the same thing. The unit of work is the joule (J), which is equal to N m.

The amount of energy transferred is:

$$\frac{\text{work}}{\text{done}} = \frac{\text{force}}{\text{applied}} \times \frac{\text{distance moved along the}}{\text{line of action of force}}$$

$$W = Fs$$

If the direction of motion is different to the direction of the force:

$$W = Fs\cos\theta$$

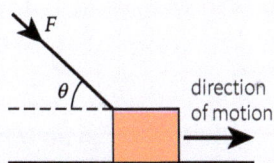

These equations can only be used if the force is constant. If the force F and the distance s are at right angles to each other, no work is done.

The area under a **force–displacement graph** represents the total work done.

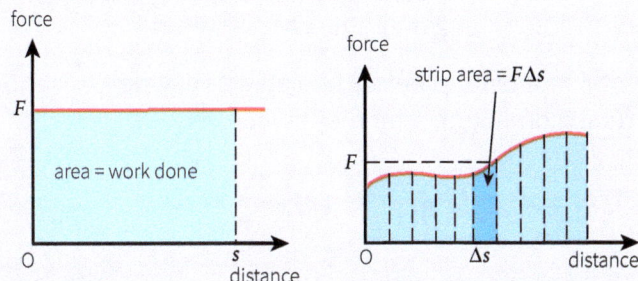

Power

Power is the rate of energy transfer or the rate of doing work:

$$\text{power} = \frac{\text{energy transferred or work done}}{\text{time}}$$

$$P = \frac{\Delta W}{\Delta t}$$

Combining this with $W = Fs$ gives:

power = force × velocity

$$P = Fv$$

Efficiency

Efficiency is a measure of how much energy is wasted in an energy transfer:

$$\text{efficiency} = \frac{\text{useful output energy transfer}}{\text{total input energy transfer}}$$

or

$$\text{efficiency} = \frac{\text{useful power output}}{\text{total power input}}$$

To give efficiency as a percentage, multiply the result by 100.

Potential and kinetic energy

The gravitational potential energy store of an object increases when it moves up and decreases when it moves down. The change in the gravitational potential energy store ΔE_p of an object can be calculated using:

$$\Delta E_p = \text{mass } m \times \frac{\text{gravitational}}{\text{field strength } g} \times \frac{\text{change of}}{\text{height } \Delta h}$$

$$\Delta E_p = mg\Delta h$$

The quantity of energy in an object's kinetic energy store E_k depends on its mass and how fast it is going. It can be calculated using:

$$E_k = \frac{1}{2} \times \text{mass } m \times \text{speed}^2 v^2 \qquad E_k = \frac{1}{2}m^2$$

The quantity of elastic potential energy E_e stored in a stretched spring or rubber band of length l (or extension e) can be calculated using:

$$E_e = \frac{1}{2} \times \text{spring constant } k \times \text{extension}^2 e^2$$

$$E_e = \frac{1}{2}k\Delta l^2 \text{ or } E_e = \frac{1}{2}ke$$

Conservation of energy

The principle of conservation of energy states that energy cannot be created or destroyed, only transferred between different stores and objects. This means the total amount of energy in a closed system is always the same. This can be used to solve problems when energy is transferred between stores.

If an object is dropped and air resistance is ignored, its kinetic energy after falling height Δh is equal to its loss of gravitational potential energy:

$$\frac{1}{2}mv^2 = mg\Delta h$$

Work done against resistive forces such as friction and air resistance means that some energy is always transferred to the thermal energy stores of an object and its surroundings.

Learn the answers to the questions below, then cover the answers column with a piece of paper and write as many as you can. Check and repeat.

	Questions	Answers
1	Define inertia.	tendency of an object to remain in a steady state at rest, or moving in a straight line at a constant speed
2	State Newton's first law of motion.	velocity of an object will only change if a resultant force is acting on it
3	State Newton's second law of motion.	acceleration of an object is proportional to the resultant force on the object, and inversely proportional to the mass of the object
4	Why do forces always occur in pairs?	Newton's third law – whenever two objects interact with each other, they exert equal and opposite forces on each other
5	What are the properties of a pair of forces related by Newton's third law?	same size; act on separate objects; act in opposite directions; along the same line; same type
6	What is the principle of conservation of momentum?	total momentum before two objects collide = total momentum after they collide
7	Give the equation for the principle of conservation of momentum for two objects which collide.	$m_1 u_1 + m_2 u_2 = m_1 v_1 + m_2 v_2$
8	What is conserved in an elastic collision?	kinetic energy
9	Write the relationship between force and momentum.	$F = \dfrac{\Delta(mv)}{t}$
10	State Newton's second law in terms of momentum.	force acting on an object is equal to rate of change of momentum of the object
11	State the relationship between Impulse and momentum.	impulse = change of momentum $(mv - mu)$
12	What does the area under the line of a force–time graph represent?	impulse of the force (or object change in momentum)
13	How do many everyday safety features reduce the force of an impact?	increase time taken for a change in momentum, reducing impact force
14	Define power.	rate of energy transfer or rate of doing work
15	State the relationship between power, work done, and time.	$P = \dfrac{\Delta w}{\Delta t}$
16	Define efficiency.	measure of how much energy is wasted in an energy transfer
17	State the principle of conservation of energy.	energy cannot be created or destroyed, only transferred between different stores and objects

Put paper here

18 Give the equation used to calculate the change in an object's gravitational potential energy store.

$$E_p = mg\,\Delta h$$

19 Give the equation used to calculate the amount of energy in an object's kinetic store.

$$E_k = \frac{1}{2}mv^2$$

20 Give the equation used to calculate the amount of energy in an object's elastic potential store.

$$E = \frac{1}{2}k\,\Delta l^2$$

Put paper here   Put paper here

Maths skills

Practise your maths skills using the worked example and practice questions below.

Rates of change	Worked example	Practice

Rates of change

The rate of change of momentum of an object is equal to the net force applied:

$$F = \frac{\Delta(mv)}{\Delta t}$$

where:

F = force (N)

m = mass of object (kg)

v = velocity of object (m s^{-1})

t = time (s)

You can solve differential equations by plotting a graph or modelling with a spreadsheet. To find the impulse, $\Delta(mv) = F\,\Delta t$, plot a graph of F against t. The area under the graph is equal to the impulse.

Worked example

Question

A ball is pushed towards a wall, hits the wall, and comes to rest. The graph shows the force acting on the ball during the motion.

Calculate the impulse during this motion.

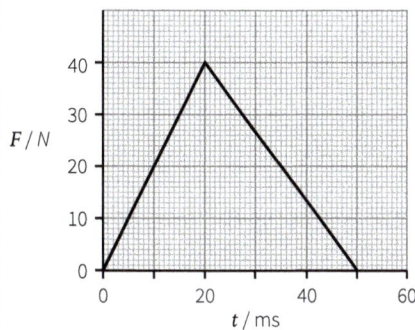

Answer

Impulse = area under the graph.

area of a triangle = $\frac{1}{2}$ base × height

base = 50 ms (5×10^{-2} s); height = 40 N

area = $\frac{1}{2}$ × 5×10^{-2} s × 40 N

impulse = 1 N s

Practice

1 This graph shows the impact of a car crash.

Calculate the impulse during the crash.

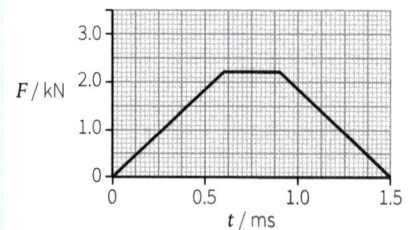

2 A tennis racket exerts a force of 420 N on a ball for 0.01 s.

Calculate the change in momentum of the ball.

3 When catching a it is less painful if you move your hands backwards with the ball, rather than holding them still and letting the ball land in them.

Explain why.

Exam-style questions

01 A roller coaster accelerates along a horizontal track. It climbs to a maximum height of 62.5 m, then comes hurtling down. The whole ride lasts 15 seconds.

01.1 The roller coaster starts from rest and is accelerated to 130 km h^{-1} in 1.9 seconds. Calculate the initial acceleration of the roller coaster.

[2 marks]

Synoptic links

3.4.1.3   3.7.5.4

acceleration = _____ m s^{-2}

01.2 Calculate the distance the roller coaster travels while accelerating.

[2 marks]

distance = _____ m

01.3 Calculate the velocity of the roller coaster as it reaches the highest point of the track.

[3 marks]

> **! Exam tip**
>
> If you are not sure whether or not acceleration is constant, then you cannot use *suvat* equations. Think about conservation of energy in these situations.

velocity = _____ m s^{-1}

01.4 On occasion, the roller coaster experiences some rollback. In case this happens, the launch side is fitted with copper fins that can be deployed in conjunction with the magnets attached to the cars of the roller coaster.

Explain how this braking system works.

Comment on whether this is a safe braking system.

[4 marks]

> **! Exam tip**
>
> Copper is not a magnetic material. How else could a conductor and a magnet be used as a braking mechanism?

_____

_____

_____

_____

_____

02 The assumptions made by the kinetic theory model include that all the collisions between molecules are elastic and that the time of collisions is much shorter than the time between collisions.

Synoptic links

3.4.1.1   3.6.2.3

02.1 State what is meant by an elastic collision. **[1 mark]**

_____

_____

02.2 A single molecule with a mass of 2.0×10^{-26} kg is within a cube with sides of 2 cm. The molecule collides with wall **X** when moving at a velocity of $500 \, \text{m s}^{-1}$.

Figure 1

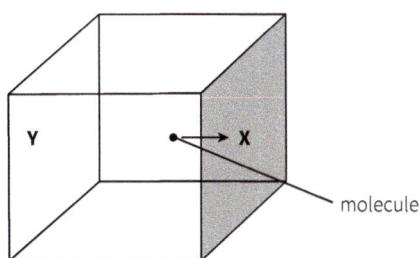

Calculate the change in momentum of the molecule, assuming the collision is elastic. Give an appropriate unit for your answer.

[3 marks]

change in momentum = _____ unit = _____

02.3 The molecule rebounds off wall **Y**, then collides with wall **X** again. Calculate the time between collisions with side **X**. **[1 mark]**

time = _____ s

02.4 Determine the average force exerted by the molecule on side **X**.

[2 marks]

force = _____ N

02.5 Calculate the number of molecules required to exert a pressure of $101\,000 \, \text{N m}^{-2}$. **[2 marks]**

Exam tip

If you have forgotten how to calculate pressure, look at the units N m^{-2} for a clue.

number of molecules = _____

03 **Figure 2** shows a pumped storage system in a hydroelectric power station. Water is pumped to the top reservoir at night when there is low demand for electricity. A website says the system has an 'energy storage capacity of 11 GWh'.

Figure 2

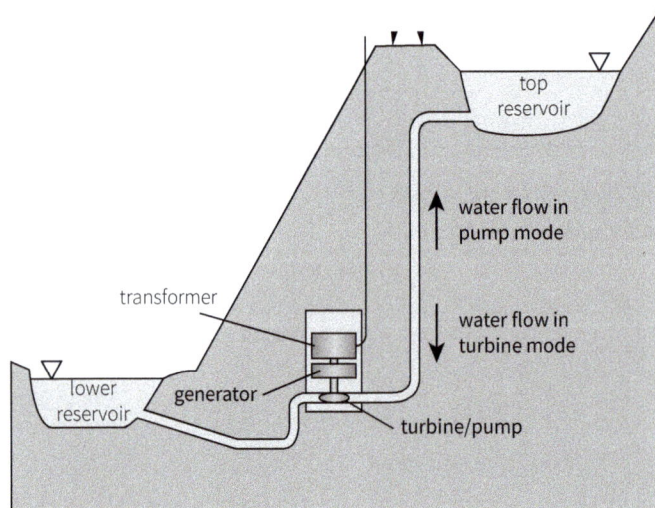

Synoptic links

3.1.1   3.4.2.1

03.1 Sugget what is meant by an 'energy storage capacity of 11 GWh'.

[2 marks]

_____

_____

Exam tip

If you are unsure, look carefully at the units. The value is given in watts × hours. What quantities do those units represent?

03.2 Water from the top reservoir falls 500 m to the turbines. At peak output, the water flows at a rate of 390 $m^3 s^{-1}$.

Calculate the power generated in W at peak output.
density of water = 1000 $kg\,m^{-3}$ **[3 marks]**

power generated = _____ W

03.3 Determine how many minutes the hydroelectric power station can run for. **[2 marks]**

minutes = _____

03.4 The power available is 1.7 GW.

Determine the efficiency of the generators in the power station.

[1 mark]

efficiency = _____

03.5 The generators and turbines used to generate electricity are also used to pump the water back up in the storage system. The overall efficiency for the 'round-trip energy' is quoted at 80%.

Suggest why this figure is lower than the value calculated in **03.4**.

[1 mark]

_____

_____

04 **Figure 3** shows the force–time graph for a snooker ball when it is struck by the cue.

Figure 3

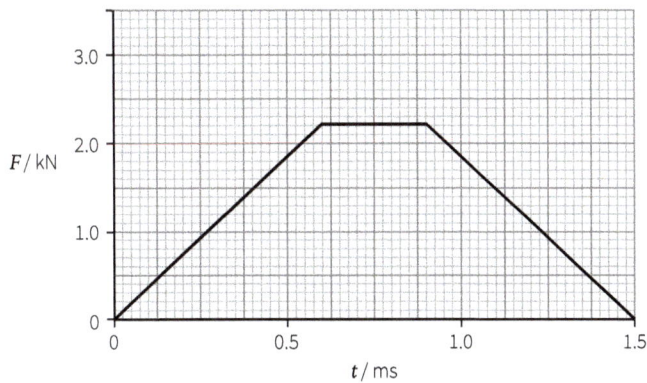

04.1 Sketch the graph showing how the force varies with time for the cue. [2 marks]

04.2 Determine the change in momentum in N s of the snooker ball using values from **Figure 3**. [2 marks]

_____

_____

04.3 The snooker ball has a mass of 140 g.

Calculate the velocity in m s^{-1} with which the snooker ball left the cue. [2 marks]

velocity = _____ m s^{-1}

04.4 Explain how your answer to question **04.3** would change if the ball had been rolling towards the cue when it was struck.

Assume the ball received the same impulse. [2 marks]

_____

_____

> ⚠ **Exam tip**
>
> If you are finding the explanation difficult you might want to try a rough calculation first. Imagine the ball has an initial velocity towards the cue of 1 m s^{-1}. What would the new velocity be?

05 Newton's second law states that, for a constant mass, force is proportional to acceleration.

05.1 Design an experiment to find the relationship between the force acting on a trolley and its acceleration.

Include a labelled diagram and explain how you will ensure your results are accurate. **[4 marks]**

05.2 **Table 1** shows some results from an experiment to test the relationship between force and acceleration.

Table 1

Falling mass / g	Acceleration of trolley / m s^{-2}
100	0.69
200	1.38
300	2.07

Show that these results are consistent with force being proportional to acceleration. **[2 marks]**

05.3 Newton's second law is better expressed as force being proportional to the rate of change of momentum.

$$F = \frac{\Delta m\, v}{\Delta t}$$

Show how this can be expressed as $F = m\,a$ for constant mass. **[1 mark]**

05.4 Explain how Newton's second law relates to car safety features such as crumple zones and seatbelts. **[2 marks]**

> **! Exam tip**
>
> It's important that you include key words in your answer to **05.4**.

06 A student makes a simple catapult using a piece of elastic. **Figure 4** shows the force–extension graph for the elastic used.

> **⚙ Synoptic link**
>
> 3.4.2.1

Figure 4

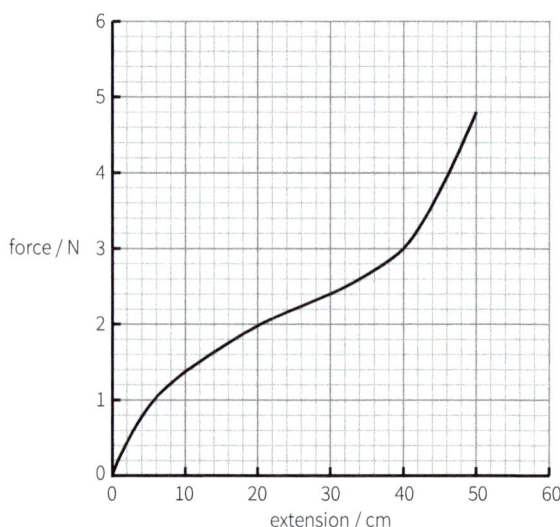

06.1 State whether the elastic obeys Hooke's law.
Explain your answer. **[3 marks]**

06.2 Determine the work done in J in stretching the spring 20 cm. **[3 marks]**

06.3 The catapult is pulled back 20 cm and used to fire a 10 g pellet. Calculate the velocity in m s^{-1} at which the pellet leaves the catapult, assuming energy is conserved. **[2 marks]**

06.4 **Figure 4** shows the loading graph for the elastic. Explain why the energy transferred to the pellet may be less than the value determined in **06.2**. **[2 marks]**

> **! Exam tip**
>
> What would the unloading graph be like for an elastic band?

07 A stationary nucleus of thorium-226 undergoes alpha decay to form radium.

07.1 Complete the equation for this decay. **[2 marks]**

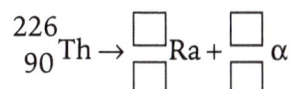

$$^{226}_{90}\text{Th} \rightarrow {}^{\square}_{\square}\text{Ra} + {}^{\square}_{\square}\alpha$$

07.2 The thorium nucleus was stationary before it decayed. After the decay, both the alpha particle and the radium nucleus are moving. Describe, compare, and explain their movements after the decay. **[3 marks]**

07.3 This decay is part of the decay series for uranium. Later in this decay series, polonium $^{214}_{84}\text{Po}$ becomes $^{210}_{84}\text{Po}$.

Describe the decays that must have happened for this to be possible. **[2 marks]**

> **Synoptic link**
>
> 3.2.1.2

> **! Exam tip**
>
> Only alpha decay can change the mass number, so there must have been at least one alpha decay.

08 A cyclist pedals a bike on a horizontal road. The combined mass of bike and cyclist is 95 kg, and the output power of the cyclist is 250 W.

08.1 When the cyclist cycles at 5 m s^{-1}, the total force resisting the forward motion is 10 N.

Calculate the work done in J each second against the total resistive force. **[1 mark]**

08.2 Calculate the forward force in N on the bike when the bike moves at 5 m s^{-1}. **[1 mark]**

08.3 State whether the bike will accelerate under these conditions. Explain your answer. **[2 marks]**

08.4 The cyclist now cycles up a hill.

Draw a free-body diagram to show the forces acting on the cyclist as they cycle up the hill. **[4 marks]**

08.5 Determine the steepest hill in degrees that the power-assisted bike can go up at a constant speed of 5 m s^{-1}. **[4 marks]**

> **Synoptic link**
>
> 3.4.1.1

> **! Exam tip**
>
> Don't forget to include all of the forces acting on the bike for **08.5**.

8 Materials

Density

The density of a substance, unit $kg\,m^{-3}$, is its mass per unit volume:

$$\text{density} = \frac{\text{mass}}{\text{volume}} = \rho = \frac{m}{v}$$

Hooke's law

Hooke's law states that the extension of a spring is directly proportional to the force exerted on it:

$$F = k\,\Delta L$$

where:

F = force extending the spring (N)

k = spring constant ($N\,m^{-1}$)

ΔL = extension of spring (m).

Hooke's law:

- only applies up to a certain point – the **limit of proportionality**
- is obeyed by wires and solid objects made of most materials
- can also be applied to the compression of a spring, where ΔL is the amount the spring gets shorter.

Elastic deformation occurs when a material can return to its original shape and size once any forces on it are removed.

Plastic deformation occurs when a material remains permanently deformed, so does *not* return to its original shape and size once any forces on it are removed.

The **elastic limit** of a spring or material is the point beyond which it becomes permanently deformed.

The Young modulus 🧪

The Young modulus E, unit Pa or $N\,m^{-2}$, is a measure of the stiffness of a material.

$$E = \frac{\sigma}{\varepsilon} = \frac{\frac{F}{A}}{\frac{\Delta L}{L}} = \frac{FL}{A\Delta L}$$

The stress and strain of a material are proportional to each other up to the limit of proportionality.

Stress and strain

Work must be done to stretch or compress an object.

When an object is elastically deformed, the energy transferred when it is stretched or compressed is stored as elastic strain energy (or **elastic potential energy**) E_p, unit J:

$$E_P = \frac{1}{2}\,F\,\Delta L$$

Since $F = k\,\Delta l$, E_p can also be calculated using:

$$E_p = \frac{1}{2}\,k\,\Delta L^2$$

An object is under **tension** if the forces acting on it are stretching it. An object is under **compression** if the forces acting on it are squashing it.

The **tensile stress** σ on an object is the force stretching F it divided by the cross-sectional area A:

$$\sigma = \frac{F}{A}$$

The unit of stress is the pascal (Pa). $1\,Pa = 1\,N\,m^{-2}$.

The **tensile strain** ε on an object is the ratio of its extension ΔL over its original length L:

$$\varepsilon = \frac{\Delta L}{L}$$

Stress–strain graphs show the general behaviour of a material.

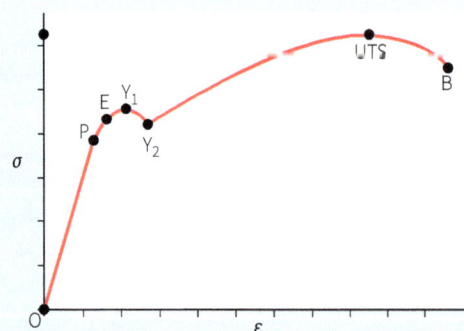

On the stress–strain graph:

- OP: stress and strain proportional to each other; material obeys Hooke's law; gradient constant and equal to Young modulus; area under section = energy stored in the material per unit volume

- P: limit of proportionality – stress and strain no longer proportional; deformation still elastic
- E: elastic limit – deformation plastic from this point
- Y_1: yield point 1 – stress at which material weakens and stretches plastically without additional force
- Y_2: yield point 2 – stress at which material undergoes **plastic flow**, where small stress leads to large strain because cross-sectional area of material decreasing rapidly
- UTS: ultimate tensile stress – maximum stress the material experiences; measure of the material's strength
- B: breaking point – stress in material at which it breaks.

Properties of a material:

- Stiff materials have steep initial gradients, large Young modulus.
- Strong materials have high UTS and breaking point.
- Brittle materials break without much plastic deformation before breaking.
- Ductile materials undergo high plastic deformation before breaking.

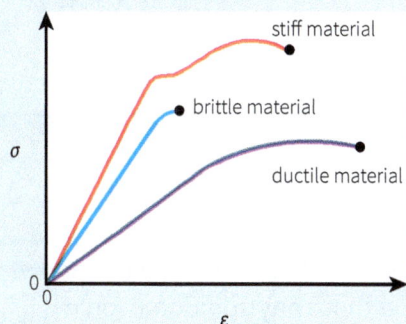

Force-extension graphs

Force-extension graphs show the behaviour of a sample of a material, with particular shape and size.

Metal wire stretched beyond its elastic limit (Graph A)

- unloading line does not go through origin – wire is permanently stretched
- loading and unloading lines parallel because Young modulus is constant
- area between loading and unloading lines is the work done to permanently deform the wire

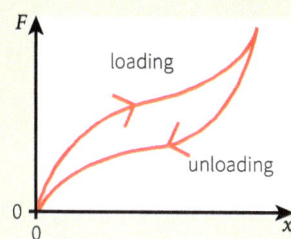

Rubber band (Graph B)

- unloading line goes through origin – rubber band returns to original length
- area under loading curve is the work done to stretch rubber band
- area under unloading curve is the work done by rubber band when unloaded
- area between loading and unloading curves is the difference between energy stored in stretched rubber band and energy recovered when unstretched
- difference between energy stored and energy recovered is because some energy transferred to internal energy of the molecules

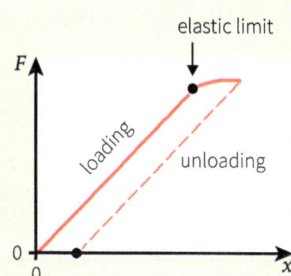

Polythene strip (Graph C)

- unloading line does not go through origin – polythene is permanently stretched
- area between loading and unloading curves is the sum of the work done to deform the strip and the energy transferred to the internal energy of the molecules

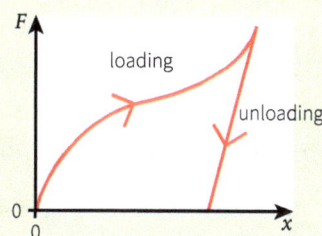

Area under the line of an object's force–extension graph represents work done and total elastic strain energy stored.

When the force is removed from an elastically deformed object, the stored elastic strain energy is transferred to other stores, such as kinetic or gravitational potential.

When an object is plastically deformed, work is done to rearrange the atoms of the material and energy is transferred to the thermal energy store of the object and its surroundings.

Crumple zones are a safety feature in many vehicles – they plastically deform during a crash so less energy is transferred to the passengers.

Retrieval

Learn the answers to the questions below, then cover the answers column with a piece of paper and write as many as you can. Check and repeat.

	Questions	Answers
1	Define the density of a substance.	$\rho = \dfrac{m}{v}$
2	State Hooke's law.	extension (or compression) of a spring is directly proportional to the force exerted on it
3	Define elastic deformation.	a material returns to its original shape and size once any forces on it are removed
4	Define plastic deformation.	a material remains permanently deformed so does not return to its original shape and size once any forces on it are removed
5	What is the elastic limit of a material?	point beyond which it becomes permanently deformed
6	Give the equations to calculate the elastic potential energy.	$E_\mathrm{p} = \dfrac{1}{2} F \, \Delta L$ and $E_\mathrm{p} = \dfrac{1}{2} k \, \Delta L^2$
7	What does the area under the line of a force–extension graph for an object represent?	$\dfrac{\text{work done on object}}{\text{total elastic potential energy stored by it}}$
8	What happens to the elastic potential energy of an elastically-deformed object when the force on it is removed?	transferred to other stores
9	What causes an object to be under tension?	forces acting on it are stretching it
10	Define the tensile stress on an object.	$\sigma = \dfrac{F}{A}$
11	Define the tensile strain on an object.	$\varepsilon = \dfrac{\Delta L}{L}$
12	What is the Young modulus of a material?	measure of the material's stiffness in $\mathrm{N\,m^{-2}}$ or Pa
13	Define the limit of proportionality of a material.	point up to which its stress and strain are proportional
14	How can the Young modulus of a material be found?	gradient of the initial straight-line section of its stress–strain graph
15	What does the area under the straight-line section of a material's stress–strain graph represent?	energy stored in the material per unit volume
16	What does the first yield point on a material's stress–strain graph represent?	stress at which the material suddenly weakens and stretches plastically without any additional force

Put paper here

Practical skills

Practise your practical skills using the worked example and practice questions below.

Determination of Young modulus

Apply a mass m to stretch a metal wire of length L, and measure the extension ΔL.

Substitute $F = mg$ into the equation for the Young modulus and rearrange:

$$\Delta L = \frac{L}{AE} mg$$

Plot a graph of mean extension ΔL against load m. The gradient $= \frac{L}{AE}$, which can be used to calculate the Young modulus.

Practice

1 In an experiment to measure the Young modulus of copper wire of length $2.82\,\text{m} \pm 0.01\,\text{m}$ and diameter $0.32\,\text{mm} \pm 0.01\,\text{mm}$, a student obtained the following values: gradient of line of best fit $= 0.281 \times 10^{-3}\,\text{m N}^{-1}$; worst gradient $= 0.315 \times 10^{-3}\,\text{m N}^{-1}$

Use these values to determine the Young modulus of copper, and the uncertainty in the measurement.

2 A student carried out the loading and unloading process with a wire, obtaining the following results.

State and explain whether the elastic limit of the wire has been exceeded.

Mass / kg	Extension (loading) / mm	Extension (unloading) / mm
1	2.5	4.0
2	4.0	5.5
3	6.0	8.0
4	10.5	12.0

Worked example

Question

The graph shows results for a copper wire that is $2.47\,\text{m} \pm 0.01\,\text{m}$ long with a mean diameter of $0.38\,\text{mm} \pm 0.01\,\text{mm}$. There is a line of best fit and a line of worst fit (red line).

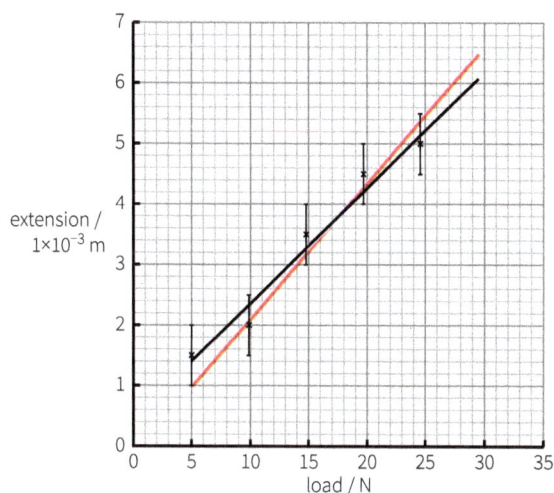

1 Calculate the gradient of the graph and the uncertainty in the gradient.

2 Determine a value for the Young modulus of copper.

Answer

1 gradient $= \dfrac{(5.8 - 2.2) \times 10^{-3}}{(28 - 9)} = 0.189 \times 10^{-3}\,\text{m N}^{-1}$

worst gradient $= \dfrac{(6.5 - 1.0) \times 10^{-3}}{(29.4 - 4.9)} = 0.224 \times 10^{-3}\,\text{m N}^{-1}$

% uncertainty $= \dfrac{(\text{worst gradient} - \text{gradient})}{\text{gradient}} \times 100\%$

$= 18.5\%$

2 gradient $= \dfrac{l}{AE}$ therefore $E = \dfrac{l}{A \times \text{gradient}}$

$$E = \frac{2.47\,\text{m}}{\pi \left(\dfrac{0.380 \times 10^{-3}}{2}\right)^2 \text{m}^2 \times 0.189 \times 10^{-3}\,\text{m N}^{-1}}$$

$= 1.2 \times 10^{11}\,\text{Pa}$

Practice

Exam-style questions

01 Spider silk is a material that enables spiders to make webs that stretch over long distances.

Figure 1

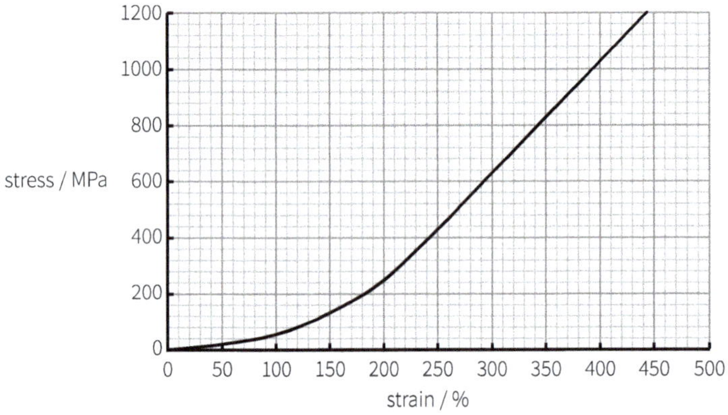

01.1 Calculate the Young modulus for spider silk for strains over 250% using data from **Figure 1**. **[2 marks]**

> **Exam tip**
>
> Always check the prefixes of units when you take measurements from a graph.

Young modulus = ____________ MPa

01.2 Describe the difference between the stiffness of the silk below 250% strain with the stiffness above 250% strain. **[1 mark]**

____

____

01.3 **Table 1** compares the ultimate tensile stress, yield stress, and density of spider silk and steel.

Table 1

Material	Ultimate tensile strength / MPa	Yield stress / MPa	Density / $kg\,m^{-3}$
steel	500–2000	250	7800
spider silk	1000	1650	1300

A cable is needed to lift a car of mass 1200 kg.

Calculate the diameters in m of both steel cable and spider silk that would lift the car without yielding.

Comment on your answer. **[4 marks]**

____

____

01.4 Calculate the ratio of the weight of the cable made of silk to the weight of the cable made of steel. Assume the length of cable is the same. **[3 marks]**

_____

02 A student stretches a length of material. They obtain the graph of stress against strain shown in **Figure 2**.

Figure 2

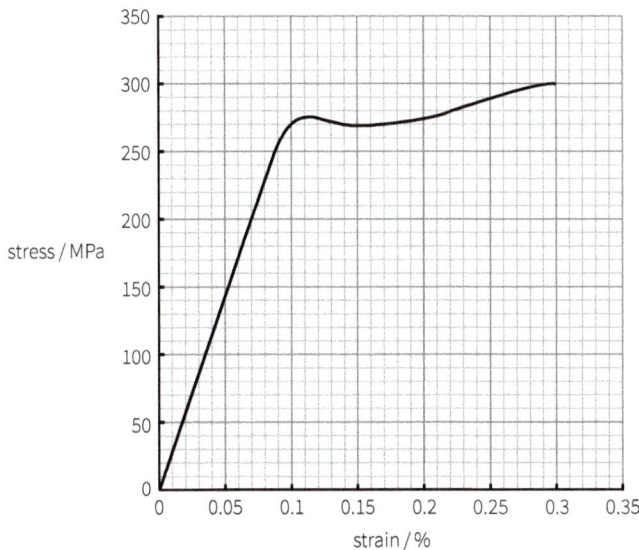

02.1 Suggest how the student measured strains of this magnitude.

[1 mark]

_____

_____

02.2 Write the limit of proportionality and the yield point for the material.

[2 marks]

limit of proportionality = _____ N

yield point = _____ N

02.3 Describe what is physically happening to the material between strains of 0.1% and 0.15%.

Using your answer, explain why the Young modulus is calculated from the initial section of the graph. **[2 marks]**

_____

_____

_____

> **! Exam tip**
>
> The graph in **Figure 2** has many different features. Make sure that you understand the meaning of each section of the graph in terms of the physical changes to the material.

02.4 State how you know that this is a ductile material. **[1 mark]**

_____

_____

02.5 Draw a line on the graph in **Figure 2** to show the behaviour of a brittle material with a Young modulus twice that of the ductile material shown. **[2 marks]**

03 A student stretches a spring with a spring constant of $150\,\text{N m}^{-1}$ and an elastic limit of $6\,\text{N}$.

03.1 Draw a line on the axes in **Figure 3** to show how the spring would behave if stretched with a force of up to $6\,\text{N}$. **[2 marks]**

Figure 3

03.2 Sketch a line on **Figure 3** to show what would happen if the student continued to load the spring up to a force of $8\,\text{N}$, and then completely unloaded the spring. **[2 marks]**

03.3 Calculate the total energy in J stored in the spring when it is stretched by a force of $6\,\text{N}$. **[1 mark]**

03.4 Sketch a line on **Figure 3** to show the loading graph for two identical springs in parallel loaded up to $6\,\text{N}$. Label the line **P**. State a reason for your placement of the line.

Compare the energy stored in this arrangement with your answer to **03.3**. **[3 marks]**

> **! Exam tip**
>
> It might be helpful to work out a ratio to answer the comparison in **03.4**.

04 A student uses the apparatus shown in **Figure 4** to collect data to investigate the relationship between the length of a spring and its spring constant. They choose a long spring and can attach the load to different parts of the spring.

Figure 4

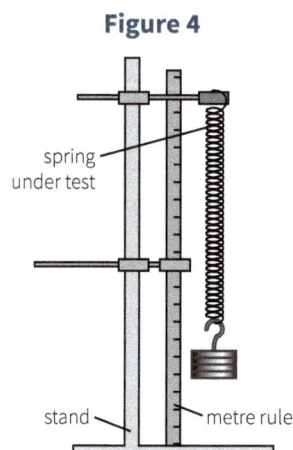

04.1 Describe in detail how the student could collect data to investigate this relationship using the apparatus shown in **Figure 4**. **[5 marks]**

04.2 Sketch the graph that you would expect to obtain by performing this experiment. **[2 marks]**

04.3 Explain the shape of the graph you sketched in **04.2**. **[2 marks]**

04.4 Suggest how this experiment demonstrates the need to use strain and not extension when determining the stiffness of materials using the Young modulus. **[2 marks]**

05 A nylon fishing line of length 3.6 m has a diameter of 0.22 mm. The graph of stress against strain for the nylon is approximately linear below the elastic limit. A fish, of weight 36 N, extends the line by 66 cm, which is below the elastic limit.

05.1 Calculate the Young modulus of the nylon fishing line. **[4 marks]**

05.2 Calculate the speed in m s^{-1} of the end of the line if it snaps. The density of the nylon is 1.15 g cm^{-3}. State clearly any assumptions you make. **[4 marks]**

05.3 Suggest the impact on the speed calculated in **05.2** if your assumptions are incorrect. **[1 mark]**

05.4 Suggest and explain what would happen, quantitatively, to your answer to **05.2** if the line were to break for a fish of *half* the weight. **[2 marks]**

06 A student making a model suspension bridge has a choice of cords made from two materials. Material **A** has a high Young modulus and material **B** has a low Young modulus.

06.1 Suggest which material would be a better choice, and give a reason for your answer. **[1 mark]**

06.2 The student plots the force against extension graph for one of the cords, as shown in **Figure 5**.

Figure 5

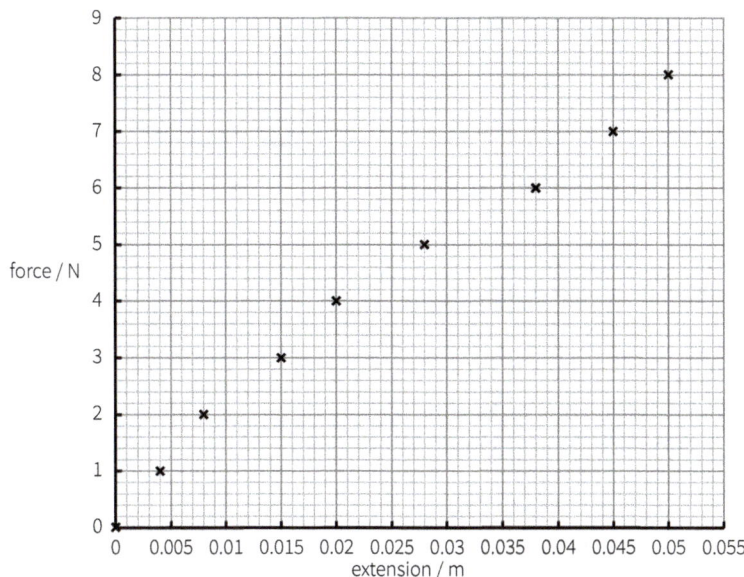

Calculate the energy in J stored in the cord when it is stretched using data estimates from **Figure 5**. **[2 marks]**

06.3 Calculate the speed in $m\,s^{-1}$ of the end of the cord if it breaks at the maximum extension, using data estimates from **Figure 5**.
cord dimensions: length = 1 m; diameter = 1 mm; density = $1.15\,g\,cm^{-3}$
[3 marks]

Exam tip

Be consistent with your use of units within calculations by paying attention to prefixes.

06.4 The student twists two cords together to make the cords for the model suspension bridge. Sketch the force–extension graph for the combined cord on the graph in **Figure 5**. **[2 marks]**

06.5 Calculate the Young modulus in $N\,m^{-2}$ of the cord using the initial part of the graph in **Figure 5**. **[3 marks]**

06.6 The student builds a simple bridge by balancing a 1 m ruler on two bricks, **X** and **Y**. They model a lorry using two masses, 100 g and 200 g, a distance of 10 cm apart. They put the model lorry at the centre of the bridge so that the 100 g mass is closer to brick **X**.

Calculate the reaction forces in N in each of the bricks. The mass of the ruler is 50 g. **[3 marks]**

07 A student used the equipment in **Figure 6** to collect data to measure the Young modulus.

Figure 6

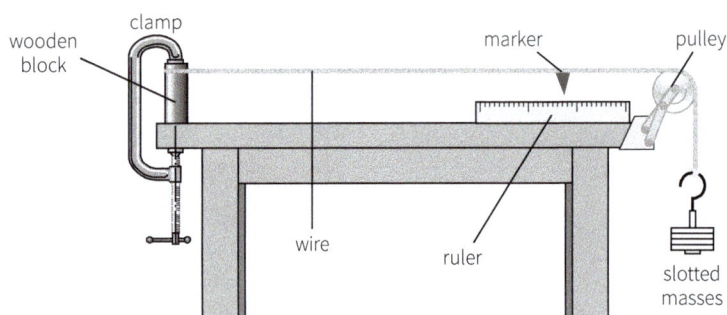

Synoptic links

3.1.2   3.4.1.3   3.4.1.4

07.1 The student measured the original length of the wire from the edge of the wooden block to the marker.

Describe the additional measurements the student would need to make to find the Young modulus of the material of the wire. Describe the graph the student could plot so that the gradient would be equal to the Young modulus. **[4 marks]**

07.2 Another student measured the original length from the edge of the wooden blocks to the hook on the slotted masses.

Describe and explain the effect this error would have on their calculated value of the Young modulus. **[2 marks]**

07.3 The original length of the wire was 60.0 cm.

Estimate the uncertainty in the measurement of the original length of the wire. Calculate the percentage uncertainty. **[2 marks]**

Exam tip

There are various ways of estimating uncertainty – for example, using the resolution of instruments, or the spread of data. An 'estimate' question means that there are a range of possible answers.

07.4 A length of wire of this material is used to suspend a window-cleaning platform. The strain must not exceed 0.1% for the cleaners to use the platform safely. The Young modulus of the material of the wire is 1.5×10^{11} Pa.

Calculate the diameter in m of the wire that would produce a strain of 0.1% in the wire, for an applied force of 1 kN. **[3 marks]**

07.5 The platform is suspended by two of these wires. The average weight of a person on the platform is 700 N.

Suggest and explain the number of people that could safely use the platform. **[2 marks]**

07.6 Calculate the speed in m s^{-1} with which a paintbrush will hit the ground if it falls, from rest, a distance of 12 m from the platform. State clearly any assumptions that you have made. **[3 marks]**

07.7 A drop of paint leaves the brush just as it is dropped.

Suggest and explain whether the drop of paint will be travelling at a higher, lower, or the same speed as the brush if your assumptions in **07.6** are not correct. **[3 marks]**

08 A tree frog can jump about 10 times its height. The mass of a tree frog is 7 g, and it is 2 cm high.

08.1 Assuming that the tendons in the tree frog's legs behave like springs, calculate the spring constant in N m^{-1} of the tendon.

Make your reasoning clear. State any assumption that you make. **[4 marks]**

08.2 Sketch the graph for extension against force for a rubber band, a spring, and a strip of polythene on the same axes. Assume they are stretched by the same force, and sketch the loading force only. Annotate your graphs clearly. **[3 marks]**

08.3 Suggest whether tendons behave more like rubber bands or polythene. Give a reason for your answer. **[2 marks]**

08.4 Suggest why humans can only jump a fraction of their height. **[2 marks]**

08.5 A human-like robot can jump to a height of 1.5 m, which is approximately the height of a human.
Calculate the extension of the springs in m that would move a robot with the mass of a human through this distance. Assume that the spring constant of the spring in the robot is the same as that of the frog. Comment on your answer. **[2 marks]**

08.6 The robot makes this manoeuvre in about 1.2 s.

Calculate the power in W of the robot. Comment on your answer by comparing it to a household appliance. **[2 marks]**

Synoptic links

3.1.4.7   3.4.1.8

! Exam tip

Make clear any assumptions or estimations you make in your answer by including 'Assuming…' and 'Estimating…'.

⚙ Knowledge

9 Current electricity

Basics of electricity

Electric current is the rate of flow of **charge**:
$$I = \frac{\Delta Q}{\Delta t}$$
An electric current of 1 ampere is a rate of flow of charge of 1 coulomb per second.

Current is measured using an **ammeter**, which must be connected in series in a circuit.

Potential difference (p.d.) is the work done per unit charge:
$$V = \frac{W}{Q}$$
1 volt = 1 joule per coulomb, $1\,V = 1\,J\,C^{-1}$

A p.d. of 1 volt across a component, or between two points, in a circuit means 1 joule of energy is transferred per coulomb of charge that moves through the component or between the points. The p.d. between two points in a circuit is measured using a **voltmeter**, which must be connected in parallel with the relevant part of a circuit.

Resistance is a measure of how much a component opposes the flow of current through it. Resistance is measured in ohms (Ω):
$$R = \frac{V}{I}$$

Resistivity 🌡

The resistivity ρ (Ω m) of a material is a measure of how difficult it is for a current to flow through it, and is defined as the resistance of a 1 metre length of the material with a cross-sectional area of 1 square metre.

For a sample of material with resistance R, cross-sectional area A, and length L, the resistivity can be calculated using:
$$\rho = \frac{RA}{L}$$
The resistivity of a material is not constant and can depend on environmental factors such as temperature and light intensity.

Circuit components

Each type of electrical component has a symbol which is used to represent it in a circuit diagram.

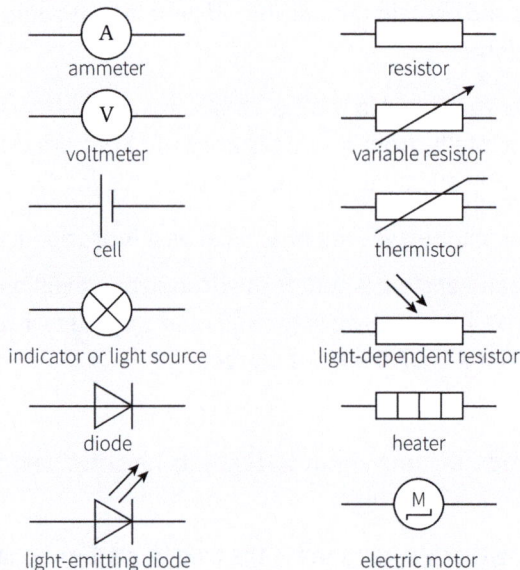

ammeter — resistor
voltmeter — variable resistor
cell — thermistor
indicator or light source — light-dependent resistor
diode — heater
light-emitting diode — electric motor

Current-voltage *I–V* characteristics

Ohm's law states that the current through a metallic conductor is directly proportional to the p.d. across it, provided the physical conditions do not change. Conductors that obey Ohm's law have a constant resistance and are known as **ohmic conductors**.

An *I–V* characteristics graph shows how the current through a component varies with the p.d. across it. The data to plot an *I–V* graph for any component can be collected using a circuit like that below.

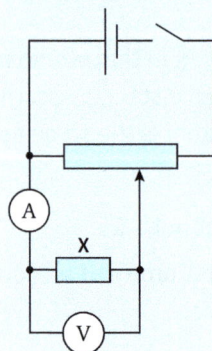

Using a potential divider   Using a variable resistor

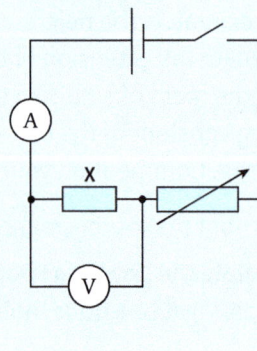

I–V graphs can also be plotted with p.d. on the *y*-axis and current on the *x*-axis.

I–V characteristics

Ohmic conductor (Graph A)

- Current is proportional to p.d., so the I–V graph is a straight line through the origin.

- Resistance is constant and can be found by $\dfrac{1}{\text{gradient}}$ of the I–V graph.

Filament lamp (Graph B)

- The gradient decreases, showing that the resistance of the filament lamp increases.

- As more current flows through the filament, its temperature increases. The atoms and ions in the wire vibrate more, and collide more often with electrons flowing through it, so there is more resistance. This means there is less current per volt increase in p.d., so the gradient of the graph decreases.

- Resistance of the filament lamp changes and its value at any point can be calculated from $R = \dfrac{V}{I}$ at that point.

Diode (Graph C)

A diode only lets current flow in one direction, known as the forward direction.

The p.d. must be above a certain threshold voltage before any current can flow.

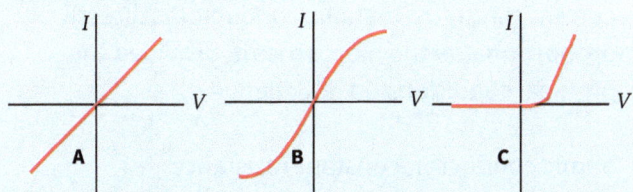

Resistance and temperature

A metal has a positive temperature coefficient because the resistance of a metal increases with temperature.

At higher temperatures, the atoms and ions in a metal vibrate more, increasing the chance of collision with the charge carriers (conduction electrons). This makes it more difficult for them to pass through the metal, so there is less current per volt of p.d. across the metal and, therefore, a greater resistance (graph A).

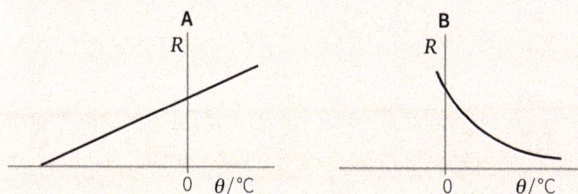

The resistance of a negative temperature coefficient (NTC) **thermistor** decreases as temperature increases (graph B).

The number of charge carriers (conduction electrons) increases with temperature, so there is more current per volt of p.d. across the thermistor, and, therefore, a smaller resistance.

The change in resistance per kelvin for an NTC thermistor is much greater than for a metal, making them useful in temperature sensing circuits.

Superconductivity

A **superconductor** is a material which has zero resistivity at and below a critical temperature, which depends on the particular material.

When a current flows through the material in the superconducting state, there is no p.d. across it because $V = IR$ and R is zero. This means the current has no heating effect and no energy is lost.

Superconductors are used to produce strong magnetic fields in devices such as MRI scanners and particle accelerators.

Superconducting transmission cables would allow much more efficient transfer of electrical energy because none of it would be lost to heating. Superconducting wires would also increase the efficiency of electric motors and generators.

Retrieval

Learn the answers to the questions below, then cover the answers column with a piece of paper and write as many as you can. Check and repeat.

	Questions	Answers
1	Define electric current.	rate of flow of charge
2	What does an electric current of one amp represent?	rate of flow of charge of one coulomb per second
3	Name the instrument used to measure current.	ammeter
4	Define potential difference.	work done per unit charge $$V = \frac{W}{Q}$$
5	Name the instrument that measures the pd between two points in a circuit.	voltmeter
6	What is the resistance of a component and how is it defined?	measure of how much a component opposes the flow of current $$R = \frac{V}{I}$$
7	State the unit of resistance.	ohm (Ω)
8	How can the resistance of an electrical component be found?	measure the current through it and the p.d. across it and use $R = \frac{V}{I}$
9	State Ohm's law.	current through a metallic conductor is directly proportional to the p.d. across it, provided the physical conditions do not change
10	What is the name for a conductor that obeys Ohm's law?	ohmic conductor; constant resistance
11	What is the I–V characteristic of a component?	graph which shows how the current through the component varies with the p.d. across it
12	Describe the I–V characteristic for an ohmic conductor.	straight line graph through the origin
13	Describe the I–V characteristic for a filament lamp.	curve with a decreasing gradient
14	Describe the I–V characteristic for a diode.	with forward bias, graph starts at (0, 0) and increases gradually; with reverse bias, graph extends along the negative p.d. with almost zero current
15	What is a negative temperature coefficient thermistor?	component for which the resistance decreases when temperature increases
16	What is a superconductor?	material with zero resistivity at and below a critical temperature
17	Why are superconductors useful?	current in a superconductor has no heating effect so no energy is lost

Put paper here

Practical skills

Practise your practical skills using the worked example and practice questions below.

Resistivity of a wire

To measure the resistivity of a wire, you need to know its diameter and length. Use a micrometer to measure the diameter for at least three points along the wire, then calculate the mean diameter.

Clip the wire onto a ruler using crocodile clips 0.1 m apart, connected to a circuit with an ammeter and voltmeter.

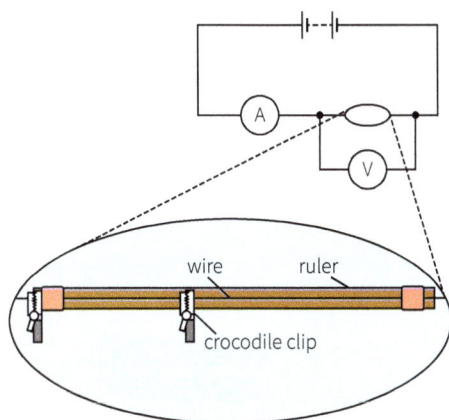

Set the p.d. to 0.5 V. Measure the current. Increase the length to 0.2 m. Increase the p.d. by 0.5 V. Measure the current. Repeat for lengths up to 1.0 m.

Worked example

Question

A student measured the diameter of a graphite rod. The diagram shows the reading on the micrometer scale.

1 Write the diameter of the rod.

2 Calculate the cross-sectional area of the graphite rod. Give your answer in m.

Answer

1 The barrel of the micrometer is slightly beyond the 6 mm mark on the main scale. This gives the first digit of the diameter. The central line of the main scale lines up with the line representing 34 on the barrel of the micrometer, so the diameter of the rod is 6.34 mm.

2 cross-sectional area $= \pi r^2 = \pi \dfrac{d^2}{4}$

diameter $= 6.34 \times 10^{-3}$ m

area $= \pi \dfrac{(6.34 \times 10^{-3})^2}{4} = 3.16 \times 10^{-7}$ m

Practice

1 The diagram shows two measurements a student made of the diameter of a constantan wire.

Write the diameter of the wire for each reading.

2 The diameter of a different constantan wire is measured at five points along the wire. The values obtained were: 0.314, 0.351, 0.315, 0.316, and 0.315 mm.

  a Calculate the mean diameter of the wire, and explain your calculation.

  b Calculate the cross-sectional area of the wire.

3 Explain how you could check a micrometer for a zero error.

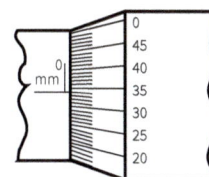

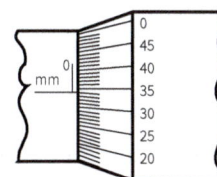

Exam-style questions

01 A student made the following measurements in an experiment to determine the resistivity of a wire.

Synoptic links

3.1.2   MS3.1
MS 3.2   MS 3.4

Table 1

Length of wire / m	Potential difference / V	Current / A	Resistance / Ω
0.80	1.46	0.14	10.43
0.70	1.32	0.15	8.80
0.60	1.32	0.17	7.76
0.50	1.33	0.19	
0.40	1.21	0.23	5.26
0.30	1.16	0.28	4.14

01.1 Complete **Table 1** by calculating the resistance for the 0.50 m length wire.   **[1 mark]**

01.2 State the resolution of the ammeter used in the experiment.   **[1 mark]**

_____

01.3 **Figure 1** shows a graph of the results.
Plot the missing data point and draw a line of best fit.   **[1 mark]**

Figure 1

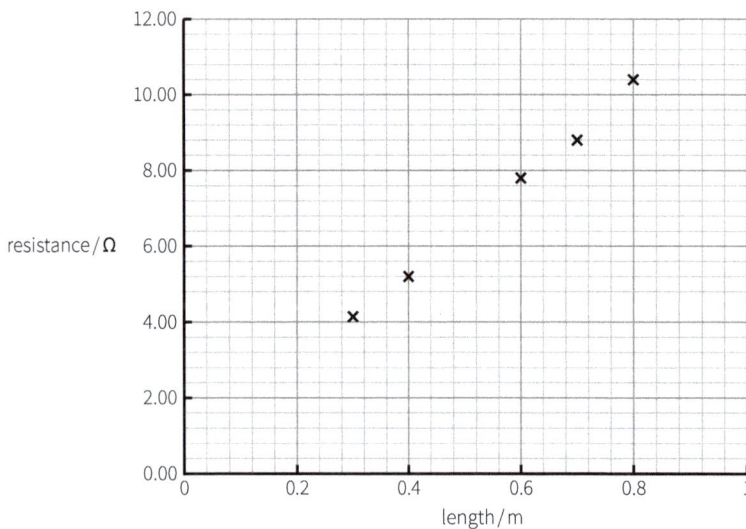

01.4 The graph suggests an error in the experiment.
State the type of error and suggest a possible cause.   **[2 marks]**

_____

_____

_____

_____

> **Exam tip**
>
> Never force a line through the origin even if that is what you are expecting.

01.5 Calculate the gradient of your line of best fit. **[2 marks]**

gradient = _____

01.6 The diameter of the wire is 0.22 mm.
Determine the resistivity of the wire using your answer to **01.5**.
[3 marks]

resistivity = _____ Ω m

01.7 The actual resistivity of the constantan wire used is $4.9 \times 10^{-7}\,\Omega$ m.
Comment on the accuracy of your result. **[2 marks]**

_____

_____

_____

02 A lightning strike is a discharge of electricity between the
atmosphere and the ground.

02.1 The current in a lightning strike is 30 kA and delivers 15 C of charge.
Calculate the time that the lightning strike lasts. **[1 mark]**

Synoptic link

3.6.2.1

time = _____ s

02.2 Calculate the number of electrons that flowed during the strike,
using data from the data sheet. **[2 marks]**

number of electrons = _____

02.3 The potential difference during a lightning strike is 40 MV.

Calculate the energy that is transferred during the lightning strike.

[1 mark]

energy = _____ J

02.4 When lightning strikes sand, the sand can melt to form a glass tube called a fulgurite. The average mass of a fulgurite is 580 g.

Calculate the energy that is required to melt this much sand, assuming the temperature rise needed is 1800 °C.

State whether the lightning strike in **02.3** could produce a fulgurite if it struck sand.

specific heat capacity of sand = 830 J kg^{-1} K^{-1}

latent heat of fusion of sand = 156 kJ kg^{-1} [3 marks]

> **! Exam tip**
>
> Total energy required will include the energy required to heat sand to melting point and the energy required to melt it.

energy = _____ J

_____

03 **Figure 2** shows the current–voltage characteristics of two components, **X** and **Y**.

> **Synoptic link**
>
> 3.5.1.4

Figure 2

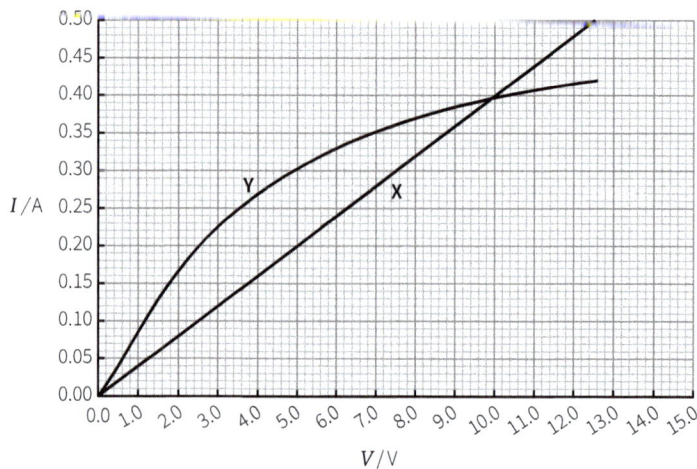

03.1 State the names the components **X** and **Y** in **Figure 2** and give a reason for your answers.   [4 marks]

03.2 Calculate the resistance in Ω of component **Y** when the voltage is 5.0 V. **[2 marks]**

03.3 The two components are connected in series.

Determine the e.m.f. in V of the power supply needed if the current is to be 0.20 A. **[2 marks]**

03.4 The two components are then connected in parallel and the e.m.f. is 5.0 V.

Calculate the total current in A in the circuit. **[2 marks]**

04 An overhead cable for the National Grid consists of 27 strands of aluminium wire of diameter 3.33 mm.

04.1 Calculate the resistance in Ω of 1 km of the overhead cable. resistivity of aluminium = $2.82 \times 10^{-8}\ \Omega\,m$ **[3 marks]**

04.2 The power losses for the transmission cables are stated as 30 W m^{-1}. Determine the current in A in the power cable. **[2 marks]**

04.3 In Germany, a 1 km length of high temperature superconductor supplies 10 000 households.

Explain what a high temperature superconductor is. **[2 marks]**

04.4 Suggest whether high temperature superconductors are a technology that more countries might adopt. **[1 mark]**

05 A thermistor can be used as a temperature sensor.

05.1 **Figure 3** shows a thermistor in a simple temperature-sensor circuit.

Figure 3

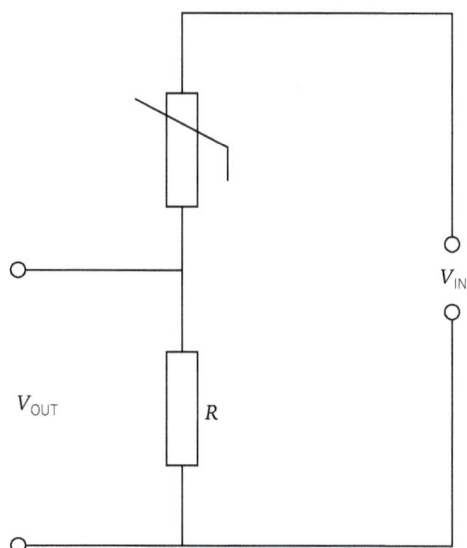

> **! Exam tip**
>
> Take care with resistance calculations – always use the values of current and voltage *at the point* to determine the resistance.

> **Synoptic link**
>
> 3.5.1.4

> **! Exam tip**
>
> Read the question for **04.2** carefully – the power loss is per metre. Your resistance calculated in **04.1** is per km.

> **Synoptic link**
>
> 3.5.1.5

> **! Exam tip**
>
> Under what conditions will the output (V_{OUT}) be high?

05.2 **Figure 4** shows how the resistance of the thermistor changes with temperature. Explain the shape of the graph. **[1 mark]**

Figure 4

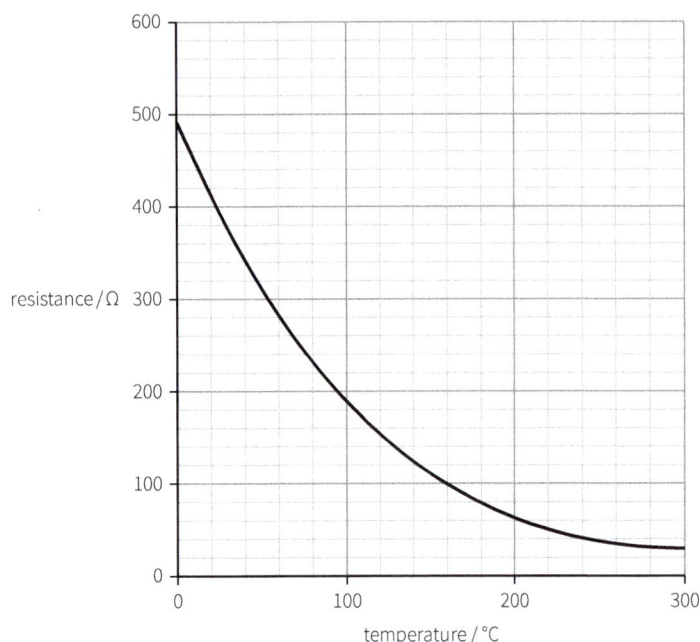

05.3 State the values of resistance in Ω at 60°C and 100°C. **[2 marks]**

05.4 The temperature sensor circuit needs to be adjustable.

The circuit needs to switch on when the temperature is either 60°C or 100°C. V_{IN} is 9.0 V and V_{OUT} is 5.0 V when the circuit is triggered. Explain, using calculations, the type of resistor and the values of resistance the resistor should have. **[3 marks]**

06 A piece of conducting paper was cut into a shape and then covered so that a group of students could not see the conducting paper. The students were asked to use a multimeter to determine the shape by taking readings of each centimetre section, as shown in **Figure 5**.

Figure 5

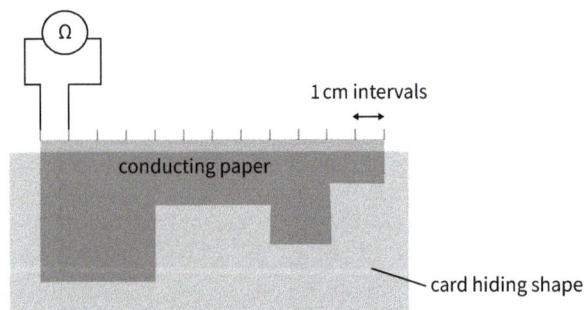

06.1 The students were asked to plot a graph of resistance against distance and use it to determine the shape of the conducting paper. Sketch the shape of this graph. **[3 marks]**

06.2 This experiment is similar to a technique used in archaeology called a resistivity survey.

Explain the difference between resistance and resistivity. **[2 marks]**

> **! Exam tip**
>
> ρ and l are constant for each cm so how are R and A related? Use this relationship to predict the shape of the graph.

06.3 A simple archaeology resistivity meter is made of a frame supporting two metal electrodes 50 cm apart, which are pushed into the ground every metre.

A current is passed between the electrodes and the potential difference measured.

Explain what this meter is recording. **[1 mark]**

06.4 Suggest how the resistivity meter can be helpful to archaeologists.
[2 marks]

07 **Figure 6** shows the p.d.–current characteristic graph for component **X**.

Figure 6

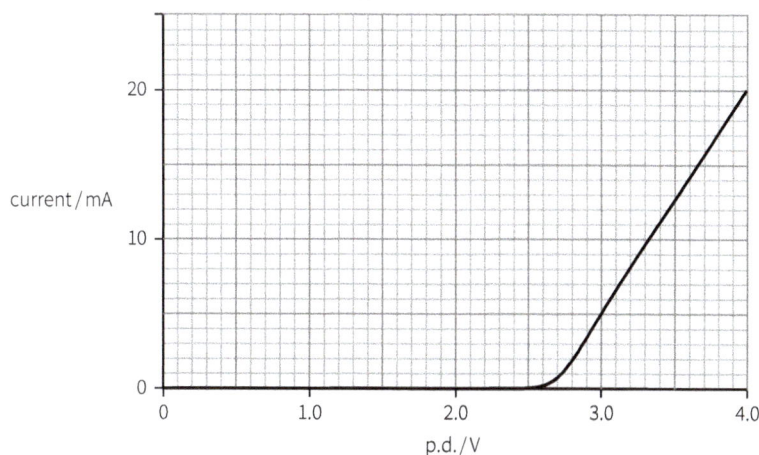

current / mA — p.d. / V

07.1 State what component **X** is and give a reason for your answer.
[2 marks]

07.2 Draw the circuit diagram for the circuit needed to produce the results in **Figure 6**. **[3 marks]**

07.3 State the resistance of the component when the potential difference is 2.0 V. **[1 mark]**

07.4 Determine the resistance of the component in Ω when the potential difference is 4.0 V. **[2 marks]**

> **Exam tip**
> You need to obtain a range of values for potential difference and current using the circuit – what component would allow you to do this?

08 Jupiter is linked to one of its moons, Io, by an electrical current of 3 MA.

08.1 The current is described as an electron beam with an energy of 20 keV.

Show that the velocity of one of these electrons is 8×10^7 m s^{-1}, ignoring relativistic effects, and using data from the data sheet.
[2 marks]

08.2 Jupiter is 4.22×10^5 km from Io.

Show that the time taken to reach Jupiter is 5 s. **[2 marks]**

08.3 Calculate the number of electrons arriving at Jupiter per second.
[2 marks]

08.4 State the direction of the current. **[1 mark]**

> **Synoptic links**
> 3.4.1.8 3.2.2.2 3.4.1.3

> **Exam tip**
> Convert the eV into J, then consider the energy changes as the electron is accelerated by the electric field.

For answers and more practice questions visit www.oxfordrevise.com/scienceanswers
Even more practice and interactive revision quizzes are available on kerboodle
9 Practice **105**

10 Circuits

Current rules

The law of **conservation of charge** states that, in a closed system, charge cannot be created or destroyed.

- Charge does not get used up or lost when it flows around a circuit.
- The same amount of charge passes through every component per second.
- The current entering a component is the same as the current leaving the component.
- The current passing through two or more components in series is the same through each component.

At any junction in a circuit, the total current entering the junction is equal to the total current leaving the junction – this is known as **Kirchhoff's first law**.

For example, in this diagram:

$$I_1 = I_2 + I_3$$

$$\Sigma I_{in} = \Sigma I_{out}$$

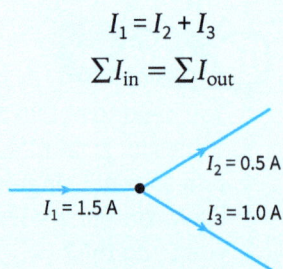

$I_2 = 0.5\,A$

$I_1 = 1.5\,A$

$I_3 = 1.0\,A$

Power and energy

Electrical power P (unit watt, W) supplied to a component can be calculated using:

$$P = IV = I^2 R = \frac{V^2}{R}$$

Total energy E (unit joule, J) transferred to the component in time t can be calculated using:

$$E = ItV$$

Potential difference rules

The potential difference (p.d.) between two points in a circuit is defined as the energy transferred per coulomb of charge that flows from one point to the other.

- If the charge carriers lose energy, the p.d. is a potential drop.
- If the charge carriers gain energy, the p.d. is a potential rise.

Due to the law of **conservation of energy**:

$$\text{energy transferred to the charge in a circuit} = \text{energy transferred from a charge in a circuit}$$

For any complete loop of a circuit, the sum of the e.m.fs around any loop in a circuit is equal to the sum of the p.d.s around the loop – this is known as **Kirchhoff's second law**.

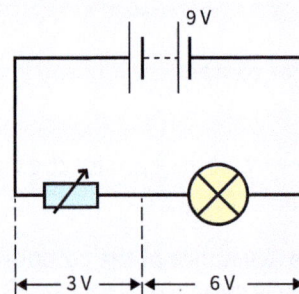

For two or more components in series, the total p.d. across all the components is equal to the sum of the p.d.s across each component.

The p.d. across components in parallel is the same.

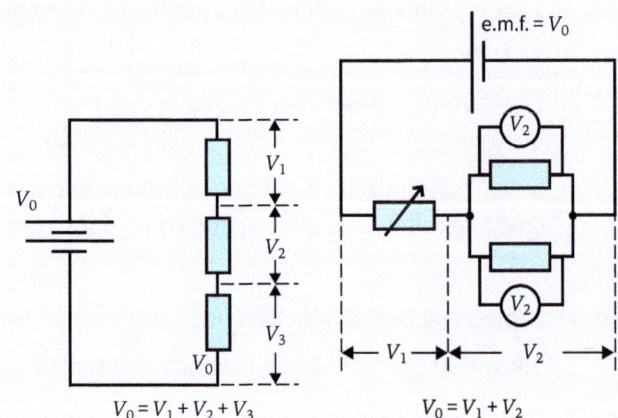

$V_0 = V_1 + V_2 + V_3$

$V_0 = V_1 + V_2$

Resistors in series

When two or more resistors are connected in a series circuit:

- current is the same through each resistor
- p.d. across any individual resistor can be calculated using $V = IR$
- p.d. is split between them in proportion to their resistances:

$$V = V_1 + V_2 + V_3 = IR_1 + IR_2 + IR_3$$

- total resistance R_{total} can be calculated using:

$$R_{\text{total}} = R_1 + R_2 + R_3$$

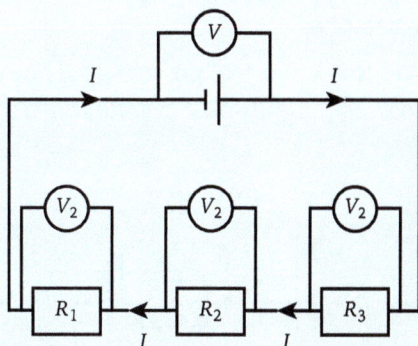

The potential divider

A potential divider circuit can be used:

- to supply a p.d. fixed at any value between zero and the pd of a source of fixed p.d.
- to supply a variable p.d.
- in sensor circuits to supply a p.d. that varies with physical conditions, such as temperature or light intensity.

A simple potential divider provides a fixed p.d. less than that of the p.d. source – it comprises two or more resistors in series with each other and the source of fixed p.d.

Resistors in parallel

When two or more resistors are connected in a parallel circuit:

- current from the supply is equal to the sum of the currents through each component:

$$I = I_1 + I_2 + I_3$$

- p.d. across components in parallel is the same:

$$V = V_1 = V_2 = V_3$$

- total resistance R_{total} can be calculated using:

$$\frac{1}{R_{\text{total}}} = \frac{1}{R_1} + \frac{1}{R_2} + \frac{1}{R_3}$$

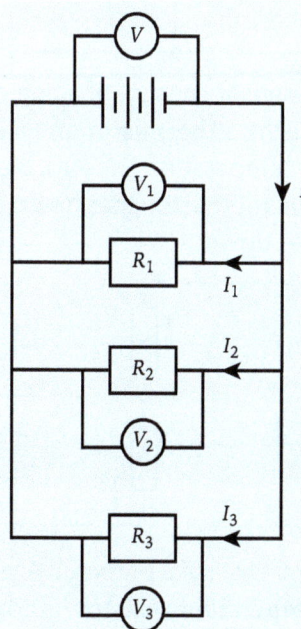

The ratio of the p.d.s across each resistor is equal to the ratio of their resistances:

$$\frac{V_1}{V_{\text{out}}} = \frac{R_1}{R_2}$$

A component connected in parallel with R_2 will have the same p.d. across it as R_2 – this is the output p.d. V_{out} which can be calculated using:

$$V_{\text{out}} = \left(\frac{R_2}{R_1 + R_2}\right) V_{\text{in}}$$

$$\frac{V_{\text{out}}}{V_{\text{in}}} = \frac{\text{output resistance}}{\text{total resistance}}$$

Variable p.d. and sensor circuits

A variable p.d. can be provided by replacing R_2 with a variable resistor.

A **potentiometer** is a variable potential divider that uses a single variable resistor connected in a way that allows V_{out} to be varied from 0 V to the maximum source p.d. This is useful in volume and light controls that need a range from zero to maximum.

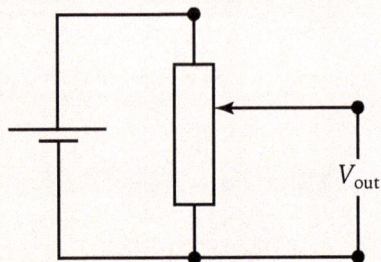

A temperature sensor consists of a potential divider made using a **thermistor** and a variable resistor. As the temperature goes up, the resistance of the thermistor goes down, so the output p.d. goes down.

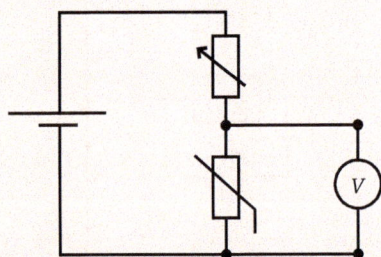

A light sensor consists of a potential divider made using a **light-dependent resistor** (LDR) and a variable resistor. As the light intensity goes up, the resistance of the LDR goes down, so the output p.d. goes down.

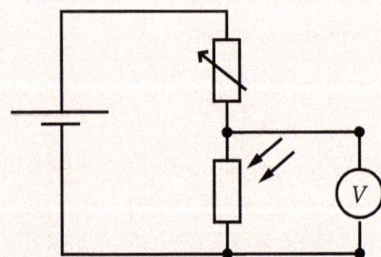

Sensors can be used to turn on connected circuits (heating or lighting) when the output p.d. goes below or above a certain p.d.

Electromotive force and internal resistance

Electromotive force (e.m.f.) ε of a power supply is the electrical energy per unit charge produced by the source. The unit of emf is the volt (V), because e.m.f. is a potential difference.

$$\varepsilon = \frac{E}{Q}$$

Internal resistance r of a power supply is the resistance to the flow of current inside the power supply due to collisions between the electrons in the current and the atoms in the supply. Power supplies can get hot when in use because of the energy transferred as a result of their internal resistance.

It is useful to think of a real power supply as a source of p.d. of magnitude ε in series with a resistor of $r\,\Omega$.

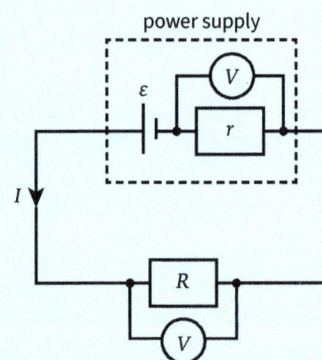

When a power supply or cell of e.m.f. ε and internal resistance r is connected to an external resistor of resistance R:

- ε = energy per coulomb supplied by the source
- I = current through the whole circuit, including the power supply
- $v = Ir$ = voltage drop across internal resistance, known as the 'lost volts'
- $V = IR$ = voltage drop across external resistor, known as the terminal p.d.

$$\text{e.m.f.} = \text{terminal volts} + \text{lost volts}$$
$$\varepsilon = IR + Ir$$

This can be rearranged to:

$$\varepsilon = I(R + r)$$

This can also be written as:

$$\text{terminal p.d.} = \text{e.m.f.} - \text{lost volts}$$
$$V = \varepsilon - Ir$$

The **terminal p.d.** V can be measured by connecting a high resistance voltmeter directly across the terminals of the power supply or cell (since I is effectively **zero**, so $V = \varepsilon - Ir$ becomes $V = \varepsilon$.

Since $V = \varepsilon - Ir$, a graph of terminal p.d. V against current I for a power supply will have:

- gradient of $-r$, the negative of the internal resistance of the cell

- y-intercept of e.m.f. ε.

gradient $= \dfrac{A-B}{B-C} = \dfrac{0.5}{-1.2}$

e.m.f. = 1.5V   So, internal resistance = 0.42 Ω

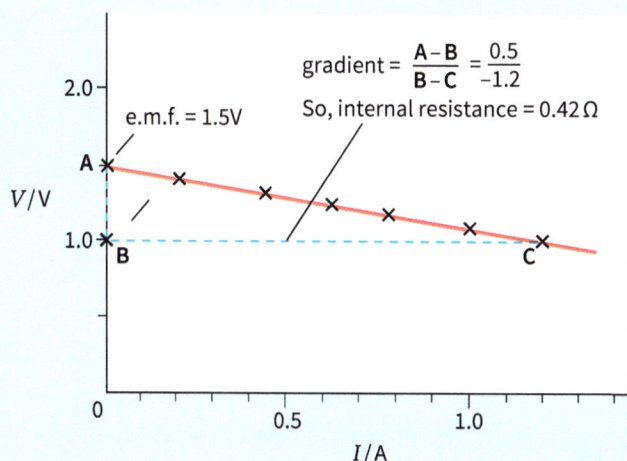

Cells in series

Connecting cells in series can provide more energy per coulomb to the charge flowing in the circuit.

If cells are connected in series in the same direction in the circuit, the total e.m.f. supplied to the circuit is the sum of the individual e.m.f.s.

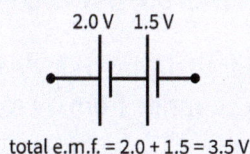

2.0 V   1.5 V

total e.m.f. = 2.0 + 1.5 = 3.5 V

If cells are connected in series in opposite directions in the circuit, the total e.m.f. supplied to the circuit is the difference between the individual e.m.f.s.

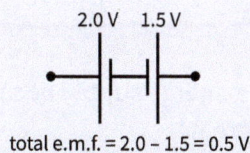

2.0 V   1.5 V

total e.m.f. = 2.0 – 1.5 = 0.5 V

The **total internal resistance** is the sum of the individual internal resistances.

Cells in parallel

Connecting cells in parallel can provide a longer lasting energy supply because the total store of energy is greater.

If the cells have no internal resistance:

- the total p.d. supplied to the circuit is equal to the p.d. of just one of the cells

- the current is the same as when there is just one cell.

If n identical cells with an internal resistance r are connected in parallel:

- the current through each cell is $\dfrac{I}{n}$

- the lost p.d. in each cell is $\dfrac{I}{n} r = \dfrac{Ir}{n}$

- the terminal p.d. across each cell is $V = \varepsilon - \dfrac{I}{n} r$

- the parallel combination of cells acts as a source of e.m.f. ε and internal resistance $\dfrac{r}{n}$.

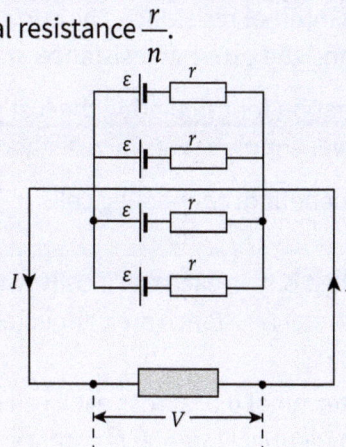

⇄ Retrieval

Learn the answers to the questions below, then cover the answers column with a piece of paper and write as many as you can. Check and repeat.

Questions	Answers
1 State the law of conservation of charge.	charge cannot be created or destroyed, so the amount of charge in a closed system remains the same
2 State Kirchhoff's first law.	total current leaving any junction in a circuit is equal to the total current entering that junction
3 Define the potential difference between two points in a circuit.	energy transferred per coulomb of charge that flows from one point to the other
4 What is a potential drop in a circuit?	p.d. between two points in a circuit if the charge carriers lose energy
5 State Kirchhoff's second law.	sum of the e.m.f.s around any loop in a circuit is equal to the sum of the p.d.s around the loop
6 Give the potential difference rule for two or more components in series.	total p.d. across all the components is equal to the sum of the potential differences across each component
7 Give the equation for calculating the total resistance for two or more resistors connected in series.	$R_{\text{total}} = R_1 + R_2 + R_3$
8 Give the equations for calculating the electrical power supplied to a component.	$P = IV = I^2R = \dfrac{V^2}{R}$
9 What can a potential divider circuit do as part of a sensor circuit?	supply a p.d. that varies with physical conditions
10 What is a potentiometer?	variable potential divider that uses a single variable resistor to allow V_{out} to be varied from $0\,\text{V}$ to the maximum source p.d.
11 Define the electromotive force of a power supply and give its unit.	electrical energy per unit charge produced by the source, in volts
12 Give the equation relating the e.m.f. of a power supply to its internal resistance, the current in the circuit, and the external resistance.	$\varepsilon = I(R + r)$
13 Give the equation for calculating the lost p.d. across a power supply's internal resistance.	lost p.d. $= Ir$
14 What is the benefit of connecting cells in parallel?	providing longer-lasting energy supply because the total store of energy is greater
15 What is the lost p.d. in each cell if cells with identical internal resistance are connected in parallel?	$\dfrac{I}{n}r = \dfrac{Ir}{n}$
16 What is the terminal p.d. across each cell if cells with an identical internal resistance are connected in parallel?	$V = \varepsilon - \dfrac{I}{n}r$

Put paper here

Practical skills

Practise your practical skills using the worked example and practice questions below.

Determining e.m.f. and internal resistance

The internal resistance r of a cell or battery means that the potential difference V across the terminals of the cell is less than the e.m.f. E:

$$V = E - Ir$$

This equation has the form of the equation for a straight line:

$$y = mx + c$$

where V is plotted on the y-axis and I on the x-axis, so $-r$ is the gradient and E is the intercept with the y-axis.

Worked example

Question

A student measures the potential difference and current through a standard AA battery, and obtains the following results.

p.d. / V	I / mA
1.34	39.9
1.32	45.2
1.31	50.1
1.30	54.6
1.29	58.0
1.28	62.8
1.27	67.3
1.26	72.5
1.25	79.0
1.23	83.8

Determine the e.m.f. and internal resistance of the battery.

Answer

Plot a graph of V against I.

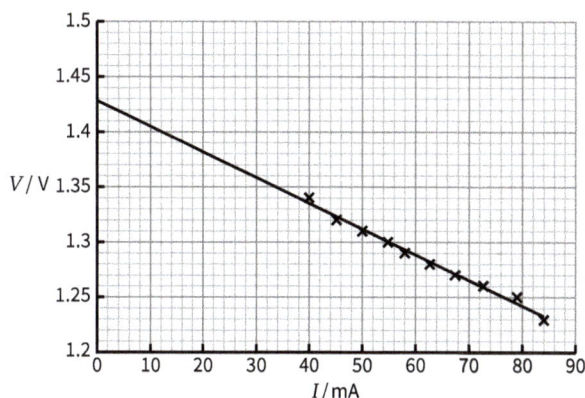

The intercept on the y-axis = 1.43 V; the gradient = -0.0023 V mA^{-1} = -2.3 V A^{-1}

Therefore, the e.m.f. of the battery = 1.43 V and the internal resistance = 2.3 Ω

Practice

A student measures the p.d. and current through a standard lithium cell.

1 Sketch a diagram of the circuit that could be used to measure the p.d. and current.

2 When taking measurements, the student connects the circuit for as short a time as possible. Explain why.

3 The student plots a graph of the data and obtains the values: y-axis intercept = 3.15 V; gradient = -0.015 V mA^{-1}.

Determine the e.m.f. and internal resistance of the lithium cell.

Practice

Exam-style questions

01 A student investigates the e.m.f. and internal resistance of a solar cell.

01.1 **Figure 1** shows the symbol for a solar cell.
Draw a diagram using **Figure 1** to show the circuit the student needs to set up to find the e.m.f. of the solar cell. **[2 marks]**

Figure 1

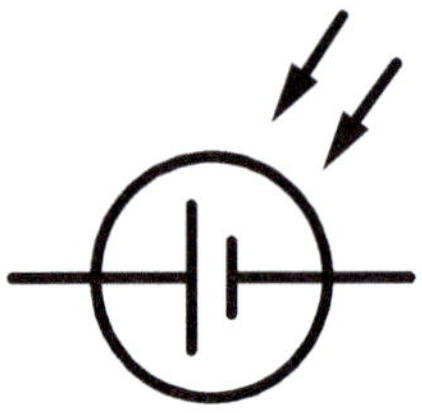

01.2 The student read that the e.m.f. of a solar cell depends on the intensity of the incident light, and that solar cells are temperature sensitive. Describe how the student could control the light intensity and temperature when taking readings. **[2 marks]**

Exam tip

This is probably an unfamiliar investigation, but use your practical knowledge to suggest how you could control the light intensity incident on a cell without changing its temperature.

01.3 Describe the readings the student would need to take from their circuit, and how they could use the readings to find the e.m.f. and internal resistance of the solar cell. **[4 marks]**

01.4 The data for the solar cell states: open circuit potential difference = 8.2 V; typical current = 100 mA; typical potential difference = 5.5 V.

Suggest why the data states a typical potential difference and current. **[1 mark]**

01.5 Determine the internal resistance of the solar cell, using the typical values of potential difference and current. **[3 marks]**

internal resistance = _____ Ω

02 Two lamps, **A** and **B**, are connected in series in a circuit.

02.1 State how the conservation of charge can be applied to circuits. **[1 mark]**

02.2 State how the conservation of energy can be applied to circuits. **[1 mark]**

02.3 Lamp **A** glows more brightly than lamp **B**.
Explain which lamp, **A** or **B**, has the higher resistance. **[3 marks]**

02.4 The two lamps are now placed in parallel.
State which lamp is brightest and explain your answer. **[3 marks]**

> ⓘ **Exam tip**
>
> The brightness of a lamp is a measure of the energy transferred per second or the power of the lamp. How can you relate power to resistance?

03 A solar-powered water feature for a garden is described as able to pump 191 litres per hour at a height of 30 cm.

03.1 Show that the water feature pumps water at 0.05 kg s^{-1}.
density of water = 1000 kg m^{-3} **[2 marks]**

03.2 Calculate the power output in W of the pump. **[2 marks]**

03.3 Under optimum conditions, the solar cell is described as having 1.2 W available and a potential difference of 5.0 V across the pump.
Calculate the maximum current in A through the pump. **[1 mark]**

03.4 The e.m.f. of the solar cell is 6.0 V.
Calculate the internal resistance in Ω of the cell. **[3 marks]**

> ⊗ **Synoptic links**
>
> 3.4.1.7   3.4.1.8   3.4.2.1

> ⓘ **Exam tip**
>
> To convert litres per hour to kg s^{-1}, first convert into cm^3 s^{-1}, then to m^3 s^{-1}.

04 Three lamps, **A**, **B**, and **C**, have different ratings.

04.1 Calculate the current in each lamp if they have the following ratings:
A is 0.7 W and 3.5 V; **B** is 1.95 W and 6.5 V; **C** is 0.3 W and 1.5 V. **[3 marks]**

04.2 The three lamps are connected in a circuit with a power supply of e.m.f. 9.0 V and negligible internal resistance. Three resistors, R_1, R_2, and R_3, are chosen so that each lamp works at its correct operating voltage, as shown in **Figure 2**.

Figure 2

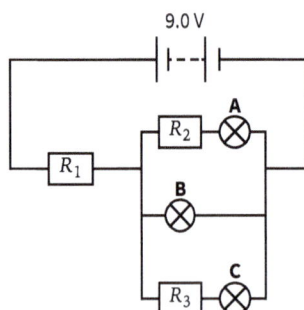

State the current in **A** flowing through R_1. [1 mark]

04.3 Calculate the resistance in Ω of R_1. [2 marks]

04.4 Calculate the resistance in Ω of R_3. [2 marks]

05 **Figure 3** shows two identical cells connected in series and in parallel.

Figure 3

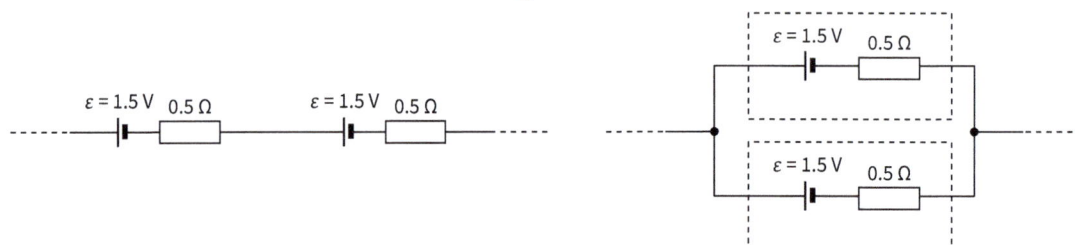

05.1 Calculate the total e.m.f. in V and internal resistance in Ω when the cells are connected in series. [2 marks]

05.2 Calculate the total e.m.f. in V and internal resistance in Ω when the cells are connected in parallel. [2 marks]

05.3 A student uses three of these identical cells to deliver an e.m.f. of 3.0 V and a maximum current of 4 A.
State the combination of cells the student must have used to achieve these values. [1 mark]

05.4 The student uses the combination of cells from **05.3** and connects them to a 2.0 Ω load resistor.
Determine the current in A flowing through the load resistor. [2 marks]

05.5 Calculate the power in W that transferred to the load resistor. [2 marks]

06 A student is given three resistors of different values, 33 Ω, 110 Ω, and 67 Ω.

06.1 Draw a diagram showing how the student should connect the resistors to give the largest possible resistance. [1 mark]

06.2 Calculate the resistance in Ω of the arrangement drawn in **06.1**. [1 mark]

> **Exam tip**
>
> If you were unsure how to answer **05.3**, you can still answer the next questions. Use the potential difference and maximum current to determine the internal resistance of the combination and carry on as if it were one cell with these values.

06.3 Draw a diagram showing how the student should connect the resistors to give the smallest possible resistance. **[1 mark]**

06.4 Calculate the resistance in Ω of the arrangement drawn in **06.3**. **[2 marks]**

06.5 Draw a diagram showing how the student could connect four of the 33 Ω resistors to give a total resistance of 33 Ω. **[1 mark]**

06.6 Suggest **one** advantage of building the circuit described in **06.5**. **[1 mark]**

> **! Exam tip**
>
> Think about the current in each branch of the circuit.

07 A 50 W heater in the rear window of a car consists of eight strips of a conducting material connected in parallel, shown in **Figure 4**.

Figure 4

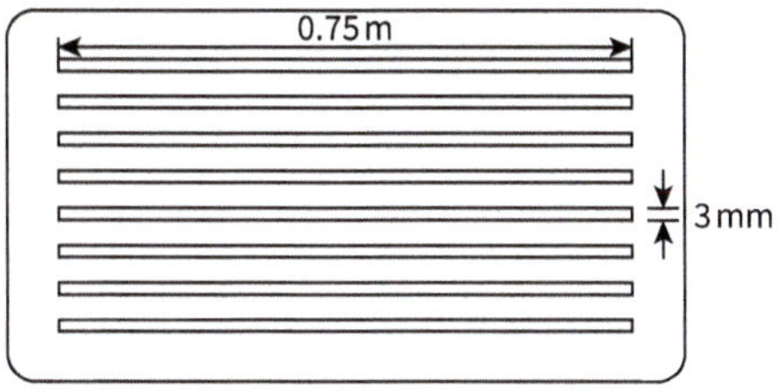

> **Synoptic links**
>
> 3.5.1.3   3.1.1   3.1.3

07.1 The potential difference across the strips is 12 V. Calculate the total resistance in Ω of the heater. **[2 marks]**

07.2 Show that the resistance of each strip is 23 Ω. **[2 marks]**

07.3 The resistivity of the material used in the strips is $1.1 \times 10^{-5}\,\Omega\,m$. Each strip is 0.75 m long and 3 mm wide. Calculate the thickness in mm of the strips. **[2 marks]**

07.4 One of the strips breaks, leaving a gap. The owner of the car replaces it by filling in the 1 cm gap with a conducting paint. The label on the conducting paint says it has a resistivity of $5\,\mu\Omega\,cm$. Determine the resistivity of the paint in Ω m. **[1 mark]**

> **! Exam tip**
>
> If in doubt, think how you would convert from cm to m – this will be the same process.

07.5 Without doing any further calculations, the effect this will have on the overall resistance of the mended strip. **[2 marks]**

08 The datasheet for a light dependent resistor (LDR) states that, in the light, the minimum resistance is 5.4 kΩ and, in the dark, the resistance is 1 MΩ.

08.1 Draw a simple potential divider arrangement, using the LDR and a fixed resistor R, whose output could be used to sense when it is light. Show this by clearly marking V_{out}. **[2 marks]**

08.2 Explain what will happen to the current in the circuit and the value of V_{out} when the light intensity increases. **[2 marks]**

> **! Exam tip**
>
> The labels are important in **08.1**– make sure you have labelled your arrangement clearly.

08.3 The sensor is intended to switch on as light level increases. There is a choice of three values for the fixed resistor, 1 MΩ, 10 kΩ, or 1 kΩ. The emf of the power supply is 6.0 V. Explain, using appropriate calculations, which resistor is the best choice for the fixed resistor. **[4 marks]**

⚙ Knowledge

11 Circular motion

Uniform circular motion

An object is in uniform circular motion if it is moving in a circle at a constant speed.

Key properties of an object moving in uniform circular motion

Period T = time taken to complete one rotation in s.

Frequency f = number of revolutions per second in Hz.

Period and frequency are related by the formula:

$$f = \frac{1}{t}$$

Radius r = distance from object to centre of circle in m.

Distance travelled in one revolution in m = $2\pi r$.

Linear speed v = distance travelled per second in $m\,s^{-1}$:

$$v = \frac{\text{circumference}}{\text{period}}$$

$$= \frac{2\pi r}{T}$$

The direction of the velocity at any instant is always at a tangent to the circle.

Angular speed ω = angle turned through per second in $rad\,s^{-1}$:

$$\omega = \frac{\theta}{t} = \frac{2\pi}{T} = 2\pi f$$

Magnitude of the angular speed is related to the linear speed by:

$$\omega = \frac{v}{r}$$

In questions involving uniform circular motion, angles are expressed in radians (rad), where $360° = 2\pi$ rad

Centripetal acceleration

An object in uniform circular motion is always accelerating even though it is travelling at a steady speed because it is constantly changing direction, so is constantly changing velocity (and therefore acceleration).

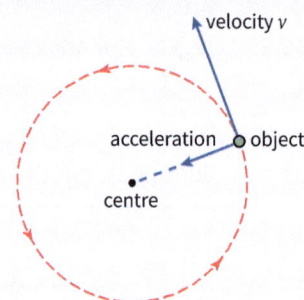

The acceleration of an object in circular motion is called the **centripetal acceleration** because the change in direction of the velocity is always towards the centre of the circle:

$$\text{centripetal acceleration } a = \frac{v^2}{r} = \omega^2 r$$

Centripetal force

Since an object in circular motion is accelerating, it must have a resultant force acting on it. This is called the centripetal force, and it always acts towards the centre of a circle.

$$\text{centripetal force } F = \frac{mv^2}{r} = m\omega^2 r$$

It is important to note that centripetal force is not a *type* of force, but the name given to the *resultant* of the forces acting on an object moving in uniform circular motion.

The centripetal force can be due to a single force or the vector sum of two or more forces.

Key examples of centripetal force

For a vehicle going round a bend, the centripetal force is friction between the car's tyres and the road surface.

For a vehicle going over a hill or curved bridge, the centripetal force is the resultant of the weight mg and the support force from the road S:

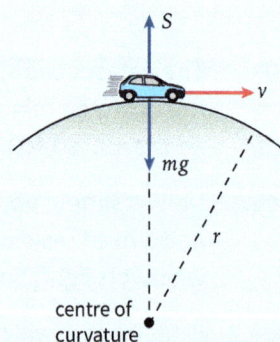

$$\frac{mv^2}{r} = mg - S$$

If the vehicle is going too fast, it will lose contact with the road, at which point there is no support from the road and $\frac{mv^2}{r} = mg$, so the maximum speed at which a vehicle can stay in contact with the road is given by:

$$v = \sqrt{(gr)}$$

For a vehicle on a banked track, the centripetal force is the horizontal component of the normal reaction to the surface:

$$\frac{mv^2}{r} = N_H = N\sin\theta$$

Since $mg = N\cos\theta$, it can be shown that for a vehicle on a banked track:

$$v^2 = gr\tan\theta$$

This can also be applied to the tension in the string of a conical pendulum or the lift force of a turning aircraft.

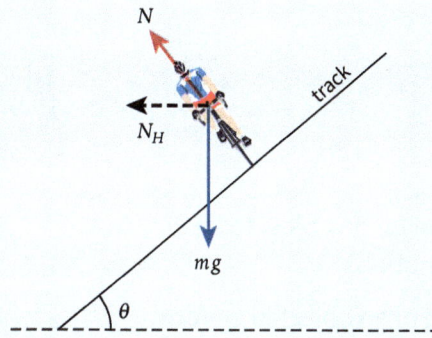

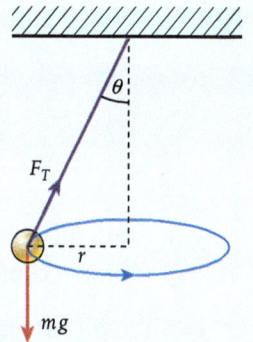

Circular motion in a vertical circle

If an object is moving in uniform circular motion in a vertical circle, the magnitude and direction of its weight must be considered when calculating the centripetal force acting on it.

For an object suspended by a string

At the top of the circle:

$$\text{centripetal force } \frac{mv^2}{r} = T + mg$$

$$T = \frac{mv^2}{r} - mg$$

At the bottom of circle:

$$\text{centripetal force } \frac{mv^2}{r} = T - mg$$

$$T = \frac{mv^2}{r} + mg$$

The tension is less at the top of the circle than at the bottom.

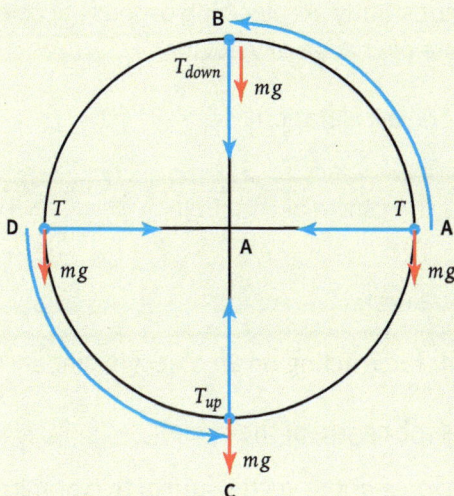

For an object supported by a surface

For example, a person on a rollercoaster going round a vertical loop.

At the top of the loop:

$$\text{centripetal force } \frac{mv^2}{r} = N + mg$$

$$N = \frac{mv^2}{r} - mg$$

At the bottom of the loop:

$$\text{centripetal force } \frac{mv^2}{r} = N - mg$$

$$N = \frac{mv^2}{r} + mg$$

The reaction force is greater at the bottom of the loop than at the top, so the person 'feels' heavier at the bottom of the ride than the top.

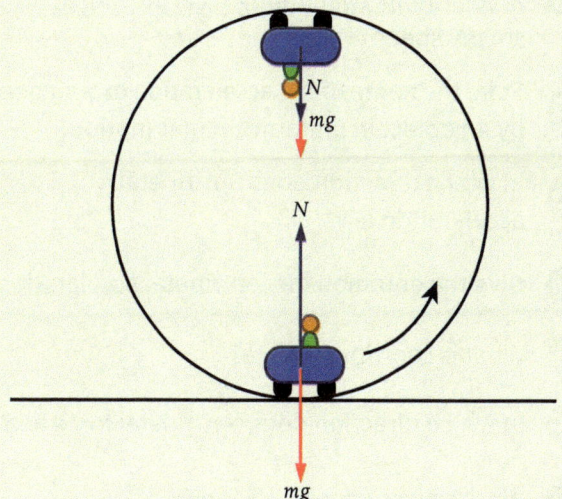

Retrieval

Learn the answers to the questions below, then cover the answers column with a piece of paper and write as many as you can. Check and repeat.

	Questions		Answers
1	State the condition for an object to be in uniform circular motion.		moving in a circle at a constant speed
2	What is the period of an object in uniform circular motion?		time taken to complete one rotation
3	Give the relationship between the period and frequency of an object in uniform circular motion.		$f = \dfrac{1}{T}$
4	Give the distance travelled in one revolution by an object in uniform circular motion.		$2\pi r$
5	Give the linear speed of an object in uniform circular motion.		$v = \dfrac{\text{circumference}}{\text{period}} = \dfrac{2\pi r}{T}$
6	What is the direction of the velocity of an object in uniform circular motion?		tangent to the circle
7	Give the angular speed of an object in uniform circular motion.		$\omega = \dfrac{\theta}{t} = \dfrac{2\pi}{T} = 2\pi f$
8	Give the relationship between the angular speed and the linear speed of an object in uniform circular motion.		$\omega = \dfrac{v}{r}$ or $v = \omega r$
9	State the unit used to express angles for an object in circular motion.		radian (rad)
10	What is the conversion factor between degrees and radians?		$360° = 2\pi$ rads
11	Why is an object in uniform circular motion accelerating, even though it is moving at a steady speed?		object is constantly changing direction and velocity, so it is constantly accelerating because acceleration is the rate of change of velocity
12	State the name of the acceleration experienced by an object in uniform circular motion.		centripetal acceleration
13	In which direction does centripetal acceleration act?		towards the centre of the circle
14	Give the equation for centripetal acceleration.		$a = \dfrac{v^2}{r} = \omega^2 r$
15	What is centripetal force?		resultant force acting on an object in circular motion
16	In which direction does centripetal force act?		towards the centre of the circle
17	What causes centripetal force?		a single force or the vector sum of two or more forces acting on an object in uniform circular motion

Put paper here

18	What provides the centripetal force for a vehicle going round a bend?	friction
19	What provides the centripetal force for a vehicle going over a hill or curved bridge?	resultant of the weight $m g$ and the support force from the road S
20	What must be considered when calculating the centripetal force of an object moving in a vertical circle?	magnitude and direction of its weight

Put paper here

🖩 Maths skills

Practise your maths skills using the worked example and practice questions below.

Centripetal force

The equation that describes the centripetal force acting on any object travelling in a circular path is:

$$F = \frac{m v^2}{r}$$

where F = centripetal force (N)

m = mass of object (kg)

v = velocity of object (m s⁻¹)

r = radius of orbit (m)

angular velocity of the object

$$\omega = \frac{2\pi}{T} = \frac{v}{r}$$

so

$$F = m \omega^2 r$$

Worked example

Question

The Hubble space telescope orbits the Earth in a circular orbit with a radius of 6.9×10^6 m. It orbits once every 95 minutes. The mass of the telescope is 11×10^3 kg.

a Calculate the angular velocity of the Hubble telescope.

b Calculate the magnitude of the centripetal force that acts on the telescope.

Answer

a angular velocity $\omega = \dfrac{2\pi}{T}$

Convert the period of the orbit into seconds:

$T = 95 \times 60\,\text{s} = 5700\,\text{s}$

$\omega = \dfrac{2\pi\,\text{rad}}{5700\,\text{s}} = 1.1 \times 10^{-3}\,\text{rad s}^{-1}$

b $F = m \omega^2$

$F = 11 \times 10^3\,\text{kg} \times (1.1 \times 10^{-3}\,\text{rad s}^{-1})^2 \times 6.9 \times 10^6\,\text{m}$

$F = 9.2 \times 10^4\,\text{N}$

Practice

1 A large Ferris wheel has a diameter of 130 m and completes a full rotation once every half an hour.

Calculate the angular velocity of the wheel.

2 An electron in a hydrogen atom can be modelled as orbiting with a circular radius of 5×10^{-11} m. In this model, the period of rotation of the electron is 1.5×10^{-16} s.

Calculate the angular velocity of the electron.

3 In athletics, the hammer throw involves an athlete spinning round on the spot with a hammer. The hammer is a large lead ball of mass 4.0 kg on a steel wire of length 1.1 m. In one event the hammer was spun at a constant speed of 8.6 m s⁻¹.

Calculate the tension in the steel wire.

Practice

Exam-style questions

01 The planet Venus orbits the Sun with a nearly circular orbit. It has a 'year' of 225 days.

01.1 Write the condition for any object moving in a circle. **[1 mark]**

_____

_____

01.2 Calculate the angular velocity of Venus.
Assume that Venus moves in a circle. **[2 marks]**

angular velocity = _____ rad s^{-1}

01.3 The radius of the orbit is 67.24 million miles.
Calculate the centripetal acceleration.

1 mile = 1609 m **[3 marks]**

> **Exam tip**
>
> Remember, the centripetal force is the force required to keep an object moving in a circle, and it is provided by a real force (gravity, friction, and so on).

centripetal acceleration = _____ m s^{-2}

01.4 The centripetal acceleration is produced by a force of 5.6×10^{22} N exerted on Venus by the Sun.
Calculate the mass of Venus. **[1 mark]**

mass = _____ kg

01.5 Venus is the Earth's sister planet. It has approximately the same mass as the Earth. The Earth orbits at 1.5×10^{11} m.

Calculate the centripetal acceleration of the Earth in orbit.
Compare this with the Venus in orbit. **[3 marks]**

centripetal acceleration = _____ m s^{-2}

_____

_____

02 A car of mass 1600 kg passes over the crest of a hill that follows the arc of a circle, as shown in **Figure 1**.

Figure 1

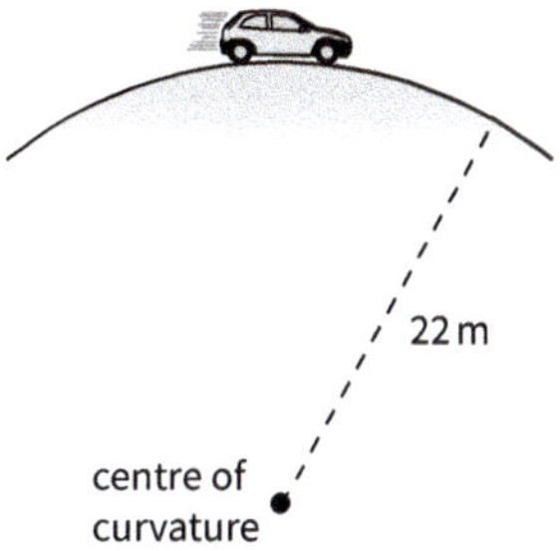

The radius of the circle is 22 m. The car is coasting over the hill, meaning the engine is no longer being used to move the car. Assume there is no friction.

02.1 Draw a free-body diagram for the car when it is on the crest of the hill shown in **Figure 1**. Label the forces. **[2 marks]**

02.2 Describe how the normal force changes as the speed of the car changes from zero. **[2 marks]**

_____

_____

02.3 The car is moving at $9\,m\,s^{-1}$.

Calculate the normal force.

Show your method. **[3 marks]**

> **① Exam tip**
>
> Remember that the normal force is what you experience as weight.

normal force = _____ N

02.4 Calculate the maximum speed at which the car can travel over the hill, and still stay on the road. **[2 marks]**

speed = _____ $m\,s^{-1}$

03 A student moves a cork on the end of a string in a horizontal circle, as shown in **Figure 2**.

Figure 2

03.1 State the force responsible for keeping the cork moving in the circle.

[1 mark]

03.2 Calculate the tension in N in the string, using estimates for any required values.

[2 marks]

03.3 The student moves the string so it is moving in a vertical circle. Compare the tension at the top and bottom of the vertical circle to the tension calculated in **03.2**.

[2 marks]

03.4 Calculate the minimum speed in $m\,s^{-1}$ that will keep the cork moving in a vertical circle.

[2 marks]

> **(!) Exam tip**
>
> You should draw free-body diagrams if you are unsure about the forces acting on an object.

04 A cyclist rides at 12 mph, or $5.3\,m\,s^{-1}$. Each bicycle wheel has a radius of 60 cm.

04.1 Calculate the angular velocity in $rad\,s^{-1}$ of one of the wheels. [1 mark]

04.2 A small, raised part of the wheel rim rubs against the bike as it rotates, producing a noise.

Calculate the frequency in Hz of the noise produced. [1 mark]

04.3 The cyclist can ride around a roundabout at 12 mph.
State the force that enables them to ride in a circle. [1 mark]

04.4 The cyclist rides on a cycling track that has a banked surface, as shown in **Figure 3**.

Figure 3

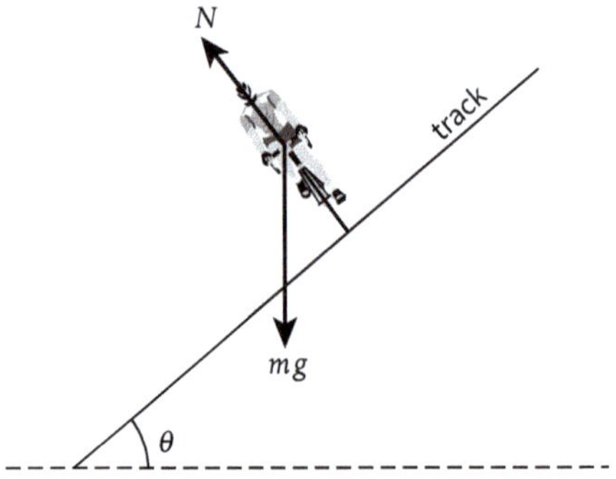

Resolve the normal force into vertical and horizontal components.

Deduce the force that provides the centripetal force keeping the cyclist moving in a circle in terms of m, g, and θ. [3 marks]

04.5 The track is banked at an angle of 15°, and the curve of the corner has a radius of 50 m.

Calculate the speed in m s^{-1} at which the cyclist could move around the track corner using your expression from **04.4**.

[2 marks]

04.6 Suggest what would happen to the frequency of the sound produced by the wheel rim when the cyclist is on the banked track.

Give a reason for your answer. [1 mark]

Exam tip

The final part of a question will often ask you to use information or values from throughout the question.

05 A theme park ride consists of a drum that can spin. People stand with their backs to the inside of the drum at the start of the ride. The drum spins until, at a certain point, the floor moves away, and the people are 'stuck' to the inside of the drum.

05.1 Explain, in terms of Newton's first law, why the people 'stick' to the inside of the drum. [2 marks]

05.2 The ride operators do not need to know the masses of the people on the ride to know when to move the floor away.

Suggest why. [3 marks]

Exam tip

Apply Newton's laws to a question by saying how they apply to the particular situation, rather than just quoting them.

05.3 The frequency of rotation of the drum is 56 revolutions per minute (rpm).

Calculate the angular velocity in rad s^{-1} and frequency in Hz of the drum. [2 marks]

05.4 Calculate the centripetal acceleration in m s^{-2} of a person rotating with the angular velocity calculated in **05.3** in a drum of radius 1.9 m.

[1 mark]

05.5 The company that makes the drum wants to market a drum with a much larger diameter. Suggest **one** benefit and **one** problem that the ride operators might have with a much wider drum. [2 marks]

06 A pilot flies a plane in a vertical loop as part of an aerobatic display, as shown in **Figure 4**.

Synoptic links

3.4.1.3   3.4.1.4

Figure 4

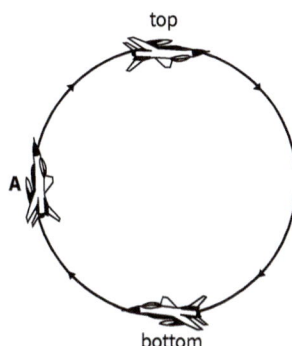

06.1 Explain why the 'weight' experienced by the pilot changes between the bottom and the top. [2 marks]

06.2 Suggest which physical force is providing the centripetal force required to move the plane in the circle. **[1 mark]**

06.3 The pilot has a mass of 70 kg, and the plane takes 12.4 seconds to complete one loop at a speed of 70 m s^{-1}.

Calculate the apparent weight in newtons of the pilot at **A**.

Assume that the pilot's head is pointed towards the centre of the circle and that the speed of the plane is constant. **[3 marks]**

06.4 In reality, the speed of the plane is not constant.

The speed at the bottom is 70 m s^{-1} and the speed decreases towards the top of the loop.

Suggest a value for the speed at **A** and describe the effect on the apparent weight you calculated in **06.3**. **[3 marks]**

06.5 As part of the display, the pilot drops a ball from the bottom of the plane when the plane is at the bottom of the loop. The ball lands in a small pool on the ground.

Describe and explain how the organisers could ensure that the pool is positioned correctly.

State any relevant equations.

06.6 Suggest and explain **one** reason why the ball may not hit the pool. State whether the ball land in front of or behind the pool.

Air resistance can be ignored. **[6 marks]**

> **!** **Exam tip**
>
> In a synoptic question, you will be asked to apply knowledge from all parts of the course. For this question, you should think of approaches that use ideas of forces, momentum, energy, etc.

07 A toy on a string attached to the ceiling moves in a circle with a constant speed. It can be modelled as a point particle as shown in **Figure 5**.

The length of the string is l, and the string makes an angle θ with the vertical. The radius of the circle is r.

> **Synoptic links**
>
> 3.4.1.1   3.1.2

Figure 5

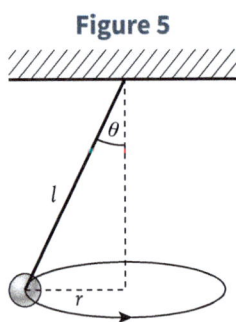

07.1 Draw a free-body diagram for the toy.

Label the forces clearly. **[2 marks]**

07.2 A similar toy also makes a circle when connected to the ceiling by the same length of string, but moves faster, and the angle θ is bigger.

Explain why the second toy moves in a circle with a bigger radius, and why you can draw no such conclusion about the mass of the second toy.

Use appropriate equations in your answer. **[4 marks]**

07.3 A student collects the following data: radius of orbit for the toy = 17 cm; length of string = 43 cm; time for one orbit = 1.3 s. Suggest how the radius and time for one orbit could be measured and estimate the percentage error in each measurement. **[6 marks]**

07.4 The strings break at the same moment for both toys. Suggest whether the toys could land at the same distance from the centres of their orbits. Explain your answer. **[2 marks]**

08 A student swings across a river using a rope attached to a tree, shown in **Figure 6**.

Figure 6

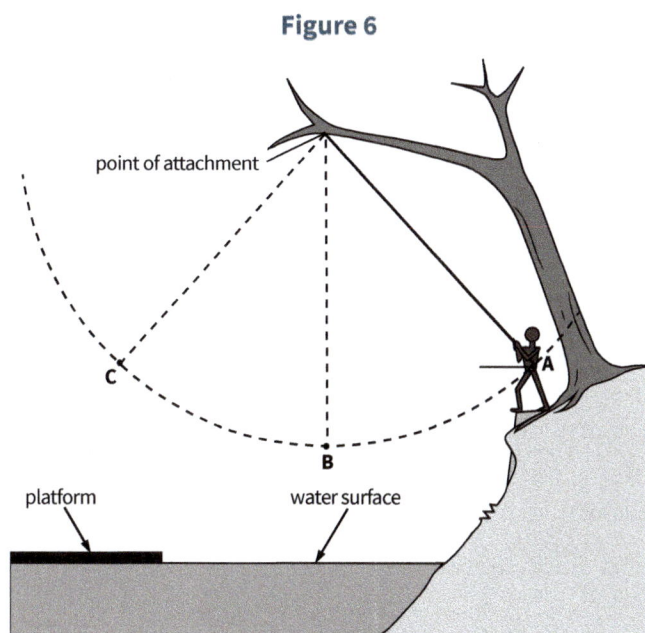

At point **A**, the student's centre of mass is a vertical distance of 2.7 m above the surface of the water. The length of the rope is 3.6 m, and point **B** is 1.4 m above the surface of the water.

08.1 Explain why the path of the student is the arc of a circle while they are holding the rope. **[1 mark]**

08.2 Calculate the speed in m s⁻¹ of the student when they reach point **B**. Ignore air resistance and assume that the rope does not stretch. Show your working. **[2 marks]**

08.3 The student lets go of the rope at point **B**. The platform on the other side of the river is a horizontal distance of 1.2 m from point **B**. Calculate the distance in m that the student travels, and state whether they will land on the platform.

Assume the platform is level with the surface of the water. **[3 marks]**

08.4 Over time, the rope stretches.

Suggest whether this would make it more or less likely that the student would reach the platform.

Explain your answer. **[2 marks]**

Synoptic links

3.4.1.3   3.4.1.8

Knowledge

12 Simple harmonic motion

Simple harmonic motion (SHM)

SHM is an oscillating motion in which:

1 the magnitude of the acceleration is proportional to the displacement from the equilibrium position.

2 the direction of the acceleration is always in the opposite direction to the displacement, directed back towards the equilibrium position.

The condition for SHM is:

$$a \propto -x$$

For an object moving with simple harmonic motion
Equilibrium position is the midpoint of its motion.

Displacement x is the distance and direction from the equilibrium position in m.

Amplitude A is the maximum displacement from the equilibrium position in m.

Time period T is the time taken in s for one complete cycle of oscillation (for example, from maximum positive displacement to maximum negative displacement and back).

Frequency f is the number of complete oscillations per second in Hz.

Frequency and period are independent of the amplitude, so they remain the same even if amplitude decreases due to energy loss.

angular frequency $\omega = \dfrac{2\pi}{T} = 2\pi f$

Phase difference between two objects oscillating with the same frequency is the fraction of a cycle by which their positions are separated.

Phase difference between two objects oscillating with the same frequency in rad $= \dfrac{2\pi \, \Delta t}{T}$, where Δt is the time between successive instants when the two objects are at maximum displacement in the same direction.

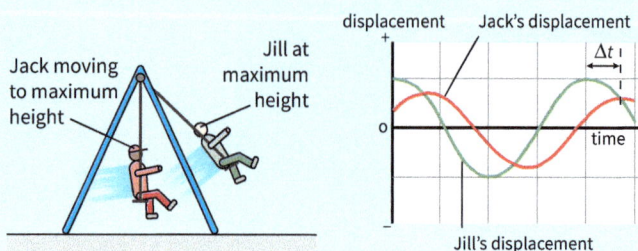

displacement   Jack's displacement

Jack moving to maximum height

Jill at maximum height

Jill's displacement

Graphs of SHM

Graphs of displacement, velocity, and acceleration against time can be produced for objects in SHM using a datalogger.

Displacement against time

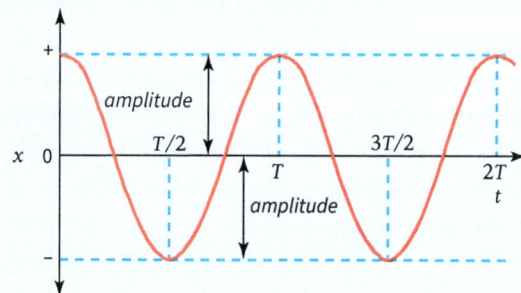

Velocity against time
Velocity is the rate of change of displacement.

$$v = \dfrac{\Delta s}{\Delta t}$$

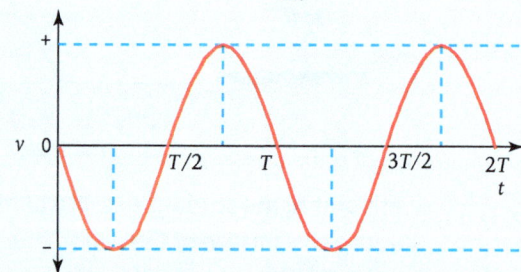

Acceleration against time
Acceleration is the rate of change of velocity.

$$a = \dfrac{\Delta v}{\Delta t}$$

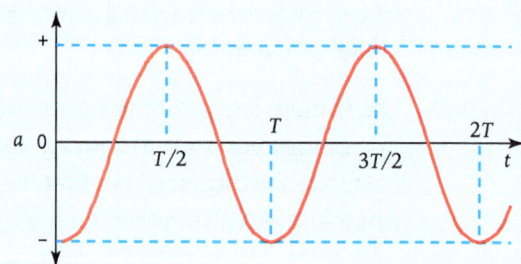

Equations of SHM

The defining equation for SHM is:

$$\text{acceleration } a = -\omega^2 x$$

If the object has an initial ($t = 0$)
velocity of zero and an initial displacement of $+A$:

- displacement x at time t is $x = A \cos \omega t$
- velocity $v = \pm\omega^2\sqrt{A^2 - x^2}$ ($\pm$ means the velocity can be in two opposite directions)
- maximum speed $= \omega A$
- maximum acceleration $a_{\max} = \omega^2 A$

The simple pendulum

Small angle approximation

A simple pendulum will be in SHM if the angle θ it makes with the vertical is <10°.

If θ is <10°, the small angle approximation ($\sin\theta = \theta$ rad) can be used to show that the acceleration of the bob is proportional to the displacement from the equilibrium and always acts towards the equilibrium.

Calculating the period

The period of a simple pendulum in SHM is:

$$T = 2\pi\sqrt{\frac{l}{g}}$$

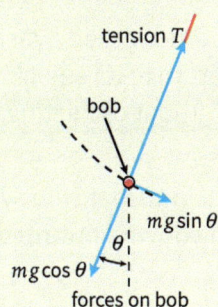

forces on bob

Mass–spring systems

Calculating the restoring force

The resultant force F due to the stretched spring is known as the **restoring force** because it always acts towards the equilibrium position.

For a spring with spring constant k stretched by a displacement x, the restoring force is:

$$F = -k x$$

Calculating the period

The period of a mass–spring system in SHM is:

$$T = 2\pi\sqrt{\frac{m}{k}}$$

Mass–spring systems can be set up in different ways to be in SHM.

Horizontal arrangement

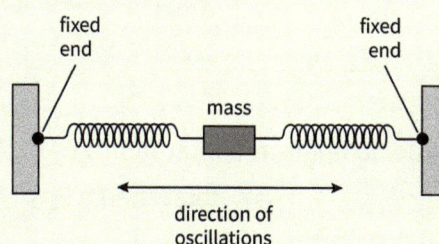

Vertical arrangement

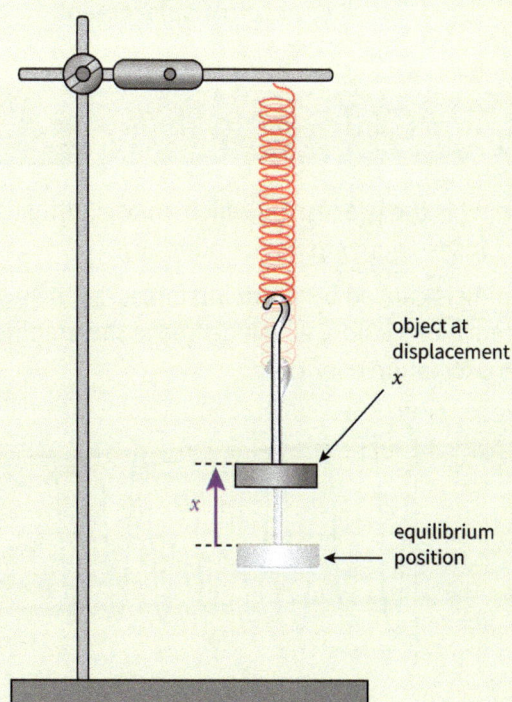

Energy of an object in SHM

There is a continuous and repeated transfer of energy from the potential energy store E_p to the kinetic energy store E_k and back again for an object in SHM.

The potential energy store can be gravitational potential or elastic potential, depending on the forces acting on the object.

Graph showing the variation with distance, x of the energy during SHM

Graph showing the variation with time, t of the energy during SHM

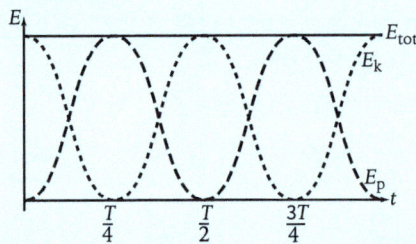

If there is no damping, the total energy E_T of the object in SHM is constant:

$$\text{total energy } E_T = E_p + E_k$$

and

$$\text{total energy} = E_p \text{ at max. displacement}$$
$$= E_k \text{ at zero displacement}$$

Damping

Damping is the process by which an oscillating object loses energy to its surroundings.

Damping is caused by 'damping forces' such as air resistance or friction, which decrease the amplitude of the oscillation over time.

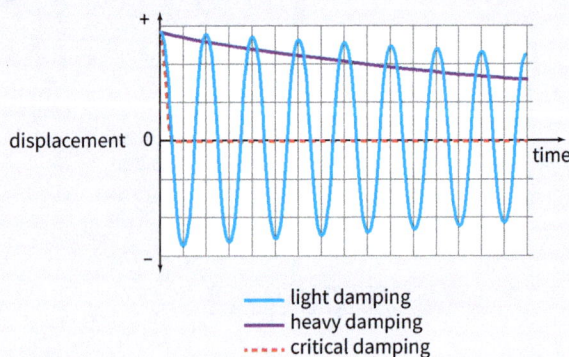

light damping
heavy damping
critical damping

Light damping
Light damping involves a small damping force, leading to a gradual decrease in amplitude.

Critical damping
Critical damping causes the oscillation to stop in the shortest possible time, returning the object to its equilibrium position without overshooting it.

Heavy damping
Heavy damping, sometimes called over damping, causes the object to return to its equilibrium position very slowly, without oscillating.

Uses of damping
Damping can be useful in suspension systems in vehicles and for cutting down unwanted sounds in recording studios.

Free and forced vibrations

Free vibration

A free vibration (oscillation) is one which occurs with no transfer of energy to or from the surroundings.

- The only forces acting on a freely oscillating object are the ones providing the resultant restoring force.
- A freely oscillating object will have a constant amplitude.

Forced vibration

A forced vibration is one where a periodic external driving force is applied.

Natural frequency

The natural frequency of an object is the frequency at which it would vibrate if left to vibrate freely, with no damping or driving forces acting on it.

Resonance

Resonance is the process by which a maximum in the amplitude of a forced oscillation is produced.

In the absence of damping, resonance occurs when driving frequency = natural frequency of an object.

The amplitude when resonance occurs depends on the degree of damping.

Uses of resonance

The lighter the damping, the larger the maximum amplitude at resonance, and the closer the resonant frequency is to the natural frequency.

Many musical instruments, microwave ovens, and MRI scanners rely on resonance to work.

Resonance can be a problem in mechanical structures and needs to be taken into consideration when constructing things such as bridges or tall buildings that can be made to resonate by high winds.

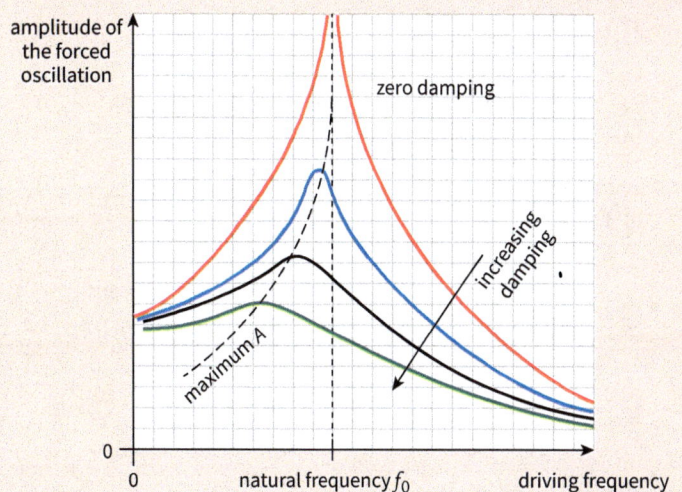

Retrieval

Learn the answers to the questions below, then cover the answers column with a piece of paper and write as many as you can. Check and repeat.

	Questions		Answers
1	State the **two** conditions for an object to be in simple harmonic motion (SHM).		magnitude of acceleration is proportional to displacement from equilibrium position; direction of acceleration is always in opposite direction to displacement
2	State the equilibrium position for an object moving with SHM.		midpoint of its motion
3	What is the displacement for an object moving with SHM?		distance and direction from equilibrium position
4	What is the time period for an object moving with SHM?		time taken for one complete cycle of oscillation
5	Define angular frequency ω.		$\omega = \dfrac{2\pi}{T} = 2\pi f$
6	Give the defining equation for SHM.		acceleration $a = -\omega^2 x$
7	Give the equation for the displacement x of an object in SHM at time t.		$x = A\cos\omega t$
8	Give the equation for the velocity of an object in SHM.		$v = \pm\omega\sqrt{A^2 - x^2}$
9	Give the equation for the maximum speed of an object in SHM.		ωA
10	Give the equation for the maximum acceleration of an object in SHM.		$a_{max} = \omega^2 A$
11	State the restoring force for a spring with spring constant k stretched by displacement x.		$F = -kx$
12	State the period of a mass–spring system in SHM.		$T = 2\pi\sqrt{\dfrac{m}{k}}$
13	What is the maximum angle a simple pendulum can make with the vertical if it is to be in SHM?		$10°$
14	State the period of a simple pendulum in SHM.		$T = 2\pi\sqrt{\dfrac{l}{g}}$
15	Define damping.		process by which an oscillating object loses energy to its surroundings, caused by 'damping forces'
16	Define a free vibration.		oscillation that occurs with no transfer of energy to or from the surroundings
17	What are the only forces acting on a freely oscillating object?		forces providing the resultant restoring force

Put paper here

18 What is a forced vibration?

Put paper here

oscillation where a periodic external driving force is applied

19 State the natural frequency of an object.

frequency at which it would vibrate if left to vibrate freely, with no damping or driving forces acting on it

20 Define resonance.

process by which a maximum in the amplitude of a forced oscillation is produced

Practical skills

Practise your practical skills using the worked example and practice questions below.

Reducing measurement errors	Worked example	Practice

Reducing measurement errors

This system used for investigating SHM is a simple pendulum.

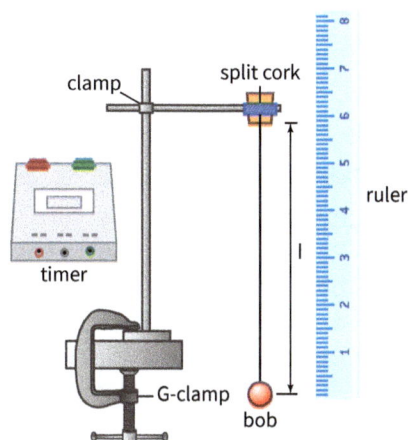

For small angles, the period of the pendulum is:

$$T = 2\pi\sqrt{\frac{l}{g}}$$

Plotting T^2 against l will give a straight-line graph with a gradient equal to $\frac{4\pi^2}{g}$.

Measurements made when investigating harmonic motion are subject to errors produced by the equipment used or human action, creating uncertainty in the results.

Worked example

Question

Explain how to reduce uncertainties in the measurement of T and l for a simple pendulum.

Answer

Estimate the delay caused by reaction time with a reaction timer. Time several oscillations, then dividing the measured time and estimated uncertainty by the number of oscillations.

A fiducial marker makes clear when the bob passes the equilibrium position.

The oscillation of the pendulum should make a small angle ($\theta < 10°$), otherwise the small-angle approximation will be invalid and the period will incorrect.

Measure the length of the pendulum from the point the string leaves the cork to the centre of mass of the bob. To reduce error, measure the diameter of the bob using Vernier callipers. Measure the string to the top of the bob, and add the radius of the bob to this. The longer the pendulum, the smaller the percentage uncertainty.

Practice

1 A mass oscillating vertically on a spring is a simple harmonic system.

Describe the practical precautions that should be taken to reduce the uncertainty in the measurement of the period of the system.

01 A student uses a datalogger to produce a graph of displacement against time for a simple harmonic oscillator, as shown in **Figure 1**.

Figure 1

01.1 Calculate the frequency of the oscillator. **[2 marks]**

frequency = _____ Hz

01.2 Calculate the maximum velocity of the oscillator using the graph and the frequency from **01.1**. **[2 marks]**

maximum velocity = _____ $m\,s^{-1}$

> **Exam tip**
> Remember that there is a link between simple harmonic motion and circular motion in terms of angular velocity.

01.3 Suggest another method that the student could use to find the maximum velocity from the graph. **[1 mark]**

01.4 Sketch the graph of acceleration against time, with maximum acceleration on the y-axis, and time on the x-axis. **[3 marks]**

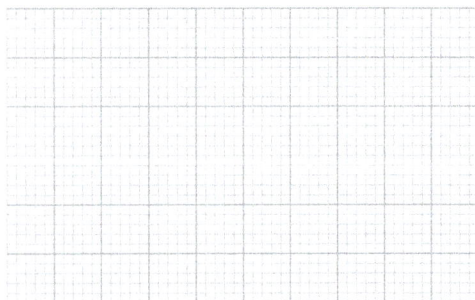

01.5 Explain how the graph in **Figure 1** and the graph in **01.4** demonstrate the condition for simple harmonic motion. **[1 mark]**

02   A student makes homemade springs by winding thick wire around a pencil. They hang up each spring, load it with a mass, and pull the mass down. The spring oscillates with a time period T.

02.1   Describe how to take the necessary measurements to find the relationship between the diameter, d, of the wire and the time period of oscillation of the mass–spring system.

State the variables, including at least **two** control variables.

Describe how you would collect sufficient reliable data to plot a graph.

[4 marks]

> **(!) Exam tip**
>
> It should be clear from any method how many measurements you need to take, and how repeat measurements can be used to calculate a mean.

_____
_____
_____
_____
_____
_____
_____
_____

02.2   Suggest how the student can reduce the uncertainty in the measurement of the time period.   **[1 mark]**

_____
_____

02.3   **Table 1** shows the measurements recorded.

Table 1

Diameter / $\times 10^{-3}$ m	1.0	1.5	2.0	2.5	3.0	3.5	4.0
Mean period / s	1.50	0.91	0.72	0.51	0.47	0.38	0.36

The student hypothesises that the spring constant of the spring is proportional to d^2.

Suggest **one** physical reason why.   **[1 mark]**

_____
_____

02.4   Suggest how the student could use the data in **Table 1** to plot a graph to confirm their hypothesis.   **[2 marks]**

_____
_____
_____
_____
_____
_____

03 A student connects a spring to an object of unknown mass m and connects the other end to a stand. When the object is displaced, it oscillates about an equilibrium position. The student has a range of different springs of known spring constants k and a stopwatch.

03.1 Describe how the student could collect data to plot a graph to find the mass of the object. You should indicate how to measure the quantities needed with accuracy and precision. **[4 marks]**

03.2 Describe **one** method that the student could use to state the uncertainty in the measurement of the time period using the stopwatch. **[3 marks]**

03.3 Suggest how the student could use light gates to measure the time period of the mass. **[2 marks]**

03.4 The student picks a spring of spring constant k. They connect an identical spring from the bottom of the mass to the stand so that both springs are under tension.

Assume that the springs obey Hooke's law under both tension and compression.

Deduce the ratio of the time period of the new arrangement to that of the old arrangement. Explain your reasoning. **[3 marks]**

Exam tip

There are different methods that can be used to estimate the uncertainty in a measurement. You should be able to choose the most appropriate for a given measurement.

04 A student wants to measure the acceleration due to gravity using a pendulum. They attach a length of string to a 1 cm diameter bob and suspend it so that the bob travels through a light gate when it is at the bottom of its swing. The light gate timing starts when the beam is broken for the first time, and ends when the beam breaks for the second time.

The student pulls the pendulum to an angle of 20° to the vertical and releases it. They record data in order to calculate the time period. They repeat the experiment with pendulums of different lengths.

Table 2 shows the student's recorded data.

Table 2

Length of pendulum / m	0.45	0.55	0.65	0.75	0.85	0.95
Time period / s	1.345	1.488	1.625	1.729	1.952	1.964

04.1 Suggest how the student could use the times recorded by the light gate to obtain an accurate value for the time period of the pendulum for a given length. **[3 marks]**

04.2 The student manipulated the data to plot a straight-line graph. They used the gradient to find a value of the acceleration due to gravity. **Figure 2** shows the graph that they plotted.

Annotate the axes in **Figure 2**, and use the graph to deduce a value for g in m s^{-2}. **[4 marks]**

Exam tip

Practice using algebra to produce equations of the form $y = mx$ from given equations such as $T = 2\pi\sqrt{\dfrac{l}{g}}$ and $T = 2\pi\sqrt{\dfrac{m}{k}}$.

Figure 2

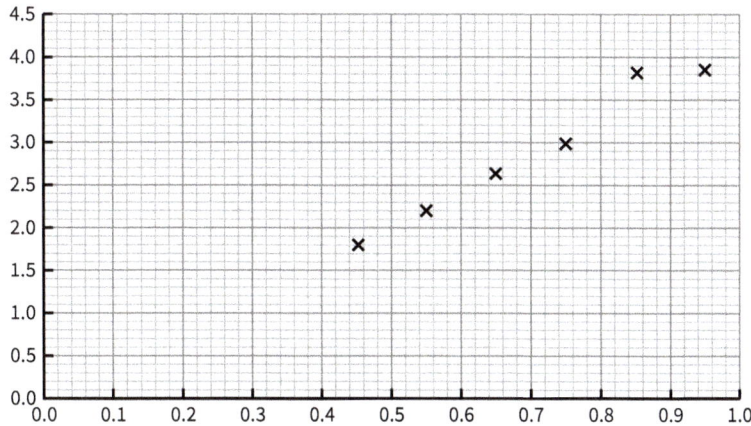

04.3 Suggest the effect on the calculated value of g of the student pulling the pendulum back to an angle that is much bigger or much smaller than 20°. **[2 marks]**

04.4 Another student carried out the same experiment but found that the straight line did not go through the origin on their graph. Suggest why. **[1 mark]**

05 A container of sand suspended above a long piece of paper is displaced from its equilibrium position. As it oscillates, sand leaks from the bottom, as shown in **Figure 3**. The paper is pulled perpendicular to the oscillation.

Figure 3

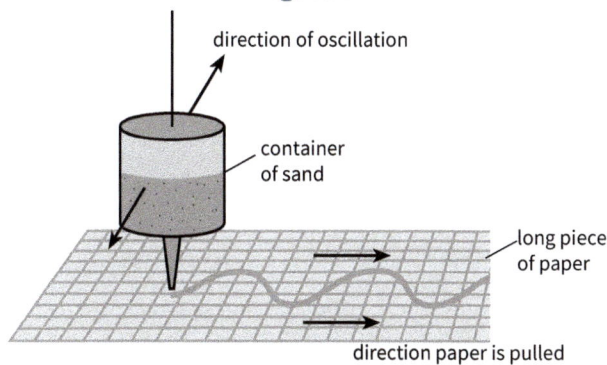

direction of oscillation

container of sand

long piece of paper

direction paper is pulled

05.1 Describe the condition for the time period of the pendulum to be independent of the mass of sand in the container. **[2 marks]**

05.2 The time period is 1.4 s, and the amplitude of the oscillation is 3.2 cm.

Deduce the equation for the displacement of the pendulum as a function of time. **[2 marks]**

05.3 The mass of sand for one oscillation stays approximately constant at 260 g.

Sketch a graph of the kinetic energy of the pendulum as a function of displacement from the equilibrium position. Write numerical values on the x- and y-axes. **[3 marks]**

> **! Exam tip**
>
> There is a condition for simple harmonic motion of a pendulum to occur; you should be able to describe this requirement.

05.4 Explain how to deduce the graph for potential energy against position on the same axes.

State any assumptions that you make. **[2 marks]**

05.5 When the pendulum has swung for many oscillations, the mass changes significantly over the course of one oscillation.

Suggest how this change would affect your graph from in **05.3**.

[1 mark]

06 Astronauts on the International Space Station need to monitor their weight while they are on missions.

06.1 Explain why the astronauts cannot use the normal bathroom scales that they would use on Earth. **[2 marks]**

Synoptic links

3.4.1.1   3.6.1.4

06.2 The astronauts attach themselves to a weighing machine that contains a seat connected to a spring. The astronaut displaces themself from the equilibrium position, and oscillates. The machine converts the time period of oscillation to a readout of mass in kg. The machine can do this because the astronaut–spring system executes simple harmonic motion.

The astronaut uses the weighing machine on Earth. Their mass is 68.62 kg and the periodic time is 2.084 s.

Calculate the spring constant in $N\,m^{-1}$ of the spring system. **[2 marks]**

06.3 To stay healthy in space, the astronaut should not lose more than 10% of their body mass.
Calculate the minimum allowable periodic time in s for this astronaut, and explain why it is a minimum. **[3 marks]**

> **Exam tip**
>
> Notice the number of significant figures in this question is larger than normal. Remember to use the correct number of significant figures when you are doing calculations.

06.4 The maximum displacement of the astronaut decreases slightly with time.

Suggest why. **[2 marks]**

06.5 Suggest and explain whether this change in maximum displacement affects the measurement of the mass of the astronaut. **[2 marks]**

07 A weighted test tube is floating in a large container of water. A student pushes the tube down, and it oscillates about an equilibrium position. A diagram of the arrangement is shown in **Figure 4**.

Synoptic link

3.4.2.1

Figure 4

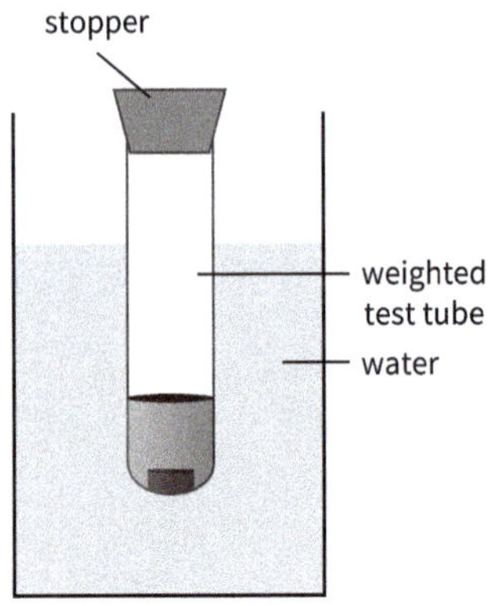

07.1 The restoring force on the test tube is equal to the additional weight of water displaced when the tube is pushed down.

Show that, if the student pushes the tube down by an additional 1.0 cm, the resultant force on the tube is 7.4×10^{-3} N.
cross-sectional area of tube = 0.75 cm²; density of water = 1 g cm⁻³

[2 marks]

Exam tip

In extended calculations, it is important to be consistent with your use of units.

07.2 State how the condition for simple harmonic motion has been met in the arrangement shown in **Figure 4**. **[1 mark]**

07.3 The mass of the tube is 12 g.

Calculate the period of oscillation of the tube in s. **[3 marks]**

07.4 A company that manufactures liquid shampoo wants to use the oscillation to measure the density of the liquid.

Describe a graph involving the time period and density of liquid that would give a straight line if the motion of the tube was simple harmonic. **[3 marks]**

07.5 To measure the time period, a light is shone on one side of the tube and is detected by a light dependent resistor (LDR) connected in a circuit on the other side of the tube.

Draw a circuit diagram for the LDR that would produce a potential difference that varies with time period. **[2 marks]**

08 A car contains springs and has a mass, so it can oscillate.

08.1 A person of weight 700 N measures a deflection of 3.0 cm when they sit on the front of the car on the driver's side.

Calculate the spring constant of the spring in N m⁻¹. **[1 mark]**

Synoptic links

3.3.1.3   3.4.2.1

08.2 The mass of the car is 1200 kg.

Calculate the natural frequency of oscillation of the car in Hz, and the time period in s. **[2 marks]**

08.3 Suggest why, in normal driving, the car might oscillate. **[1 mark]**

08.4 A student measures the period for different masses of the car plus its passengers.

Describe what the student needs to plot to obtain a straight-line graph. Deduce what can be calculated from the gradient. **[2 marks]**

08.5 When traveling in most modern cars, passengers do not experience excessive oscillations.

Suggest why. **[1 mark]**

Exam tip

A straight-line graph has the form $y = mx$. Use algebra to rearrange the relevant equation to give an equation of this form.

13 Thermal physics

Internal energy

Internal energy is the sum of the randomly distributed kinetic energies and potential energies of the particles in a body.

The internal energy of an object can be:

- increased by heating it, or by doing work on it
- reduced by cooling it or if work is done by it.

According to the **first law of thermodynamics**, when work is done on or by an object and/or energy is transferred to or from it by heating:

$$\begin{array}{ccc} \text{change in internal} \\ \text{energy of an} & = & \text{total energy transfer} \\ \text{object} & & \text{due to work done} \\ & & \text{and heating} \end{array}$$

Changing state

When a substance changes state:

- its temperature stays constant
- the kinetic energy of its particles does not change
- its internal energy changes because the potential energy of its particles changes as the forces between them change and bonds between them are made or broken.

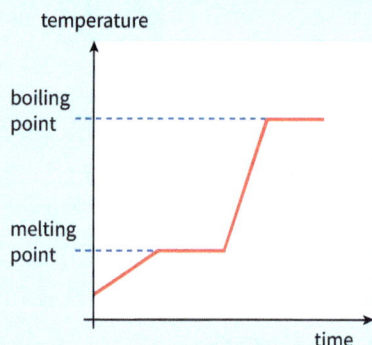

Specific heat capacity

The specific heat capacity of a substance c is the amount of energy needed to raise the temperature of 1 kg of the substance by 1 K (or °C).

For a substance of mass m, the energy transferred Q for a change in temperature $\Delta\theta$, can be calculated using:

$$Q = m c \Delta\theta$$

If a substance is heated using an electrical heater, the energy transferred to it can be calculated using:

$$E = I V t$$

Where:

E = electrical energy supplied in J

I = current in A

V = p.d. in V

t = heating time in s

Continuous flow heating

Continuous flow heating is the process of heating a fluid by having it flow continuously over a heater.

Calculations involving continuous flow heating can be simplified by the following, assuming energy transfer from the heater is 100% efficient:

energy transfer per second = power of heater

$$= \begin{array}{c}\text{mass flow} \\ \text{per second}\end{array} \times \begin{array}{c}\text{specific heat} \\ \text{capacity}\end{array} \times \begin{array}{c}\text{change in} \\ \text{temperature}\end{array}$$

Specific latent heat

Specific latent heat of fusion of a substance l_f is the amount of energy needed to change 1 kg of it from solid to liquid without a change in temperature.

Specific latent heat of vaporisation of a substance l_v is the amount of energy needed to change 1 kg of it from liquid to gas without a change in temperature.

Calculating specific latent heat

The energy transferred during a change of state is given by:

$$Q = m l$$

where l is the specific latent heat of fusion or vaporisation.

Moles and masses

The Avogadro constant

The Avogadro constant $N_A = 6.02 \times 10^{23}$ mol^{-1}. It is defined as the number of atoms in exactly 12 g of the isotope carbon-12.

The mole

One mole of any substance contains 6.02×10^{23} particles (atoms or molecules) of that substance.

Molar mass

The molar mass of a substance is the a mass of one mole of that substance in kg mol^{-1}.

The **molar mass of an element** in g equal to its relative atomic or relative molecular mass, depending on whether their particles are atoms (most elements) or molecules (a few elements). For most elements, the relative atomic mass is equal or approximately equal to its mass (nucleon) number.

The **molar mass of a compound** in g can be calculated by adding up the relative atomic masses of its constituent atoms shown in its chemical formula.

Equations

The total number of molecules N in a sample of a substance is:

$$N = n \times N_A$$

where n is number of moles of the substance in the sample. The number of moles n of substance in a sample of known mass is:

$$n = \frac{\text{mass of sample}}{\text{molar mass}}$$

The number of molecules of a substance in a sample of known mass is:

$$N = \frac{\text{mass of sample} \times N_A}{\text{molar mass}}$$

Absolute zero of temperature

Absolute zero is defined as 0 K on the absolute temperature or Kelvin scale. A temperature difference of 1 K is equal to a temperature difference of 1 °C.

To convert between K and °C:

temperature in °C = temperature in K – 273.15

A substance at absolute zero will have its minimum internal energy and its particles will have virtually zero kinetic energy.

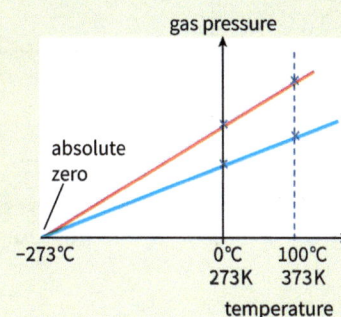

The experimental gas laws 🌡️

Boyle's law

For a fixed mass of gas at constant temperature the pressure p and volume V are inversely proportional:

$$p \propto \frac{1}{V} \text{ or } pV = \text{constant or } p_1 V_1 = p_2 V_2$$

Plotting data for p against $\frac{1}{V}$ gives a straight line through the origin.

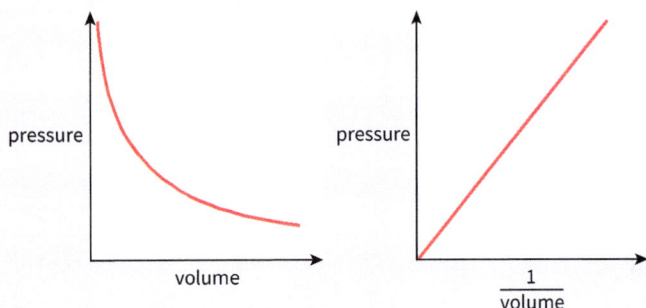

The pressure law

For a fixed mass of gas at constant volume the pressure p and absolute temperature T are directly proportional:

$$p \propto T \text{ or } \frac{p}{T} = \text{constant or } \frac{p_1}{T_1} = \frac{p_2}{T_2}$$

Plotting data for p against T gives a straight line through the origin.

Charles' law

For a fixed mass of gas at constant pressure, the volume V and absolute temperature T are directly proportional:

$$V \propto T \text{ or } \frac{V}{T} = \text{constant or } \frac{V_1}{T_1} = \frac{V_2}{T_2}$$

Plotting data for V against T gives a straight line through the origin.

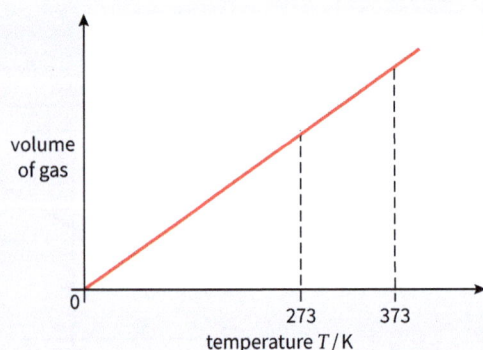

The gas laws

The gas laws describe the behaviour of gases using the following quantities and units:

- volume V in m^3
- pressure p in $N\,m^{-2}$ or pascals Pa
- temperature T in K
- number of atoms or molecules N (no unit)
- amount of gas n in mol.

The ideal gas equation

An ideal gas is a theoretical gas that obeys the experimental gas laws at all pressures and temperatures. The internal energy of an ideal gas is only dependent on the kinetic energy of its particles; the particles do not have any potential energy.

Combining Boyle's law, Charles' law, and the pressure law gives:

$$\frac{pV}{T} = \text{constant or } \frac{p_1V_1}{T_1} = \frac{p_2V_2}{T_2}$$

For n moles of gas, this can be written as the **ideal gas equation**:

$$pV = nRT$$

where $R = 8.31\,J\,K^{-1}\,mol^{-1}$, known as the molar gas constant.

For N molecules of gas, the ideal gas equation can be written as:

$$pV = NkT$$

where $k = 1.38 \times 10^{-23}\,J\,K^{-1}$ and is known as the Boltzmann constant.

Work done by an expanding gas

The energy transferred (work done) to increase or decrease the volume of a gas at a constant pressure is given by:

$$W = p\,\Delta V$$

where:

W = energy transferred in J

p = pressure in $N\,m^{-2}$ or Pa

ΔV = change in volume in m^3

Brownian motion

Brownian motion is the random motion of small particles suspended in a fluid.

Brownian motion provided evidence for the existence of atoms. Albert Einstein showed mathematically that the jiggling motion of particles suspended in a fluid could be caused by random collisions with much smaller particles (atoms and molecules) that are in constant random motion.

Scientific knowledge and understanding of the behaviour of gases has changed over time.

The three gas laws were discovered between the late 1600s and early 1800s, but the idea that the behaviour of gases could be explained by the constant random motion of molecules was not proposed until the late 1800s and not widely accepted until the early 1900s.

Boyle's law, **Charles' law**, and the **pressure law** are known as the experimental gas laws because they are based on experimental results. The gas laws are described as being empirical because they are based on observations, not theory.

Molecular kinetic theory model of gases

The behaviour of gases can be explained using the molecular kinetic theory model in which:

- pressure is caused by the change in momentum of the molecules colliding with the container walls or surfaces
- temperature is proportional to the average kinetic energy of the molecules.

Molecular kinetic theory can be used to explain the gas laws.

Boyle's law

The pressure of a gas at constant temperature is increased by reducing its volume because the gas molecules travel a smaller distance between collisions with the container walls, so there are more collisions per second and a greater pressure.

The pressure law

The pressure of a gas at constant volume is increased by raising its temperature because this raises the average kinetic energy of the molecules, so their collisions with the container wall are harder and more frequent, resulting in a greater pressure.

Charles's law

The volume of a gas at constant pressure is increased by raising its temperature because this raises the average kinetic energy of the molecules, so the particles will spread out more and collide more frequently and harder with the container walls.

The kinetic theory equation

According to kinetic theory, an ideal gas consisting of N identical molecules, each of mass m, in a container of volume V, the pressure of the gas is given by the kinetic theory equation:

$$p V = \frac{1}{3} N m (c_{rms})^2$$

c_{rms} is the root mean square speed of the gas molecules and is a kind of average of the speeds given by:

$$c_{rms} = \left[\frac{c_1^2 + c_2^2 \ldots + c_N^2}{N} \right]^{\frac{1}{2}}$$

where $c_1, c_2, c_3, \ldots c_N$ represent the speeds of the individual molecules and N is the number of molecules in the gas.

The molecules in an ideal gas do not all travel at the same speed, but have a continuous distribution of speeds that depends on the temperature:

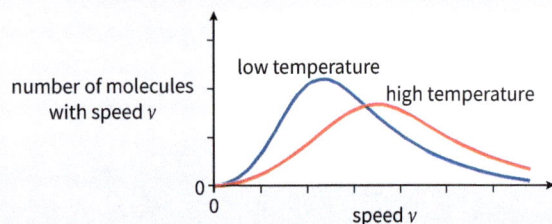

This is the **Maxwell–Boltzmann distribution** and was predicted by the kinetic theory model and later confirmed experimentally.

The speed of individual molecules can change due to collisions, but the distribution stays the same provided the temperature does not change.

Derivation of the kinetic theory equation

To derive the equation $p V = \frac{1}{3} = N m (c_{rms})^2$, consider a single molecule of an ideal gas of mass m, moving with velocity c, inside a cubic container with sides of length L.

The velocity of the molecule can be split into three perpendicular components, c_x, c_y, and c_z, which combine to give:

$$c^2 = c_x^2 + c_y^2 + c_z^2$$

If the molecule collides with the right-hand wall:

- x-component of its velocity, c_x, reverses and becomes $-c_x$, but c_y and c_z remain the same
- x-component of its momentum, mc_x, reverses and becomes $-mc_x$, but mc_y and mc_z remain the same

Deriving the kinetic theory equation

To derive the kinetic theory equation, the following assumptions about an ideal gas are made.

- All molecules are identical and the volume of each one is negligible compared with the volume of the gas.
- Molecules are in continual, random motion.
- Newton's laws can be applied to the molecules, and there are enough molecules to apply statistical laws.
- Collisions between the particles and the walls of the container are perfectly elastic (no loss of kinetic energy).
- Molecules exert no force on each other except during collisions (no intermolecular forces of attraction).
- Time between each collision with the container surface is much longer than the duration of a collision.

- change in the momentum of the particle is $-mc_x - mc_x = -2mc_x$
- time t between successive impacts with this wall is the total distance to the opposite wall and back divided by the x-component of the velocity:

$$t = \frac{2L}{c_x}$$

- number of collisions per second $= \frac{1}{t} = \frac{c_x}{2L}$
- force on the molecule is equal to its rate of change of momentum, which is equal to the momentum change per collision multiplied by the number of collisions per second:

$$F = -2 m c_x \times \frac{c_x}{2L} = \frac{-m c_x^2}{L}$$

- according to Newton's third law, the force on the molecule is equal and opposite to the force on the wall so force on wall

$$F = \frac{m c_x^2}{L}$$

- pressure on wall $= \dfrac{\text{force}}{\text{area}} = \dfrac{m c_x^2}{L} \div L^2 = \dfrac{m c_x^2}{L^3}$
- L^3 = volume of box V, so pressure $= \dfrac{m c_x^2}{V}$

The total pressure on the right-hand wall due to N molecules moving at different velocities $(c_1, c_2, c_3, \dots c_N)$, is the sum of the pressures due to each molecule:

$$P_{total} = \frac{m c_{1x}^2}{V} + \frac{m c_{2x}^2}{V} + \frac{m c_{3x}^2}{V} + \dots + \frac{m c_{Nx}^2}{V}$$

$$P_{total} = \frac{m}{V} (c_{1x}^2 + c_{2x}^2 + c_{3x}^2 + \dots + c_{Nx}^2)$$

$$P_{total} = \frac{N m \overline{c_x^2}}{V}$$

$$N m \overline{c_x^2} = \frac{\overline{c_{1x}^2 + c_{2x}^2 + c_{3x}^2 + \dots + c_{Nx}^2}}{N} = (c_{x\,rms})^2$$

So, pressure $= \dfrac{N m (c_{x\,rms})^2}{V}$ ①

But recall that $c^2 = c_x^2 + c_y^2 + c_z^2$

So, $(c_{rms})^2 = (c_{x\,rms})^2 + (c_{y\,rms})^2 + (c_{y\,rms})^2$

Since molecules move randomly, we can assume:

$(c_{x\,rms})^2 = (c_{y\,rms})^2 = (c_{y\,rms})^2$

$(c_{x\,rms})^2 = \frac{1}{3}(c_{rms})^2$

Substituting this into ① gives:

$$\text{pressure } p = \frac{1}{3} \frac{N m (C_{rms})^2}{V}$$

Rearranging gives:

$$p V = \frac{1}{3} N m (c_{rms})^2$$

Molecular kinetic energy and temperature

The ideal gas equation and the kinetic theory equation are equivalent to each other:

$$p V = \frac{1}{3} N m (c_{rms})^2 = N K T = n R T$$

This leads to:

$$\frac{1}{2} m (c_{rms})^2 = \frac{3}{2} k T = \frac{3 R T}{2 N_A}$$

These equations show that:

- average molecular kinetic energy of particles in a gas is directly proportional to its temperature in K
- mean kinetic energy of a molecule of an ideal gas $= \frac{3}{2} k T$
- total kinetic energy of 1 mol of an ideal gas $= \frac{3}{2} R T$
- total kinetic energy of n moles of an ideal gas $= \frac{3}{2} n R T$

Retrieval

Learn the answers to the questions below, then cover the answers column with a piece of paper and write as many as you can. Check and repeat.

Questions	Answers
1 How can the internal energy of an object be increased?	heating it or doing work on it
2 How can the internal energy of an object be decreased?	cooling it or if work is done by it
3 What happens to the temperature of a substance when it changes state?	stays constant
4 Define the specific heat capacity of a substance.	amount of energy needed to raise the temperature of 1 kg of it by 1 K (or °C)
5 Give the equation for the specific heat capacity of a substance.	$Q = mc\Delta\theta$
6 Give the equation for the energy transferred to a substance if it is heated using an electrical heater.	$E = IVt$
7 Define the specific latent heat of fusion of a substance.	amount of energy needed to change 1 kg of it from solid to liquid without a change in temperature
8 Give the formula for the energy transferred during a change of state of a substance (with no change in temperature).	$Q = ml$, where l is the specific heat of fusion or vaporisation
9 What is absolute zero?	lowest possible temperature – 0 K (or – 273.15 °C)
10 Define the Avogadro constant.	6.02×10^{23} – number of atoms in 12 g of carbon-12
11 What is the molar mass of a substance?	mass of one mole of that substance
12 How can the molar mass of a compound be calculated?	by adding up the relative atomic masses of its constituent atoms
13 Give the relationship between the total number of particles in a sample of a substance and the number of moles.	$N = n \times N_A$
14 What does Boyle's law state?	for a fixed mass of gas at constant temperature the pressure and volume are inversely proportional
15 What does Charles' law state?	for a fixed mass of gas at constant pressure the volume and absolute temperature are directly proportional
16 What does the pressure law state?	for a fixed mass of gas at constant volume the pressure and absolute temperature are directly proportional
17 Give the equation for the energy transferred to increase or decrease the volume of a gas at a constant pressure.	$W = p\Delta V$

Put paper here

18 Define an ideal gas.

> theoretical gas which obeys the experimental gas laws at all pressures and temperatures

19 Give the two ideal gas equations

> $pV = nRT$ and $pV = NkT$

20 Give the kinetic theory equation.

> $pV = \frac{1}{3}Nm\,(c_{rms})^2$

Practical skills

Practise your practical skills using the worked example and practice questions below.

Pressure and temperature of gases

This equipment can be used to investigate Boyle's law.

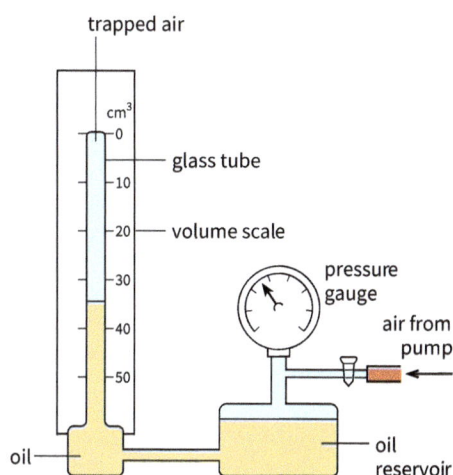

Worked example

Question

1 State the physical quantity (apart from mass) that should be kept constant during an investigation to determine the relationship between pressure and volume of a gas.

2 Sketch a graph to show the expected relationship between pressure and volume obtained using the apparatus shown.

Answer

1 The temperature of the gas should be kept constant.

2

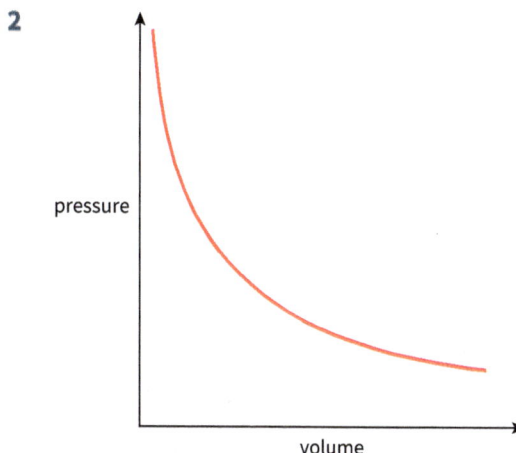

Practice

1 Suggest how the temperature of a gas can be kept constant during the experiment.

2 Sketch a graph to show the expected relationship between pressure and $\frac{1}{volume}$ obtained using the apparatus shown.

3 Charles's law states the volume of an ideal gas at constant pressure is directly proportional to the absolute temperature. Sketch the apparatus you would use to investigate this relationship.

Exam-style questions

01 An ideal gas is a theoretical gas that fits standard models.

01.1 State **two** ways that ideal gases differ from real gases. **[2 marks]**

_____

_____

_____

01.2 A factory makes dry ice (solid carbon dioxide) from carbon dioxide. A container of carbon dioxide at temperature 20 °C, and pressure 101 kPa, is compressed to 2.1 MPa. The initial volume of the gas is 3.51 m³.

Calculate the final volume of the gas.

State **one** assumption that you have made. **[2 marks]**

> **⚠ Exam tip**
>
> Remember that when doing calculations involving the ideal gas equation, the temperature is in K and not in °C.

final volume = _____ m³

_____

01.3 When carbon dioxide is compressed to the volume calculated in **01.2**, it does not exert a pressure of 2.1 MPa.

Suggest which of the assumptions of kinetic theory no longer holds, and suggest the effect on the magnitude of the pressure in comparison to the pressure of an ideal gas. **[2 marks]**

_____

_____

_____

01.4 The carbon dioxide is still in a gas state. 1.31 MJ of energy is transferred to cool the gas so that it changes to a liquid state.

Calculate the specific latent heat of fusion of carbon dioxide. The mass of a carbon dioxide molecule is 7.32×10^{-26} kg. **[3 marks]**

specific latent heat of fusion = _____ kJ kg⁻¹

02 When cooking potatoes, water in a pan is brought to a boil. In this question, energy dissipated to the surroundings should be ignored.

02.1 Show that the power required to bring 1.45 kg of water to the boil on an electric hob in 10 minutes is about 850 W. Room temperature is 20 °C

The specific heat capacity of water is 4200 J kg^{-1} °C^{-1}. The pan has a mass of 800 g and a specific heat capacity 385 J kg^{-1} °C^{-1}. **[2 marks]**

_____

02.2 When potatoes are added to the water, the temperature drops by 14 °C.

Calculate the mass in kg of the potatoes added.

Assume that the specific heat capacity of the potatoes is 3.39 kJ kg^{-1} °C^{-1}.

State any other assumptions that you have made. **[2 marks]**

mass = _____ kg

_____

_____

02.3 It takes longer to bring the potatoes and water back to the boil than is predicted using the power calculated in **02.1**.

Explain why, using ideas about the internal energy of particles. **[2 marks]**

_____

_____

_____

02.4 Some kitchens have a hot-water tap that can deliver water at a very high temperature. An element in a pipe heats water as the water passes over it.

Compare the time required to fill a pan of boiling water if the heater has the power calculated in **02.1**. The specific heat capacity of water is 4200 kJ kg^{-1} °C^{-1}. **[2 marks]**

_____

_____

> **! Exam tip**
>
> You can divide the equation for energy transferred for a temperature change by time to get an equation involving power and mass per second.

03 A student hears that the low temperatures at the South Pole affect the atmospheric pressure there. The cold air produces higher air pressure than would be expected. The student decides to investigate the prediction that a decrease in temperature produces an increase in gas pressure.

03.1 List all the factors that affect the pressure of an ideal gas. **[1 mark]**

_____

_____

_____

03.2 The student sets up an experiment to measure the gas pressure p, volume V, and temperature T, for a range of situations.

The data collected is shown in **Table 1**.

Table 1

T/°C	273	273	273	273	313	313	313	353	353	353	393	393	413
p/kPa	200	400	500	1000	459	153	229	517	862	739	576	576	1009
V/cm³	20	10	8	4	10	30	20	10	6	7	10	10	6

Select data from **Table 1** that the student could use to test their prediction, and explain your choice.

Plot the selected data on the graph paper in **Figure 1**. **[4 marks]**

Figure 1

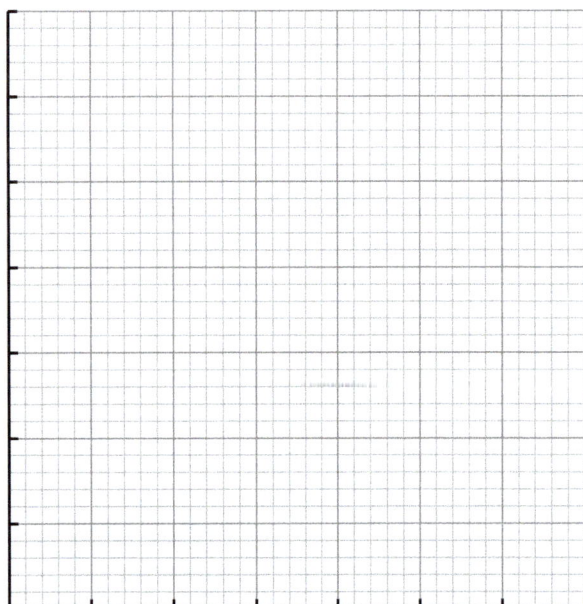

! **Exam tip**

If you are trying to find the relationship between two variables, you need to control the remaining variables.

03.3 Explain why your graph from **03.2** does not support the prediction.

Use your graph to calculate the number of moles of gas used in the experiment. **[3 marks]**

03.4 The increase in pressure is related to the change in density of the air.

Suggest the extent to which the pressure and density of a gas are proportional, using ideas about kinetic theory. **[3 marks]**

04 A student wanted to test the hypothesis that the volume of a gas and its pressure are inversely related.

04.1 Describe a procedure that the student could use to test this hypothesis.

[4 marks]

04.2 **Table 2** shows the data that the student collected at a temperature of 21.0 °C.

Table 2

p / kPa	$\dfrac{1}{p}$ / $\times 10^{-3}$	V / $\times 10^{-6}$ m³
270	3.7	5
140	7.1	10
90		15
65	15.4	20
50	18.0	25
45	22.2	30

State why the student has calculated the middle column in the table.

Complete the table by calculating the missing value. **[2 marks]**

04.3 The student plotted a graph of volume against $\dfrac{1}{\text{pressure}}$ as shown in **Figure 2**.

Figure 2

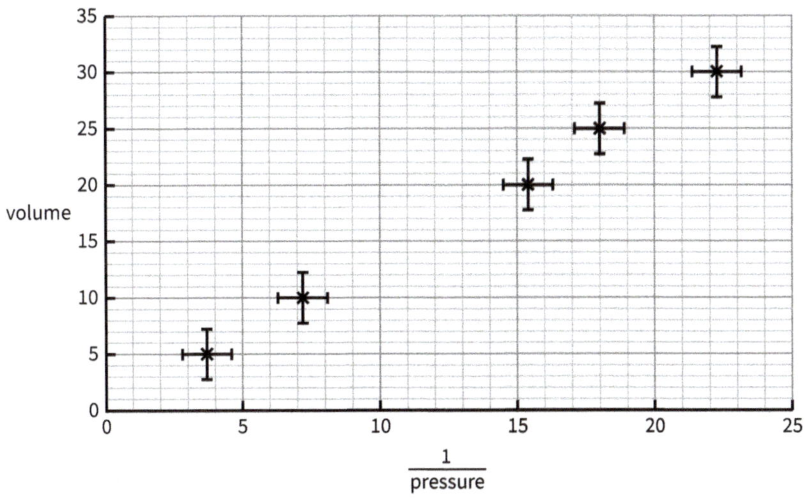

Complete the labels on the axes in **Figure 2**, and plot the missing point.

Calculate the number of moles of the gas used in the experiment using **Figure 2**, and give the uncertainty for this value. **[4 marks]**

04.4 The student drew error bars of equal length on each of the points plotted on the graph.

Comment on the extent to which these uncertainties are likely to be accurate for both horizontal and vertical error bars. **[1 mark]**

> **! Exam tip**
>
> Remember that plotting variables, or manipulated variables, to get a straight line through the origin shows that the two variables are proportional.

05 **Figure 3** shows the graph of pressure against volume as a piston in a car engine compresses a gas-containing fuel, which is then ignited, and the gas pushes the piston back out. Between points **A** and **B**, the gas is compressed.

Figure 3

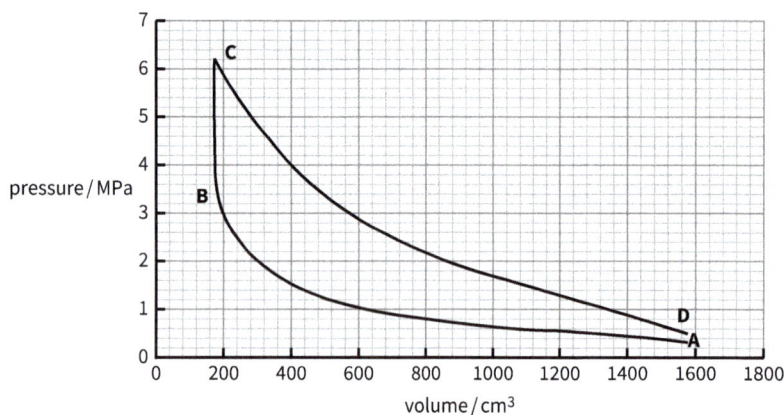

volume / cm³

05.1 Annotate the graph by shading the area that you could use to calculate the work done on the gas. Explain in terms of particles why the temperature of the gas rises. **[2 marks]**

05.2 The temperature of the gas at point **B** is 450 °C, the volume is 195 cm³, and the pressure is 3.2 MPa. Calculate the number of gas molecules in the piston, assuming it behaves like an ideal gas. **[2 marks]**

> **⚠ Exam tip**
>
> Always check the units of the data that you are using and convert to standard units.

05.3 At point **B**, the temperature of the gas mixture has risen sufficiently to ignite the fuel.

Calculate the rise in temperature in °C of the gas between points **B** and **C**. **[2 marks]**

05.4 The gas expands between points **C** and **D**.

Suggest why the work done by the gas is different from the work done on the gas. **[2 marks]**

06 A sample of gas is contained in a square box. Each side has a length of 30 cm. A gas molecule of mass 4.6×10^{-20} kg collides with a side of the container at an angle of 90°. Its speed before and after the collision is 510 m s⁻¹.

06.1 Calculate the change in momentum in kg m s⁻¹ of the molecule and the time in s before the molecule collides with another surface. **[3 marks]**

> **⚙ Synoptic link**
>
> 3.4.1.6

06.2 Calculate the pressure in Pa exerted on one surface of the box. **[2 marks]**

06.3 Calculate the number of moles of gas molecules that would need to collide with the surface to produce a pressure of 100 kPa. **[2 marks]**

> **⚠ Exam tip**
>
> Remember that force is the rate of change of momentum in any context.

06.4 Compare the temperature suggested by the equation below with that suggested by the ideal gas equation.

Suggest reasons for the difference. **[5 marks]**

$$\frac{3}{2}kT = \frac{1}{3}mc^2$$

150 **13** Thermal physics

07 The mean temperature of the air at the surface of the Earth is about 22.4 °C. The temperature on the Moon's surface varies from +125 °C to −175 °C.

07.1 Sketch a graph of the speed distribution of the same number of molecules in a gas at a low and at a high temperature. **[3 marks]**

Synoptic link

3.7.2.3

07.2 Calculate the root mean square speed in m s⁻¹ of the gas molecules at the surface of the Earth.

The molar mass of air is 28.97 g mol⁻¹. **[3 marks]**

07.3 Show that the escape velocity of an object from an astronomical object of mass M is:

$$\sqrt{\frac{2GM}{r}}$$

Exam tip

For **07.3** you should show an understanding of why gravitational potential is negative.

07.4 Explain why the Earth has an atmosphere, using the following data: $M_{Earth} = 5.97 \times 1024$ kg; $r_{Earth} = 6378$ km $= 6.378 \times 10^6$ m.

Suggest why the Moon, which has an escape velocity of 2.4 km s⁻¹, does not have an atmosphere. **[5 marks]**

08 A hot air heating system uses hot water to heat air and a fan to blow the air into a room, as shown in **Figure 4**.

Figure 4

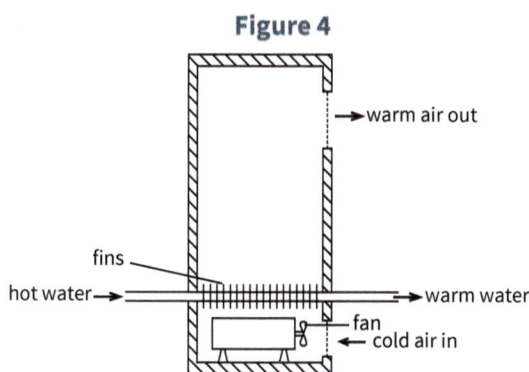

Synoptic links

3.7.5.1   3.7.5.4

08.1 The hot water enters the system at a temperature of 45 °C and leaves at 34 °C. The water moves through the pipe at a rate of 2.7×10^{-3} m³ s⁻¹. The density of water = 997 kg m⁻³, and the specific heat capacity of water = 4200 J kg⁻¹ °C⁻¹.

Calculate the energy in J that is transferred per second from the water. **[2 marks]**

08.2 The air in the heater is moved out of the heater at a rate of 2.2 m³ s⁻¹. The density of air is 1.225 kg m⁻³, and the specific heat capacity of air is 1.00 kJ kg⁻¹ °C⁻¹.

Calculate the temperature in °C of the air when it leaves the heater if it enters the heater at a temperature of 16 °C. State any assumptions that you have made. **[3 marks]**

08.3 The temperature of the air leaving the heater is closer to 50 °C. Suggest why. **[1 mark]**

08.4 The fan is connected to a mains electricity supply.

Explain why the blades of the fan turn when it is plugged in. **[2 marks]**

Exam tip

You should use ideas about the construction of a simple motor.

08.5 When the fan is turned on, the current does not immediately reach its maximum value. Explain why. **[2 marks]**

⚙ Knowledge

14 Gravitational fields

Newton's law of universal gravitation

Gravity is a universal attractive force that acts between all matter. A **point mass** is one that behaves as if all its mass is concentrated at its centre, like a sphere of uniform density. Newton's law of universal gravitation for two point masses is:

$$\text{gravitational force } F = \frac{Gm_1m_2}{r^2}$$

where G = gravitational constant, $6.67 \times 10^{-11}\,\text{N}\,\text{m}^2\,\text{kg}^{-2}$

The gravitational force between two point masses is directly proportional to the product of their masses and inversely proportional to the square of the distance between them. So, the gravitational force between any two point masses is:

- always attractive
- proportional to the product of their masses,
  $F \propto (m_1 \times m_2)$
- an inverse square law $F \propto \dfrac{1}{r^2}$

Fields

Force fields

A force field is a region of space in which an object experiences a non-contact force.

Gravitational fields

Gravitational fields exist around all objects with mass. Any mass will experience an attractive force if placed in the gravitational field of another mass. Since both masses have a gravitational field, the other mass will experience an attractive force of the same size but in the opposite direction.

Representing force fields

Force fields can be represented as vectors using field lines or lines of force on a diagram, where:

- the arrows on field lines show the direction in which the force acts
- the separation of field lines indicates the strength of the field – the closer together the lines, the stronger the field.

Radial fields are where the field lines point towards or away from the centre of the object causing the field. The separation between them increases with distance, indicating that the strength of the field decreases with distance.

Uniform fields are where the field strength and direction is the same at every point, indicated by the field lines being parallel to one another and equally spaced.

The Earth's gravitational field is radial overall, but can be considered uniform close to the Earth's surface.

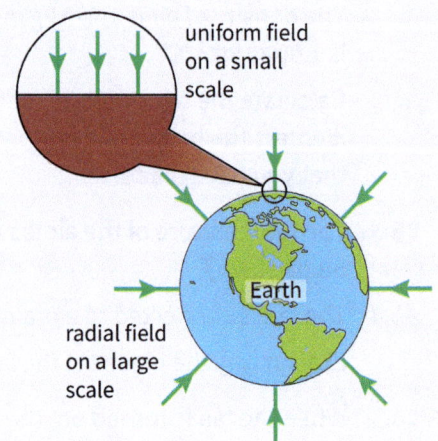

uniform field on a small scale

radial field on a large scale

Earth

Gravitational field strength

The gravitational field strength g (unit N kg^{-1}) at any point in a field is the force per unit mass on a small test mass placed at that point:

$$g = \frac{F}{m}$$

Combining:

$$F = ma$$

with:

$$g = \frac{F}{m}$$

gives:

$$a = g$$

This shows that g is also the acceleration due to gravity at that point in the field.

Combining:

$$g = \frac{F}{m}$$

with:

$$F = \frac{Gm_1 m_2}{r^2}$$

gives the field strength for a radial gravitational field at a distance r from a point mass M:

$$g = \frac{GM}{r^2}$$

For a planet of radius R, the gravitational field strength at its surface:

$$g_s = \frac{GM}{R^2}$$

Gravitational potential

The gravitational potential V (unit J kg^{-1}) at a point in a gravitational field is defined as the work done per unit mass to move a small object from infinity to that point. Gravitational potential is also the gravitational potential energy (E_p) per kg that any object would have at that point in the field.

The gravitational potential at a point in a radial field caused by a mass M is:

$$V = \frac{-GM}{r}$$

The gravitational potential energy of an object of mass m at any point in a gravitational field can be calculated using:

$$E_p = Vm = \frac{-GMm}{r}$$

The values of gravitational potential are negative because at an infinite distance away from the mass causing the field, the gravitational potential is zero. Work must be done against gravitational attraction to move masses apart, so an object gains E_p as it moves towards infinity but, since the maximum value E_p can have is zero, all the points in the field must have a negative value for gravitational potential.

Gravitational potential difference

Gravitational potential difference is the difference between the gravitational potential at two points in a gravitational field.

The work done ΔW in moving a mass m between two points in a gravitational field is:

work done = mass × gravitational potential difference

$$\Delta W = m\,\Delta V$$

The change in gravitational potential ΔV can be found from the area under the graph of g against r.

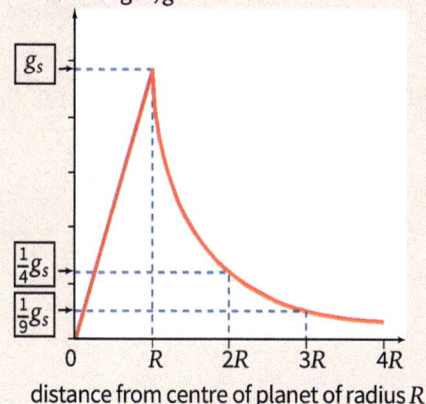

gravitational field strength, g

distance from centre of planet of radius R

Equipotential surfaces

An equipotential surface is one where the gravitational potential is the same at all points.

No work is done when moving along an equipotential surface because the potential difference between any two points on the surface is zero.

Equipotential surfaces are represented as equipotential lines on 2D diagrams.

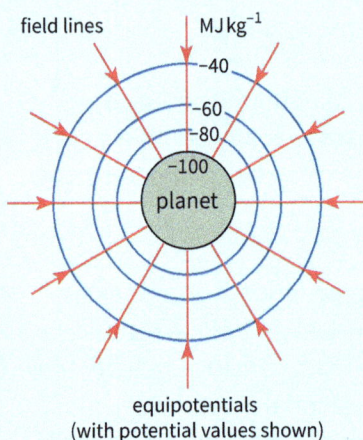

equipotentials
(with potential values shown)

Equipotential surfaces are perpendicular to field lines.

The equipotential surfaces in a radial field are spheres.

Potential gradients

The potential gradient at a point in a field is the change of potential per metre at that point:

$$\text{potential gradient} = \frac{\Delta V}{\Delta r}$$

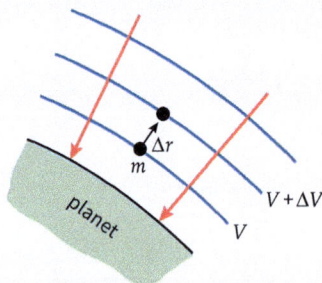

Potential gradient and gravitational field strength

Gravitational field strength is related to potential gradient by:

$$g = -\frac{\Delta V}{\Delta r}$$

The minus sign shows that g acts in the opposite direction to the gradient.

$-\dfrac{\Delta V}{\Delta r}$ is the gradient of the graph of gravitational potential against distance. The value of g at any point in a gravitational field can be found by finding the gradient of the V against r graph at that point.

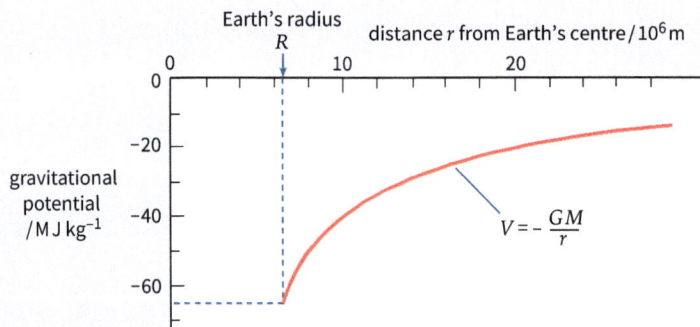

Escape velocity

The escape velocity of an astronomical body is the minimum speed an object must be given to escape the body's gravitational field when launched from the surface.

The kinetic energy of an object of mass m at the surface of a planet of radius R and mass M must be the same as the work done to move the object from from infinity to the surface:

$$\frac{1}{2}mv^2 = \frac{GMm}{R}$$

Rearranging this gives the formula for the escape velocity from the surface of a planet of radius R and mass M:

$$v_{esc} = \sqrt{\frac{2GM}{R}}$$

From this, it is clear the escape velocity is the same for all masses in the same gravitational field.

Orbits of planets and satellites

The time period T of a planet or satellite in a circular orbit is related to the radius of the orbit r by:

$$T^2 \propto r^3$$

This means:

- for any planets orbiting the same star:

$$\frac{T^2}{r^3} = \text{constant}$$

- for any two planets orbiting the same star:

$$\frac{T_1^2}{r_1^3} = \frac{T_2^2}{r_2^2}$$

Note that the radius of the orbit r is measured from the centre of the object being orbited. $\propto$

Deriving $T^2 \propto r^3$

For an object of mass m in a circular orbit of radius r around an object of mass M, gravitational attraction provides the centripetal force:

$$\frac{GMm}{r^2} = \frac{mv^2}{r}$$

Rearranging this leads to:

$$\frac{GM}{r} = v^2$$

Substituting $v = \frac{2\pi r}{T}$ into the above equation:

$$\frac{GM}{r} = \frac{(2\pi r)^2}{T^2} = \frac{4\pi^2 r^2}{T^2}$$

Rearranging this gives:

$$T^2 = \frac{4\pi^2 r^3}{GM}$$

Since $\frac{4\pi^2}{GM}$ is a constant, we can write $T^2 \propto r^3$

Types of satellite orbit

Geostationary orbits

A satellite with a synchronous orbit has a time period equal to the rotational period of the object it is orbiting.

A geostationary satellite is an example of a satellite with a synchronous orbit that:

- remains above the same point on the Earth's equator
- orbits in the equatorial plane
- has a time period of 24 hours
- moves in the same direction that the Earth rotates
- orbits ~36 000 km above the surface of the Earth with an orbital radius of about 42 000 km.

Geostationary satellites are useful as communication satellites because transmitters and receivers on Earth can point in a fixed direction.

Low orbits

Low-orbiting satellites are those that orbit less than 2000 km above the Earth.

Satellites with a polar orbit move in a plane that is 90° to the equatorial plane, and pass over every point on the Earth as it rotates beneath them.

Low-orbiting satellites with a polar orbit are useful for scanning the Earth for imaging, mapping, and weather prediction purposes.

Energy of orbiting satellites

A satellite in a circular orbit remains at the same height above the object it is orbiting and moves with a constant speed, so its kinetic and gravitational potential energies are constant.

The total energy of a satellite is the sum of its kinetic energy and its gravitational potential energy – this remains constant.

A satellite's total energy is:

$$E_{\text{total}} = E_k + E_p = \frac{GMm}{2r}$$

Retrieval

Learn the answers to the questions below, then cover the answers column with a piece of paper and write as many as you can. Check and repeat.

	Questions	Answers
1	What is gravity?	universal attractive force that acts between all matter
2	What is a point mass?	one which behaves as if all its mass is concentrated at its centre
3	Give Newton's law of universal gravitation for two point masses.	$F = \dfrac{G\,m_1 m_2}{r^2}$
4	What is a force field?	region of space in which an object experiences a non-contact force
5	Where do gravitational fields exist?	around all objects with mass
6	How can force fields be represented?	vectors using field lines or lines of force on a diagram
7	What kind of field is the Earth's gravitational field?	radial overall, but can be considered uniform close to the Earth's surface
8	Define the gravitational field strength at any point in a field.	$g = \dfrac{F}{m}$
9	Give the field strength for a radial gravitational field at a distance from a point mass.	$g = \dfrac{GM}{r^2}$
10	Define the gravitational potential at a point in a gravitational field.	work done per unit mass to move a small object from infinity to that point
11	Give the formula for the gravitational potential at a point in a radial field.	$V = \dfrac{-GM}{r}$
12	What is gravitational potential difference?	difference between the gravitational potential at two points in a gravitational field
13	Give the equation for the work done in moving a mass between two points in a gravitational field.	$\Delta W = m\,\Delta V$
14	What is an equipotential surface?	surface where the gravitational potential is the same at all points
15	What is the potential gradient at a point in a field?	potential gradient $= \dfrac{\Delta V}{\Delta r}$
16	How is gravitational field strength related to potential gradient?	$g = -\dfrac{\Delta V}{\Delta r}$
17	Give the formula for the escape velocity from the surface of a planet.	$v_{esc} = \sqrt{\dfrac{2GM}{R}}$
18	State the relationship between the time period of a planet in a circular orbit and the radius of the orbit.	$T^2 \propto r^3$

Put paper here

19 Give the equation for a satellite's total energy. $E_{total} = E_k + E_p = -\dfrac{GMm}{2r}$

20 What are low orbiting satellites?   satellites orbiting below 2000 km above the Earth

Put paper here

🖩 Maths skills

Practise your maths skills using the worked example and practice questions below.

Orbits of planets and satellites

Kepler's third law relates the time period T of a planet to the cube of its mean orbital radius r.

$$T^2 = k\,r^3$$

where k is a constant of proportionality.

However, data for this cover many orders of magnitude and are difficult to plot on a linear graph. Taking logarithms of both sides of the equation we obtain:

$$2\ln T = \ln k + 3\ln r$$

This can be rearranged into the form of the equation of a straight line:

$$\ln T = \tfrac{3}{2}\ln r + \tfrac{1}{2}\ln k$$

Plotting a graph of $\ln T$ against $\ln r$ gives a straight-line graph with a gradient of $\tfrac{3}{2}$ and an intercept of $\tfrac{1}{2}\ln k$.

Worked example

Question

Use the data in this table to show that these planetary orbits obey Kepler's third law.

Planet	T/yr	r/m
Mercury	0.241	5.83×10^{10}
Mars	1.52	2.27×10^{11}
Jupiter	11.8	7.78×10^{11}
Uranus	84.0	2.87×10^{12}
Neptune	165	4.50×10^{12}

Answer

Calculate $\ln T$ and $\ln r$ for each planet.

Planet	$\ln(T/\text{yr})$	$\ln(r/\text{m})$
Mercury	−1.42	24.8
Mars	0.41	26.1
Jupiter	2.47	27.4
Uranus	4.43	28.7
Neptune	5.11	29.1

Then plot a graph of $\ln(T/\text{yr})$ against $\ln(r/\text{m})$.

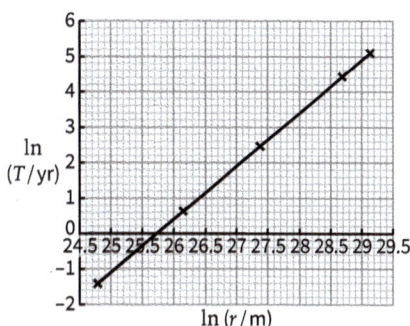

The graph is a straight line of gradient 1.5, which shows that the data follow Kepler's third law.

Practice

1 Use the data to determine whether the orbits of these major moons of Saturn obey Kepler's third law.

Satellite	T/days	$r/10^3\,\text{km}$
Mimas	0.94	185.5
Enceladus	1.37	238.0
Tethys	1.88	294.7
Dione	2.73	377.4
Rhea	4.51	527.0

2 Combining Kepler's third law with gravitation gives the equation:

$$T^2 = \frac{4\pi^2}{GM}r^3$$

Neptune's moon Triton has an orbital radius of $3.5\times10^8\,\text{m}$. Neptune has a mass of $1.02\times10^{26}\,\text{kg}$.

Calculate the orbital period of Triton.
$G = 6.67\times10^{-11}\,\text{m}^3\,\text{kg}^{-1}\,\text{s}^{-2}$

3 Using the data for Enceladus in question **1**, calculate the mass of Saturn.

01 **Figure 1** shows how the gravitational field strength g varies with distance r from the centre of the planet Mercury. The mean radius of Mercury is 2.4×10^6 m.

Figure 1

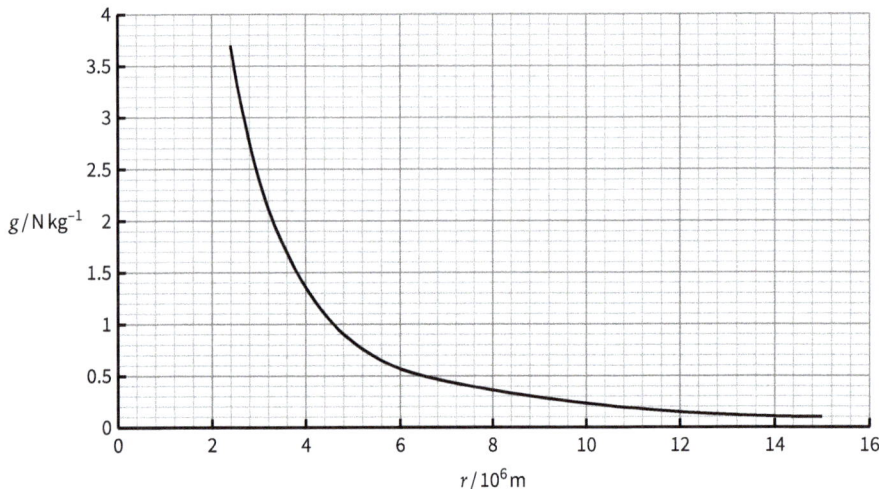

$g/\text{N kg}^{-1}$ plotted against $r/10^6 \text{ m}$

01.1 At the surface of Mercury, the gravitational field strength is 3.7 N kg^{-1}.

State the meaning of gravitational field strength. **[1 mark]**

01.2 Mercury and Earth have similar densities.

Mercury has a lower value of g than the value of g at the Earth's surface.

Determine whether Mercury is bigger or smaller than Earth. **[3 marks]**

> **! Exam tip**
>
> Substitute into the equation for g to find an expression involving density, then refer to this equation in your explanation.

01.3 Annotate **Figure 1** by shading the area that represents the gravitational potential at the surface of Mercury.

Explain your choice. **[1 mark]**

> **! Exam tip**
>
> Think about the definition for gravitational potential and then what the area under the graph represents.

01.4 Show, using **Figure 1** or otherwise, that the gravitational potential at the surface of Mercury is $9 \times 10^6 \text{ J kg}^{-1}$. **[2 marks]**

01.5 Calculate the velocity that a 1 kg mass needs to reach in order to escape from the planet Mercury. **[2 marks]**

velocity = _____ m s^{-1}

01.6 Complete the graph in **Figure 1** to show how the gravitational field strength varies inside Mercury. **[1 mark]**

02 This question is about hydrogen in the Earth's atmosphere. Assume that the height of the atmosphere is negligible compared with the radius R of the Earth.

02.1 Show that the gravitational potential at the Earth's surface can be expressed as $g R$, where R is the radius of the Earth. **[1 mark]**

Synoptic link

3.6.2.3

02.2 Show that the escape velocity of the Earth is given by $\sqrt{2 g R}$. **[2 marks]**

02.3 Calculate Earth's escape velocity. **[1 mark]**

escape velocity = _____ m s^{-1}

02.4 Calculate the temperature of the Earth's atmosphere at the point where the root-mean-square speed of the hydrogen molecules is equal to this escape velocity.

molar mass of hydrogen = 0.002 kg mol^{-1} **[3 marks]**

temperature = _____ K

02.5 The temperature of the upper atmosphere varies considerably. The maximum temperature is approximately 650 K.

Explain why hydrogen escapes from the Earth's atmosphere.
[2 marks]

_____

_____

_____

! Exam tip

Consider the velocity distribution curve for a particular temperature – will all the molecules be moving at the same velocity?

03 This question is about gravitational potential.

03.1 State what is meant by gravitational potential at a point. [1 mark]

Figure 2

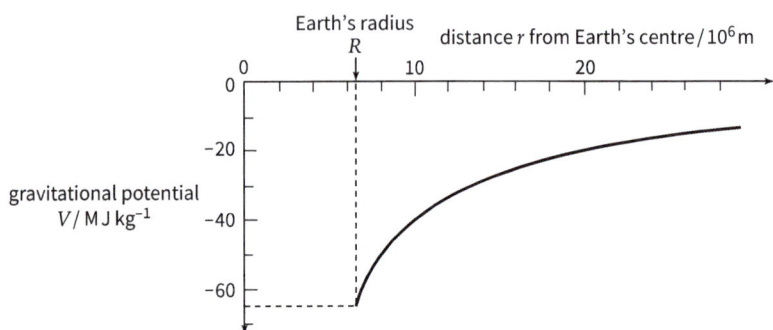

03.2 Explain how you could use data from **Figure 2** to show that gravitational potential is inversely proportional to distance from the Earth.
[2 marks]

_____

_____

_____

03.3 Determine, using **Figure 2**, the gravitational field strength in $N\,Kg^{-1}$ at a distance of $14 \times 10^6\,m$ from the centre of Earth. [2 marks]

! Exam tip

If a question asks you to use a graph, you must include how you can use the graph to produce an answer.

gravitational field strength = _____ $N\,Kg^{-1}$

03.4 Sketch a graph on the axes in **Figure 3** to show how the gravitational potential due to the Moon varies with distance along a line from the surface of the Earth to the surface of the Moon. [3 marks]

Figure 3

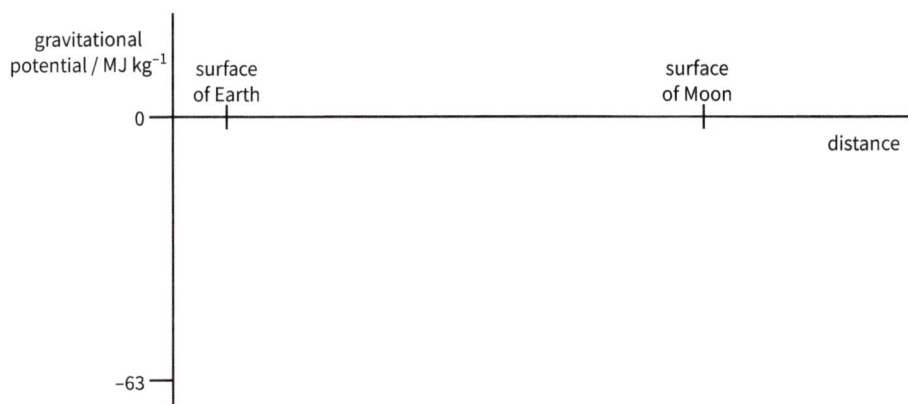

04 This question is about the Earth's gravitational field.

Figure 4

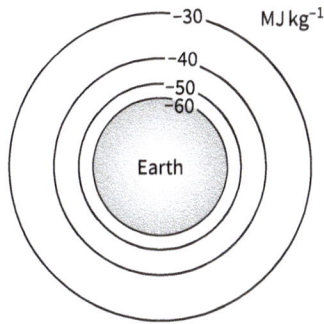

04.1 Explain how you can tell from **Figure 4** that this is not a uniform gravitational field. **[1 mark]**

_____

_____

_____

04.2 Draw the gravitational field lines on **Figure 4**. **[2 marks]**

04.3 Determine the distance in m from the centre of Earth for the 40 MJ kg^{-1} equipotential. **[2 marks]**

> **! Exam tip**
>
> Make sure you draw at least four lines using a ruler.

distance = _____ m

04.4 Describe the change in gravitational potential for a satellite orbiting at the distance calculated in **04.3**. **[2 marks]**

_____

_____

_____

_____

_____

05 In an experiment to determine the mean density of the Earth, the deflection of a pendulum due to a mountain was measured, as shown in **Figure 5**.

> **Synoptic links**
>
> 3.4.1.5 3.4.1.1

Figure 5

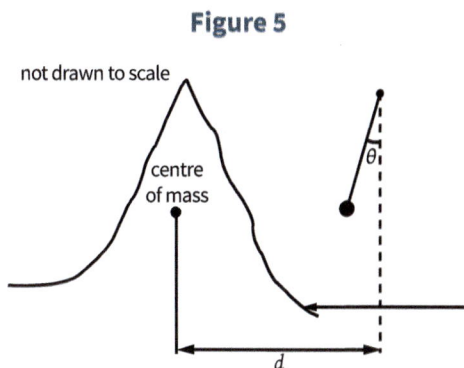

05.1 Draw a free-body diagram for the forces acting on the pendulum bob in **Figure 5**. **[3 marks]**

05.2 State Newton's law of gravitation. **[1 mark]**

05.3 The pendulum is placed at a distance d from the centre of mass of a mountain with mass M.

Show that $\tan\theta = \dfrac{MR^2}{M_E d^2}$, where M_E is the mass of the Earth and R is the distance from the centre of mass of the bob to the centre of the Earth. **[3 marks]**

05.4 The experiment yielded a mean density of the Earth of $4560\,\text{kg m}^{-3}$. The modern value is recorded as $5510\,\text{kg m}^{-3}$.

Calculate the percentage difference between these values. **[1 mark]**

Exam tip

Resolve the tension into vertical and horizontal components and equate these to the gravitational attraction of the mountain and the Earth.

06 This question is about orbits.

06.1 The relationship between the radius of an orbit r and the time period T is given by:

$$T^2 \propto r^3$$

Show how this relationship is derived. **[3 marks]**

Synoptic links

3.6.1.1   MS3.11

06.2 **Table 1** shows data for this relationship for the moons of Jupiter.

Table 1

Moon of Jupiter	T / days	r / Mm
Io	1.769	422
Europa	3.551	671
Ganymede	7.155	1070
Callisto	16.689	1883

Use the data to determine whether the moons of Jupiter follow the relationship $T^2 \propto r^3$. **[3 marks]**

06.3 Determine the mass of Jupiter using **Table 1**. **[2 marks]**

06.4 The relationship could also be proved by plotting a graph with $\log T$ on the y-axis against $\log r$ on the x-axis.

Explain how this graph would prove that $T^2 \propto r^3$. **[2 marks]**

Exam tip

Take care with units in **06.3** or you will end up with an unlikely mass for a planet.

07 A student believes it is possible for a satellite to be in geostationary orbit above London. They explain their reasoning using **Figure 6**.

Figure 6

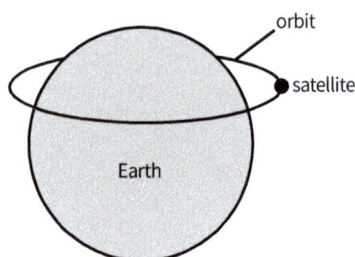

Synoptic link

3.6.1.1

07.1 Draw the gravitational force acting on the satellite in the position shown. **[1 mark]**

07.2 Explain why a satellite could not be in a geostationary orbit above London. **[2 marks]**

07.3 The height of a geostationary orbit is 36×10^6 m.

Show that the speed of a satellite in a geostationary orbit is 3 km s^{-1}.

[3 marks]

07.4 Calculate the total energy in J of a geostationary satellite of mass 282 kg in this orbit. **[3 marks]**

> **! Exam tip**
>
> Take care with gravitational potential energy – it increases with height and is zero at an infinite distance, so it must be negative in **07.4**.

08 On the International Space Station, artificial gravity could be created by rotating the space station.

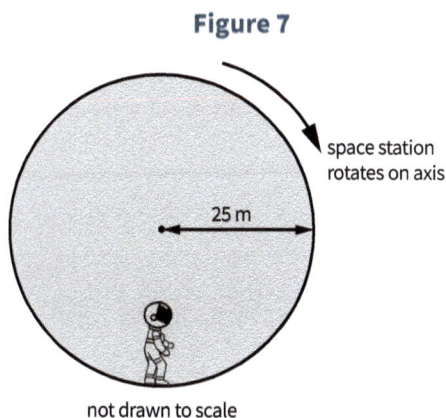

Figure 7

space station rotates on axis

25 m

not drawn to scale

> **Synoptic links**
>
> 3.6.1.1   3.1.3   3.1.2

08.1 Draw an arrow on **Figure 7** to show the direction of the resultant force acting on the astronaut in the spinning space station. **[1 mark]**

08.2 Show that the space station would have to rotate approximately once every 10 s to have an acceleration equivalent to g on the Earth's surface.

Assume that the radius of the space station is 25 m. **[2 marks]**

08.3 If there is a difference in acceleration between their head and feet, the astronaut will experience a rush of blood to their feet.

Calculate the acceleration in m s^{-2} of the astronaut's head at this rotation. Use estimated values for unknown variables. **[2 marks]**

08.4 The rush of blood could be overcome by having a space station with a radius of at least 125 m.

Explain why this would overcome the problem.

Suggest **one** issue with building a much larger space station.

[2 marks]

> **! Exam tip**
>
> For **08.3** you have to estimate the height of an astronaut; there will be a range of acceptable answers for this question.

⚙ Knowledge

15 Electric fields

Coulomb's law

Electric fields exist around all objects with charge. Any object with charge will experience a force if placed in the electric field of another charged object.

A charged sphere can be considered to have all its charge at its centre. The size of the force between two point charges in a vacuum is given by Coulomb's law:

$$F = \frac{1}{4\pi\varepsilon_0}\frac{Q_1 Q_2}{r^2}$$

where:

Q_1 and Q_2 = charges in C

r = distance between charges in m

ε_0 = permittivity of free space, $8.85 \times 10^{-12}\,\mathrm{F\,m^{-1}}$

Each charge will experience the same size force but in opposite directions.

The electric force between any two point charges:

- can be attractive or repulsive depending on the charges of the objects
- is proportional to the product of their charges, $F \propto (Q_1 \times Q_2)$
- obeys an inverse square law, $F \propto \dfrac{1}{r^2}$
- depends on the permittivity ε of the material between them (air can be treated as a vacuum).

Strength of an electric field

The strength of an electric field

The electric field strength E (unit $\mathrm{N\,C^{-1}}$) at any point in a field is the force per unit charge on a positive test charge placed at that point:

$$E = \frac{F}{Q}$$

Combining this equation with

$$F = \frac{1}{4\pi\varepsilon_0}\frac{Q_1 Q_2}{r^2}$$

gives the field strength for a radial electric field at a distance r from a point charge Q:

$$E = \frac{1}{4\pi\varepsilon_0}\frac{Q}{r^2}$$

Drawing electric fields

Drawing electric fields

Electric fields can be represented by field lines where:

- the arrows on field lines show the direction of the force acting on a positive charge placed in the field
- the separation of field lines indicates the strength of the field – the closer together the lines, the stronger the field.

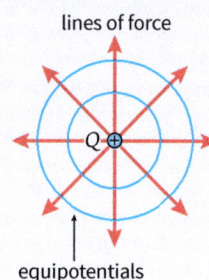

Point charges have radial fields.

Positive charge   **Negative charge**

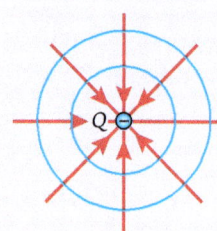

Uniform fields

A uniform electric field can be produced by applying a potential difference between two parallel metal plates.

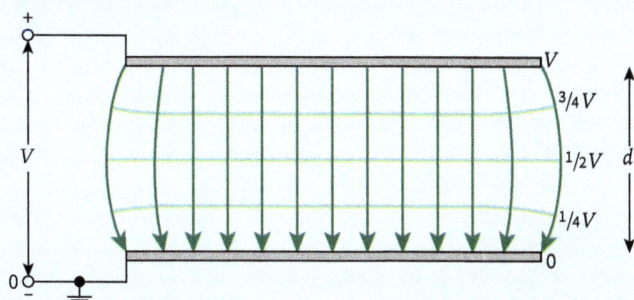

In a uniform field, the field strength and direction are the same at every point, indicated by the field lines being parallel to one another and equally spaced. The magnitude of the electric field strength at any point in a uniform field is:

$$E = \frac{V}{d}$$

where:

E = electric field strength in $V\,m^{-1}$ or $N\,C^{-1}$

V = potential difference between the plates in V

d = distance between plates in m.

This formula is derived as follows. The force F on a charge Q due to an electric field E is:

$$F = EQ \qquad ①$$

If the charge is moved by this force from the positive to the negative plate, the work done by the field on Q is:

$$\text{work done} = \text{force} \times \text{distance}$$
$$W = Fd \qquad ②$$

Combining ① and ② gives:

$$W = EQd \qquad ③$$

Work done W in moving a charge Q through a pd V is:

$$W = QV \qquad ④$$

Combining ③ and ④ gives:

$$V = \frac{W}{Q} = \frac{EQd}{Q} = Ed$$

Rearranging gives:

$$E = \frac{V}{d}$$

Moving charged particles in a uniform field

A charged particle entering a uniform field in a direction initially perpendicular to the field will experience a constant force.

- If the particle is positively charged, the direction of the force will be in the same direction as the field lines.

- If the particle is negatively charged, the direction of the force will be in the opposite direction to the field lines.

The force will result in a parabolic trajectory for the particle inside the field.

This situation is analogous to a horizontally launched projectile in a uniform gravitational field (initial vertical velocity is zero) and problems based on it can be solved using the same mathematical approach.

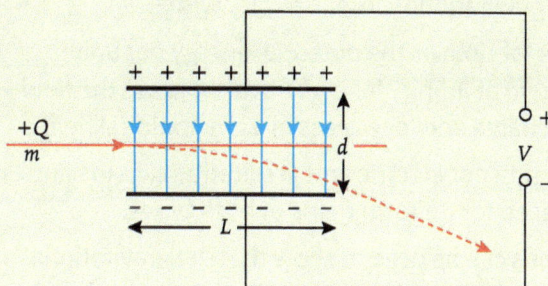

Electric potential

The absolute electric potential V at a point in an electric field is defined as the work done per unit charge to bring a small positive test charge from infinity to that point. Electric potential (unit V, equal to JC^{-1}) is the electric potential energy per coulomb that a charged object would have at that point in the field. The magnitude of the absolute electric potential at a point in a radial field caused by a charge Q is:

$$V = \frac{1}{4v\pi\varepsilon_0}\frac{Q}{r}$$

The sign of electric potential is positive for the field around a positive charge and negative for that around a negative charge.

Electric potential difference is the difference between the electric potential at two points in an electric field.

The work done ΔW in moving a charge Q between two points in an electric field is:

work done = charge × electric potential difference

$$\Delta W = Q\Delta V$$

Comparing electric and gravitational fields

Differences

Gravitational fields	Electric fields
Gravitational fields are caused by and act on objects with mass.	Electric fields are caused by and act on objects with charge.
Gravitational forces are always attractive.	Electric forces can be attractive or repulsive.

Similarities

- The size of the force in both electric fields and gravitational fields follows an inverse square law: $F - \frac{1}{4\pi\varepsilon_0}\frac{Q_1 Q_2}{r^2}$ and $F - \frac{Gm_1m_2}{r^2}$
- The field strength in both electric fields and gravitational fields follows an inverse square law: $E = \frac{1}{4\pi\varepsilon_0}\frac{Q}{r^2}$ and $g = \frac{GM}{r^2}$
- Field strength is defined similarly for both, as the force per unit charge or unit mass: $E = \frac{F}{Q}$ and $g = \frac{F}{m}$
- Both electric potential and gravitational potential vary with $\frac{1}{r}$
- Potential is defined similarly for both as the potential energy per unit charge or unit mass, and is zero at infinity for a positive charge in the field caused by another positive charge and for a mass in a gravitational field.
- Field lines can be used to represent electric and gravitational fields and are parallel and equally spaced for uniform fields of both types.

At a subatomic level, calculations can be used to show that the gravitational force between particles is many, many times weaker than the electric force.

Electric equipotentials

As with gravitational fields, an **equipotential surface** is one where the electric potential is the same at all points. No work is done when a charge moves along an equipotential surface because the potential difference between any two points on the surface is zero.

Equipotential surfaces

Equipotential surfaces are perpendicular to field lines. The equipotential surfaces in a uniform field are flat planes. The equipotential surfaces in a radial field are spheres.

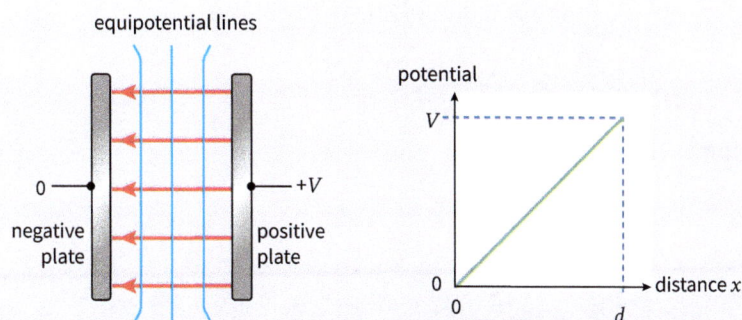

Graphs for electric potential

Graphs of electric field strength and electric potential look similar, but E is a $\frac{1}{r^2}$ line, whilst V is a $\frac{1}{r}$ line.

$$V = \frac{1}{4\pi\varepsilon_0} \frac{Q}{r}$$

The electric field strength is related to potential difference by:

$$E = \frac{\Delta V}{\Delta r}$$

So, the electric field strength at any point in an electric field can be found from the gradient of the V against r graph at that point.

The potential difference ΔV between two points in a field can be found from the area under the E against r graph between those two points.

For a positive charge

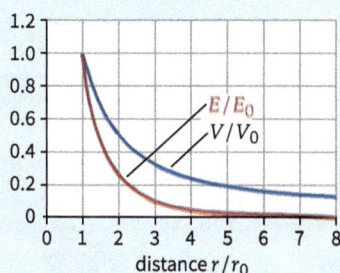

For a negative charge

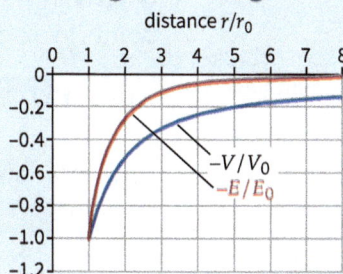

Retrieval

Learn the answers to the questions below, then cover the answers column with a piece of paper and write as many as you can. Check and repeat.

	Questions	Answers
1	What is an electric field?	region of space in which any object with charge will experience a force
2	Where do electric fields exist?	around all objects with charge
3	State Coulomb's law in symbols.	$F = \dfrac{1}{4\pi\varepsilon_0}\dfrac{Q_1 Q_2}{r^2}$
4	Give the electric field strength at any point in a field.	$E = \dfrac{F}{Q}$
5	Give the unit of electric field strength.	$N\,C^{-1}$
6	Give the field strength for a radial electric field at a distance r from a point charge Q.	$E = \dfrac{1}{4\pi\varepsilon_0}\dfrac{Q}{r^2}$
7	How can a uniform electric field be produced?	applying a potential difference between two parallel metal plates
8	Give the properties of a uniform electric field.	field strength and direction are the same at every point
9	Give the formula for the magnitude of the electric field strength at any point in a uniform field.	$E = \dfrac{V}{d}$
10	What kind of trajectory will a charged particle entering a uniform field in a direction initially perpendicular to the field have?	parabolic trajectory
11	Define the absolute electric potential at a point in an electric field.	work done per unit charge to bring a small positive test charge from infinity to that point
12	Give the formula for the magnitude of the absolute electric potential at a point in a radial field caused by a charge.	$V = \dfrac{1}{4\pi\varepsilon_0}\dfrac{Q}{r}$
13	What is electric potential difference?	difference between the electric potential at two points in an electric field
14	Give the formula for the work done in moving a charge between two points in an electric field.	$\Delta W = Q\Delta V$
15	What is an equipotential surface in an electric field?	surface where the electric potential is the same at all points
16	How are equipotential surfaces arranged relative to field lines?	perpendicular to field lines
17	What shape are the equipotential surfaces in a radial field, and a uniform field?	spheres; flat planes
18	How is electric field strength related to potential difference?	$E = \dfrac{\Delta V}{\Delta r}$

Put paper here

Put paper here

19 How can the electric field strength at any point in an electric field be found from the graph of V against r?

gradient of the graph at that point

20 How can the potential difference between two points in a field be found from the graph of E against r?

area under the graph between those two points

Maths skills

Practise your maths skills using the worked example and practice questions below.

Magnitudes of electrostatic forces

Coulomb's law allows us to calculate the magnitude of the electrostatic force between two charged objects.

$$F = \frac{1}{4\pi\varepsilon_0}\frac{Qq}{r^2}$$

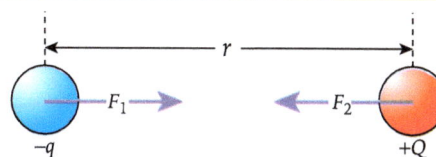

r is the position of q measured from Q
F_1 is the force on q due to Q
F_2 is the force on Q due to q
$F_1 = -F_2$

Worked example

Question

1 In the Bohr model of the hydrogen atom, the distance between the nucleus and the electron is 5.291×10^{-11} m.

Calculate the force between the proton and electron in the Bohr atom.

2 Two point charges are a distance of 60 mm apart.

State the nature of the force exerted on each charge.

$Q = +25\,\mu C$ $q = +100\,\mu C$

Answer

1 Using the equation noted:

charge on proton $Q = +1.6 \times 10^{-19}$ C

charge on electron $q = -1.6 \times 10^{-19}$ C

$\varepsilon_0 = 8.85 \times 10^{-12}$ F m^{-1}

$r = 5.291 \times 10^{-11}$ m

Substituting these values into the equation gives:

$$F = \frac{1}{4\pi \times 8.85 \times 10^{-12}}\frac{1.6 \times 10^{-19}(-1.6 \times 10^{-19})}{(5.291 \times 10^{-11})^2}$$

$$= -8.2 \times 10^{-8}\,N$$

2 The force is repulsive because the two charges have the same sign.

Practice

1 For the point charges in worked example question 2, calculate the magnitude of the force exerted on the $+25\,\mu C$ charge by the $+100\,\mu C$ charge.

2 Two point charges are placed 80 mm apart.

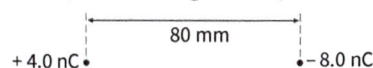

80 mm

$+4.0$ nC -8.0 nC

a State the nature of the force exerted on each charge.

b Calculate the magnitude of the force exerted on the $+4.0$ nC charge by the -8.0 nC charge.

3 Two point charges are placed a distance apart. The force on each charge at this distance is 40 N. The distance between the charges is doubled.

Determine the new force that each charge experiences.

Practice

Exam-style questions

01 **Figure 1** shows the path of an alpha particle as it approaches a gold nucleus.

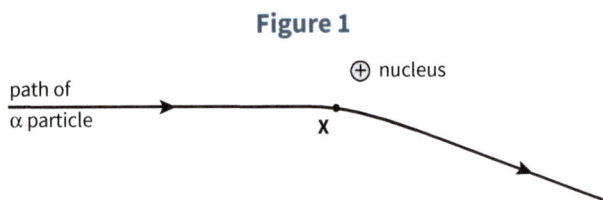

Figure 1

⊕ nucleus

path of
α particle
                        X

Synoptic links

3.7.1   3.1.1   3.2.1.2

01.1 Draw the direction of the resultant force acting on the alpha particle at point **X** on **Figure 1**. **[1 mark]**

01.2 The initial kinetic energy of the alpha particle is 6.2 MeV.

Show that the initial velocity of the alpha particle is $1.7 \times 10^7 \, \text{m s}^{-1}$.

mass of alpha particle = $6.64 \times 10^{-27} \, \text{kg}$ **[2 marks]**

01.3 Show that the closest approach r_c an alpha particle can get to the gold nucleus is given by the following expression, where Z is the atomic number for gold. **[3 marks]**

$$r_c = \frac{Z \, e^2}{\pi \, \varepsilon_0 \, m \, v^2}$$

Exam tip

Think about the energy changes of the alpha particle – the kinetic energy is transferred to potential energy.

01.4 Calculate the distance of closest approach of a 6.2 MeV alpha particle to a gold nucleus, $^{197}_{79}\text{Au}$. **[2 marks]**

distance = _____ m

02   A student uses a digital balance and two identically charged spheres to investigate Coulomb's law.

02.1 Draw the electric field lines between the two spheres in **Figure 2**.
[3 marks]

Synoptic links
PS1.1   ATc

Figure 2

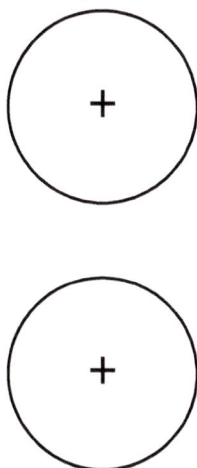

02.2 Suggest a solution to **one** problem of measuring the distances between the spheres.
[2 marks]

Exam tip

Practical skills can be tested in exam questions and will probably contain unfamiliar contexts. Use your experience to deduce why this might be a difficult measurement and how you could overcome any difficulty.

02.3 **Table 1** shows the results of this experiment.

Table 1

Distance between centres / mm	Reading on balance / g
15	147.1
20	82.7
25	53.0
30	36.8

Suggest and solve a test to show whether the data in **Table 1** supports Coulomb's law.
[3 marks]

02.4 Determine the charge on each sphere, using the data in **Table 1**. Assume it is the same on each sphere.
[2 marks]

For answers and more practice questions visit www.oxfordrevise.com/scienceanswers
Even more practice and interactive revision quizzes are available on kerboodle
15 Practice 171

03 **Figure 3** shows lines of equipotential around an electricity power cable at +500 V.

Figure 3

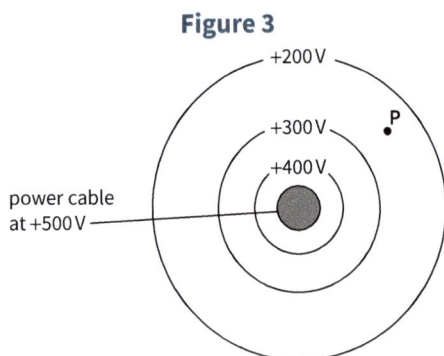

power cable at +500 V

03.1 State how can you tell from **Figure 3** that this is not a uniform electric field. **[1 mark]**

_____

_____

03.2 Draw the electric field lines on **Figure 3**. **[2 marks]**

03.3 The 300 V equipotential has a radius of 10 cm.

Determine the radius in cm of the 200 V equipotential.

Assume the power cable acts as a point charge. **[2 marks]**

! **Exam tip**

Take care with **03.4** – the question uses the command word 'determine', which suggests that you will have to use data to answer the question.

radius = _____ cm

03.4 Point **P** is at a distance halfway between the equipotential surfaces of 200 V and 300 V.

Determine the potential in V at point **P**. **[2 marks]**

potential = _____ V

04 **Figure 4** shows an electron beam entering a uniform electric field produced by two oppositely charged plates.

Figure 4

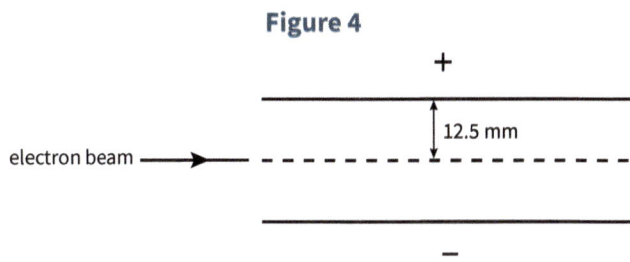

Synoptic links

3.4.1.5   3.4.1.4

04.1 Draw at least **six** lines on **Figure 4** to represent the electric field between the plates. **[2 marks]**

04.2 Sketch the path of the electron beam on **Figure 4** as it passes between the plates. **[1 mark]**

04.3 The potential difference across the plates is 1500 V.

Show that the acceleration of the electrons between the plates is approximately 1×10^{16} m s^{-2}. **[3 marks]**

acceleration = _____

04.4 The plates are 4.0 cm long. The electrons enter the electric field at a velocity of 3×10^7 m s^{-1}.

Determine the distance in mm of the electrons from the top plate as they exit the plates and the electric field. **[4 marks]**

> **Exam tip**
>
> **04.4** is really a question on projectile motion – the electron undergoes a vertical acceleration as it moves at a constant velocity in the horizontal direction. To answer this, consider the horizontal and vertical motion separately.

distance = _____ mm

05 A Van de Graaff generator is used to store a large amount of positive electrical charge. To store charge, it must have a high capacitance.

05.1 Show, by considering the definition of capacitance and the equation for electric potential, that the capacitance of the generator is given by the following equation, where R is the radius of the sphere.
$C = 4\pi\varepsilon_0 R$ **[2 marks]**

Synoptic link

3.7.4.1

_____

05.2 The radius of the sphere on one Van de Graaff generator is 0.20 m.

Calculate the capacitance of the generator and state the units.

[2 marks]

capacitance = _____

units = _____

05.3 Air breaks down when the electric field strength at the surface of the sphere is $3 \times 10^6 \, V \, m^{-1}$.

Calculate the potential of the sphere in V just before it begins to discharge by sparking across an air gap. [2 marks]

> **! Exam tip**
>
> This is not a uniform field, but think how E and V are related and use this to find a solution.

potential = _____ V

05.4 Determine the number of excess positive charges on the sphere at this potential. [2 marks]

excess positive charges = _____

06 In a simple model of a hydrogen atom, an electron orbits a proton. The average distance between an electron and proton in a hydrogen atom is 5.3×10^{-11} m.

06.1 Calculate the electrical force in N on the electron at this distance.
[2 marks]

> **⊗ Synoptic links**
>
> 3.7.4.1   3.1.1

06.2 State the electrical force in N on the proton. [1 mark]

06.3 Calculate the acceleration in $m \, s^{-2}$ of the electron. [1 mark]

06.4 Show that the total energy of the electron is 13.6 eV. [4 marks]

> **! Exam tip**
>
> Think how you would approach **06.4** if it were a question involving a gravitational field.

07 **Figure 5** shows two parallel plates 10 cm apart.

Figure 5

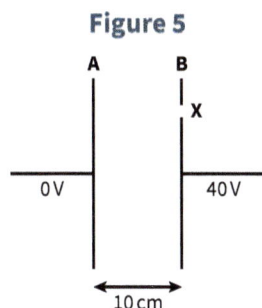

Exam tip

Remember that electric and gravitational fields are analogous. If you are unsure, consider how a ball would behave if thrown into the air – what determines how high it will go? How would you describe its motion? How does its final kinetic energy compare with the initial kinetic energy?

07.1 Draw and annotate the equipotential surfaces between the plates corresponding to 10 V, 20 V, and 30 V. **[1 mark]**

07.2 Calculate the electric field strength in $V\,m^{-1}$ between the plates if the potential difference across the plates is 40 V. **[2 marks]**

07.3 Calculate the work done in J on an electron as it moves from plate **A** to **B**. **[2 marks]**

07.4 An electron with a kinetic energy of 20 eV is injected into the field at point **X**.

Describe and explain its motion in the space between **A** and **B**. **[3 marks]**

08 **Figure 6** shows a graph of electric potential V against distance r from a positive point charge.

Synoptic link

3.1.1

Figure 6

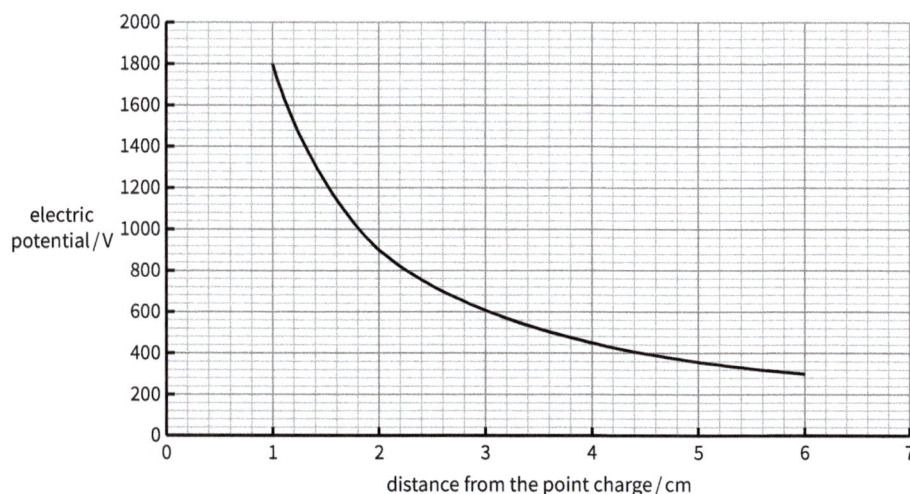

distance from the point charge / cm

08.1 Explain what is meant by electric potential. **[1 mark]**

08.2 Suggest a test and take readings from **Figure 6** to show that V is inversely proportional to r. **[3 marks]**

08.3 Calculate the value of the point charge. State your answer in nC. **[2 marks]**

08.4 Determine the field strength in $V\,m^{-1}$ at a distance of 3 cm from the point charge, using data from **Figure 6**. **[3 marks]**

Exam tip

Think carefully about the definition for electric potential.

08.5 Calculate the energy in J required to remove a second charge of −4 nC from 6 cm to an infinite distance. **[2 marks]**

Knowledge

16 Capacitance

Capacitance

A **capacitor** is a device designed to store charge. The circuit symbol for capacitor is:

$$-\!\!|\,|\!\!-$$

The capacitance C (unit farad, F, equal to $1\,C\,V^{-1}$) of a capacitor is the charged stored per unit potential difference:

$$C = \frac{Q}{V}$$

1 F is a large capacitance, so it is common to see capacitors labelled in microfarads, μF ($1 \times 10^{-6}\,F$). A simple capacitor consists of two parallel metal plates opposite each other with a gap between them.

Energy stored by a capacitor

The energy stored by a charged capacitor can be calculated using:

$$E = \frac{1}{2}QV = \frac{1}{2}CV^2 = \frac{1}{2}\frac{Q^2}{c}$$

The energy stored by a charged capacitor is equal to the area under the charge against p.d. graph or the p.d. against charge graph.

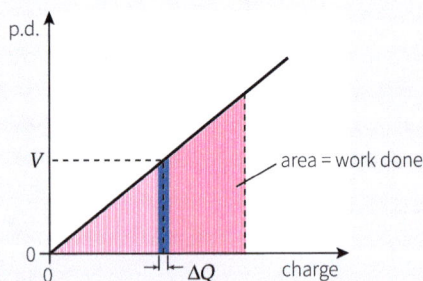

The energy supplied by a power supply to charge up a capacitor is $E = QV$.

The energy stored by the capacitor is half of this because the rest of the energy is lost to the resistance of the circuit and the internal resistance of the power supply.

Dielectrics

The charge stored on the plates of a capacitor can be increased by inserting a **dielectric** (an insulating material) between the plates.

The relative **permittivity** (or dielectric constant) of a dielectric is the ratio of the charge stored with the dielectric to the charge stored without the dielectric:

$$\varepsilon_r = \frac{C}{C_0} = \frac{Q}{Q_0} = \frac{\varepsilon_1}{\varepsilon_0}$$

where:

C = capacitance with dielectric

C_0 = capacitance when the gap between the plates is a vacuum

Q = charge stored with dielectric

Q_0 = charge stored when the gap between the plates is a vacuum

ε_1 = permittivity of dielectric

ε_0 = permittivity of 'free space' (vacuum).

Dielectrics increase the capacitance of a capacitor.

They contain polar molecules that align themselves with the electric field between the plates, which causes an electric field that partially cancels the electric field caused by the charge on the plates.

This reduces the p.d. across the capacitor, meaning more charge can be stored per volt.

Parallel plate capacitor

For a parallel plate capacitor with a dielectric between the plates, the capacitance is:

$$C = \frac{A\,\varepsilon_0\,\varepsilon_r}{d}$$

where:

A = surface area of a plate (m^2)

d = distance between the plates (m)

ε_r = relative permittivity of the dielectric.

Discharging a capacitor 🌡️

This circuit can be used to discharge a capacitor.

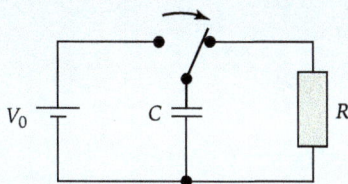

Graphs of current, p.d., or charge against time for a discharging capacitor show an exponential decay.

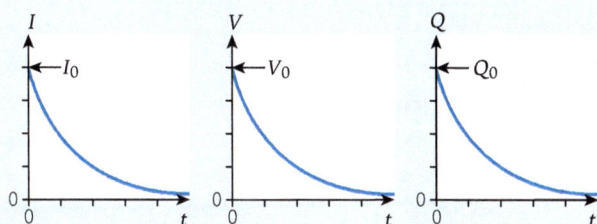

The area under the current–time graph is equal to the charge that has flowed off the capacitor (and, therefore, the charge that was stored on the capacitor). The gradient of the tangent of the charge–time graph at any point is equal to the current at that time. The equations for these graphs all have the same form:

$$I = I_0\, e^{-\frac{t}{RC}}, \quad V = V_0\, e^{-\frac{t}{RC}}, \quad Q = Q_0\, e^{-\frac{t}{RC}}$$

where:

$I / V / Q$ = current / p.d. / charge at time t

$I_0 / V_0 / Q_0$ = initial current / p.d. / charge

t = time (s)

e = constant = 2.71828…

R = resistance (Ω)

C = capacitance of capacitor (F).

The time constant of a capacitor discharge circuit is:

$$\text{time constant } (\tau) = RC$$

RC is the time taken for the charge, current, and p.d. to fall to $\frac{1}{e}$ (approximately 0.37) of its original value.

Five time constants ($5RC$) is a good approximation of the time taken to fully discharge a capacitor because the current, p.d., and charge will decrease by over 99% in this time.

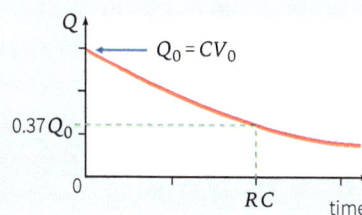

The time constant is equal to $-\dfrac{1}{\text{gradient}}$ of the graphs of $\ln I$, $\ln V$, and $\ln Q$ against time.

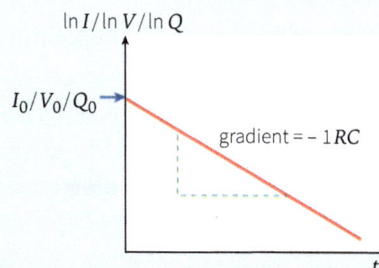

The half-life ($T_{\frac{1}{2}}$) of a capacitor discharge circuit is the time taken for the charge, current, and p.d. to halve. Half-life and time constant are related by:

$$T_{\frac{1}{2}} = 0.69\tau = 0.69RC$$

Charging a capacitor 🌡️

This circuit can be used to charge a capacitor.

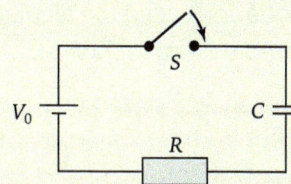

The graph of current against time for a charging capacitor has the same shape and equation as one for a discharging capacitor.

$$I = I_0\, e^{-\frac{t}{RC}}$$

The graphs of charge and p.d. against time show an exponential decrease in their rate of growth.

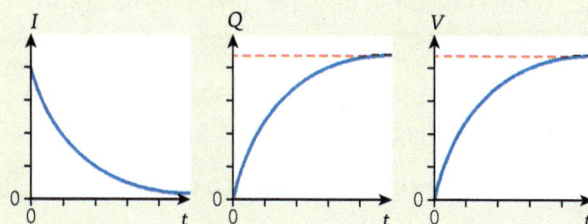

The equations for how charge and p.d. vary with time for a charging capacitor are:

$$Q = Q_0 \left(1 - e^{-\frac{t}{RC}}\right)$$

$$V = V_0 \left(1 - e^{-\frac{t}{RC}}\right)$$

Retrieval

Learn the answers to the questions below, then cover the answers column with a piece of paper and write as many as you can. Check and repeat.

	Questions	Answers
1	What is a capacitor?	device designed to store charge
2	What is the capacitance of a capacitor?	charged stored per unit potential difference, $C = \dfrac{Q}{V}$
3	What does a simple capacitor consist of?	two parallel metal plates opposite each other with a gap between them
4	What effect does inserting a dielectric between the plates have on a capacitor?	increases its capacitance
5	Give the alternative name for the relative permittivity of a dielectric.	dielectric constant
6	What is the relative permittivity of a dielectric?	ratio of the charge stored with the dielectric to the charge stored without the dielectric $\varepsilon_r = \dfrac{C}{C_0} = \dfrac{Q}{Q_0} = \dfrac{\varepsilon_1}{\varepsilon_0}$
7	Give the formula for the capacitance of a parallel plate capacitor with a dielectric between the plates.	$C = \dfrac{A\,\varepsilon_0\,\varepsilon_r}{d}$
8	Give the equations used to calculate the energy stored on a capacitor.	$E = \tfrac{1}{2}QV = \tfrac{1}{2}CV^2 = \tfrac{1}{2}\dfrac{Q^2}{C}$
9	How can the energy stored on a capacitor be found from a charge against potential difference graph?	area under the graph
10	Why is the energy stored by a capacitor only half the energy supplied by the power supply?	rest of the energy is lost to the resistance of the circuit and the internal resistance of the power supply
11	Give the equations for how current, p.d., and charge vary with time for a discharging capacitor.	$I = I_0\,e^{-\frac{t}{RC}},\ V = V_0\,e^{-\frac{t}{RC}},\ Q = Q_0\,e^{-\frac{t}{RC}}$
12	What is the area under a current–time graph equal to?	charge that has flowed off the capacitor
13	What is the gradient of the tangent of a charge–time graph at any point equal to?	current at that time
14	What is the time constant of a capacitor discharge circuit?	time taken for the charge, current, and p.d. to fall to $\dfrac{1}{e}$ (approximately 0.37 or 37%) of its original value
15	Give the formula for calculating the time constant.	time constant $(\tau) = RC$
16	How can the time constant be found from the graphs of $\ln I$, $\ln V$, and $\ln Q$ against time?	time constant $= -\dfrac{1}{\text{gradient}}$ of the graphs

Put paper here

Practical skills

Practise your practical skills using the worked example and practice questions below.

Time constants of capacitors

This circuit can be used to measure the capacitance of a capacitor.

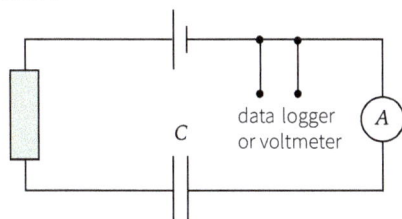

For a capacitor, the p.d. decreases as $V = V_0 e^{-\frac{t}{RC}}$

We can rearrange the equation to obtain the equation of a straight line:

$$\ln V = -\frac{t}{RC} - \ln V_0$$

Plotting a graph of $\ln V$ against t will give a straight-line graph of gradient $-\frac{1}{RC}$ and an intercept on the y-axis equal to $\ln V_0$.

Worked example

Question

A student used the equipment in the diagram (left) to measure the change in voltage over time when the switch is opened.

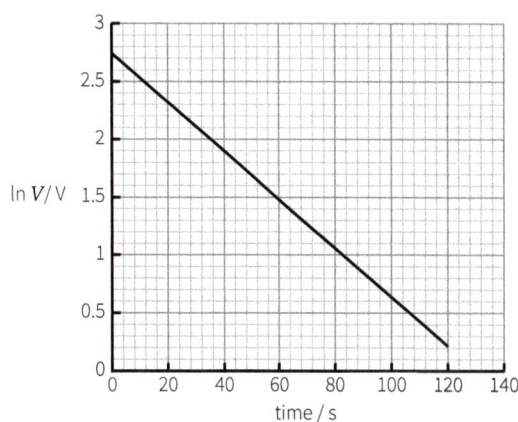

1 Use the graph to calculate the time constant RC for the circuit.

2 The resistor has a value of $10\,k\Omega$. Calculate the capacitance of the capacitor.

Answer

1 $\text{gradient} = \dfrac{2.75 - 0.2}{0 - 120} = -0.02\,125\,s^{-1}$

$RC = -\dfrac{1}{\text{gradient}} = 47.1\,s$

2 $RC = 47.06\,s$

therefore $C = \dfrac{47.06}{10 \times 10^3} = 4.71 \times 10^{-3}\,F$

Practice

1 A student measures the capacitance of a capacitor, using a digital voltmeter, a stopwatch, and a $330\,k\Omega$ resistor. The table shows their data.

Time / s	p.d. / V
0	10.3
20	6.9
40	4.6
60	3.1
80	2.1
100	1.4
120	0.9
140	0.6

a Use the data to plot a graph of $\ln V$ against time and use it to calculate the time constant for the circuit.

b Determine the capacitance of the capacitor.

2 The capacitor used is an electrolytic capacitor.

Describe what precautions the student needs to take when connecting the capacitor in the circuit.

3 Another student chooses a capacitor of $430\,\mu F$ and a resistor of $6.8\,k\Omega$. Discuss whether this is a good combination to use for this experiment.

01 A student investigates the charge stored on a capacitor using the circuit shown in **Figure 1**.

Figure 1

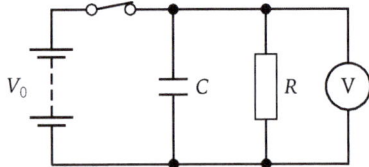

01.1 The switch has been closed for a long time. The capacitance of the capacitor is $1000\,\mu F$, the p.d. of the supply is $12\,V$ and the resistor has a resistance of $2\,k\Omega$.

Calculate the charge stored in the capacitor and the current through the resistor. **[2 marks]**

charge = _____ C

current = _____ A

01.2 The switch is opened.
Calculate the time it takes for the voltmeter reading to half.

Annotate **Figure 2** by adding numerical values to the axes.
Sketch a graph of the voltmeter reading against time during the first six seconds. **[3 marks]**

Figure 2

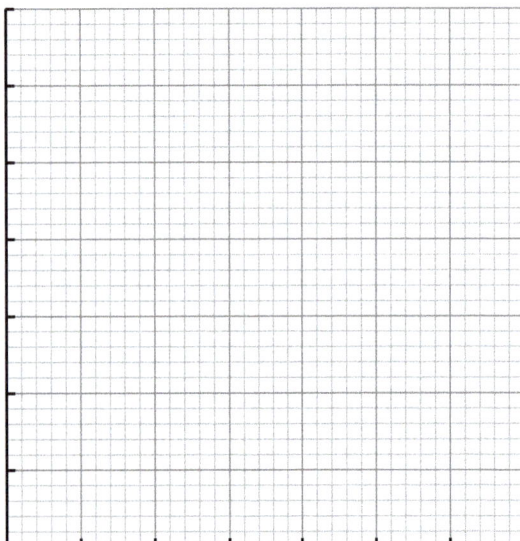

01.3 A capacitor of unknown capacitance is connected in parallel with the capacitor shown in **Figure 1**. The time constant of the circuit, as calculated from the graph of p.d. against time, is now 1.8 times the original time constant.

Calculate the value of the unknown capacitor. **[3 marks]**

capacitance = _____ μF

01.4 These calculations can be carried out if an assumption is made about the resistance of the voltmeter.

State the assumption made, and suggest the effect on your answer to **01.3** if this assumption is violated. **[2 marks]**

_____

_____

_____

02 A student connects a potential difference sensor across a 10 kΩ resistor in the charging circuit of a capacitor shown in **Figure 3**.

Figure 3a   **Figure 3b**

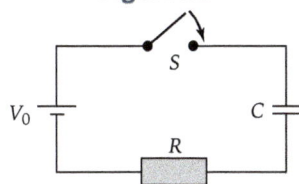

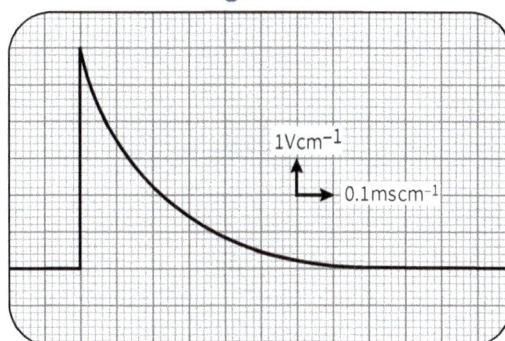

02.1 Deduce, using **Figure 3b**, the time constant of the circuit.

Show your method. **[2 marks]**

time constant = _____ s

02.2 Use your answer to **02.1** to deduce the capacitance of the capacitor in farads. **[2 marks]**

capacitance = _____ F

02.3 The student doubles the resistance in the circuit.
Sketch the graph that the student would expect to get.

Explain the values on the x-and y-axes. [3 marks]

> **! Exam tip**
>
> If a question says something is 'doubled', you can follow the consequence of that through by saying your answer 'doubled' or 'halved' as a result.

_____

_____

_____

02.4 Suggest how the student could use the graph from **02.3** to deduce the charge stored in the capacitor. [2 marks]

_____

_____

_____

_____

03 Capacitors are used in timing circuits because the charge on their plates follows an exponential relationship. The changes in quantities such as p.d. and current in capacitor circuits are predictable. A series circuit contains a charged capacitor, an open switch, an ammeter, and a resistor. A voltmeter is connected across the capacitor. The p.d. across the capacitor is shown in **Table 1**.

Table 1

Time / min	p.d. / V		
0	2.50		
10	1.62		
20	1.08		
30	0.70		
40	0.46		
50	0.30		
60	0.20		

03.1 Explain why the current in the circuit and the p.d. across the capacitor show exponential decay when the switch is closed. [3 marks]

03.2 Use the blank columns in **Table 1** to manipulate the data to produce a linear graph with quantities measured in SI units. [3 marks]

03.3 Explain what can be deduced from the gradient of the graph plotted in **03.2**. [2 marks]

> **! Exam tip**
>
> If you are plotting time on a graph, you should use seconds and not minutes.

03.4 The capacitor is used in a circuit as a back-up device. When there is a power cut, the capacitor starts to discharge. The device needs a p.d. above 1.5 V to continue to operate.

Deduce the number of minutes for which the device will continue to operate properly. [1 mark]

04 A capacitor of capacitance 470 µF is connected across a battery with a p.d. of 6.0 V.

04.1 Calculate the energy in J stored in the capacitor when it is fully charged. **[1 mark]**

04.2 The charged capacitor is connected to a single bulb. The specification of the lamp is 6 V, 0.3 A.

Determine how long in s it takes the capacitor to discharge to 37 % of its original charge through the lamp, and calculate the average power in W of the discharge.

Suggest whether the lamp would appear to light. **[5 marks]**

> **!** **Exam tip**
>
> Remember that the power of a bulb determines how bright it appears to be.

04.3 The p.d. to which the capacitor is charged is doubled.

Suggest how this doubling would affect what is observed when the capacitor discharges through the lamp. **[2 marks]**

04.4 A capacitor in a defibrillator is used to produce a p.d. that, when applied to the chest of a patient, produces a current that can cause their heart to start beating normally again. The energy that is transferred by the defibrillator is of the order 100 J.

Suggest how defibrillators produce an energy of 100 J by considering the p.d. needed to store 100 J in the 470 µF capacitor, and the capacitor needed to store 100 J with 6.0 V p.d. **[3 marks]**

05 A student wants to find out how the charge stored on a capacitor depends on its capacitance. They set up the circuit shown in **Figure 4**.

Figure 4

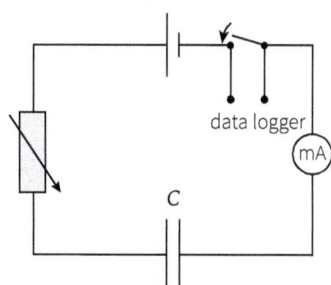

05.1 Sketch the graph that will be displayed on the computer screen from when the switch is closed to when the capacitor is fully charged. The p.d. of the battery is 12 V.

Explain the shape of the graph. **[3 marks]**

05.2 The capacitor is then discharged. In a separate experiement. When the switch is closed, the student continually adjusts the resistance of the variable resistor so that the reading on the ammeter is constant.

Describe what the student is doing to the resistance of the variable resistor to achieve this objective.

Describe the graph that will be displayed on the computer screen while the capacitor is charging and when it is fully charged. **[3 marks]**

> **!** **Exam tip**
>
> Remember to check whether a capacitor is charging or discharging before you plot a graph.

05.3 Describe how the student could use the equipment in **Figure 4** and method described in **05.2** to find the relationship between capacitance and charge stored. **[3 marks]**

05.4 Suggest the most significant source of uncertainty in this experiment, and suggest **one** way the student could reduce this uncertainty. **[2 marks]**

06 A humidity sensor is made from a capacitor that has a porous dielectric material between the plates. Water vapour can enter the sensor and penetrate the dielectric material.

Water molecules are polar molecules.

06.1 Describe what happens to a water molecule when it enters the porous dielectric material when the plates of the capacitor are connected to a potential difference. **[1 mark]**

06.2 Data relating the capacitance to the relative humidity of the air are shown in **Table 2**.

Explain the relationship between humidity and capacitance shown by the data, and relate your answer to **06.1**. **[3 marks]**

Table 2

Relative humidity / %	Capacitance / μF
10	290
30	300
53	310
75	320
98	330

06.3 Draw a graph of capacitance against humidity.

Show that the relative permittivity of the material used as the dielectric is about 470.

Assume that the relative permittivities add up, and that the relative permittivity of water is 80. **[5 marks]**

06.4 A typical humidity sensor has dimensions of $10.8\,mm \times 3.81\,mm$. Show that the thickness of the dielectric is about $0.2\,\mu m$. **[2 marks]**

06.5 The dielectric will break down when the field strength is greater than $74\,kV\,mm^{-1}$.

Calculate the current in A in the dielectric during breakdown. The resistivity of the dielectric is $10^{12}\,W\,m$.

Comment on your answer. **[4 marks]**

07 Two parallel plates are connected to a high voltage supply. Wooden retort stands are used to hold the plates a distance of 10 cm apart. There is air between the plates. The plates are disconnected from the supply at a p.d. of 4 kV. The area of each plate is $150\,cm^2$.

07.1 Calculate the charge in C on each plate. **[2 marks]**

> **Synoptic links**
> 3.7.3.3   3.5.1.3

> **Exam tip**
> Remember that polar molecules have an uneven distribution of charge.

> **Synoptic links**
> 3.5.1.1   3.6.1.3

07.2 A small, light ball is coated with conducting paint and suspended on a very light string so that it hangs between the plates.

When the ball is touched to the negative plate it oscillates between the plates. Explain why. **[2 marks]**

07.3 The length of the string is 30 cm.

Determine the time in s that it takes the ball to swing between the plates if the ball on the string behaves like a simple pendulum. **[2 marks]**

07.4 About 10% of the charge on the plate is transferred each time the ball hits it.

Calculate the current in A that flows in the wires attached to the plates. **[1 mark]**

07.5 The plates are disconnected from the supply.

Describe and explain what happens to the pd between the plates if they are moved closer together. **[2 marks]**

08 A capacitor can be used as a back-up supply in the case of a power failure to prevent loss of essential data from a system. The capacitors shown in **Table 3** have a high value of capacitance but operate at a low p.d.

Table 3

Capacitance / F	Maximum operating p.d. / V
0.5	3
1.3	3
3.3	6
10.0	6

08.1 The capacitors have a maximum operating p.d.

Suggest what would happen if this p.d. is exceeded. **[1 mark]**

08.2 A capacitor is needed that can provide a p.d. that does not fall below 1.5 V for several hours. The effective load across the capacitor will be 10 kΩ.

Suggest a value of capacitance in F that will maintain a p.d. above 1.5 V for more than 2 hours.

The capacitor can be charged to either 3 V or 6 V.
Show your working. **[3 marks]**

08.3 An alternative to a capacitor is a battery with a p.d. of 3 V and a capacity of 1400 mA h.

Compare the energy stored in the capacitors that work at a p.d. of 3 V with the energy stored in the battery. **[3 marks]**

08.4 The terminal p.d. of the battery is found to be less than 3 V.

Explain why, and describe how to calculate the minimum value of load resistance in the circuit that produces a pd above 1.5 V. **[3 marks]**

! **Exam tip**

Remember that the time period of a pendulum is the time for a complete oscillation.

Synoptic links

3.5.1.1   3.5.1.6

Knowledge

17 Magnetic fields

Magnetic field lines

A magnetic field is a region of space in which objects with magnetic properties, such as moving charges, experience a force. Magnetic fields are formed around all moving charged objects.

Magnetic fields can be represented using field lines, also known as lines of magnetic flux, where:

- the arrows on field lines show the direction of the field, always going from north to south
- the separation of field lines indicates the strength of the field – the closer together the lines, the stronger the field.

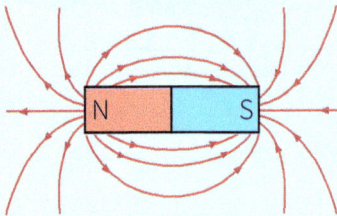

The field around a solenoid resembles that around a bar magnet. Its direction can be found using the **right-hand grip rule**, where the fingers of a right hand curl in the direction of the current, and the thumb points towards the north pole.

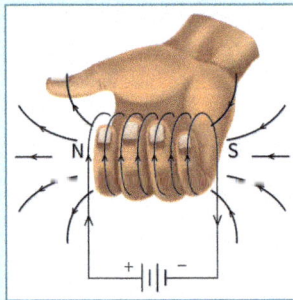

For a current-carrying straight wire, the thumb points in the direction of the current, and the fingers curl in the direction of the filed.

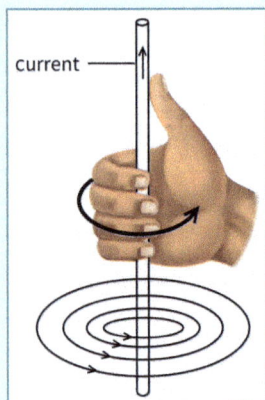

Magnetic flux density and magnetic flux

Magnetic field strength is also known as magnetic flux density, symbol B.

It is defined as the force per unit length per unit current on a current-carrying conductor at right angles to the direction of the magnetic field, and is measured in tesla (T).

1 T is the magnetic flux density that causes a force of 1 N on a 1 m length of wire at right angles to the field carrying a current of 1 A.

Magnetic flux ϕ

Magnetic flux (in weber, Wb) is defined as the number of flux lines passing through an area A (m^2) at right angles to the field the flux passes through:

$$\phi = BA$$

Magnetic flux linkage 🧪

For a coil of wire with cross-sectional area A and N turns:

$$\text{magnetic flux linkage } N\phi = BAN$$

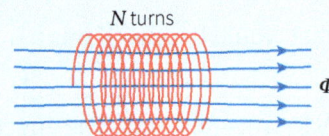

N turns

The unit of magnetic flux linkage is weber-turn, Wb turn. If the magnetic field is at an angle θ to the normal at the face of the coil, the flux linkage through the coil is:

$$N\phi = BAN\cos\theta$$

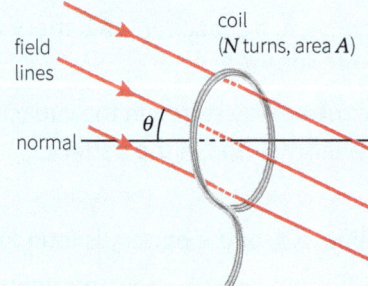

flux linkage = $BAN\cos\theta$

For a single loop of wire, $\phi = BA\cos\theta$

Force on a current-carrying wire 🧪

A current-carrying wire in a magnetic field will experience a force. This is known as **the motor effect**. The size of the force F in N of a wire of length L in m, that is perpendicular (at 90°) to the field lines, is:

$$F = BIL$$

The direction of the force can be found using **Fleming's left hand rule**, where:

- thumb points in the direction of the force
- first finger points in the direction of the magnetic field, from north to south
- second finger points in the direction of the current.

Moving charges in a magnetic field

The force on a charged particle moving at right angles to a magnetic field is:

$$F = BQv$$

The direction of the force on a:

- positive particle can be determined using Fleming's left hand rule, with the second finger representing the particle's direction
- negative particle is in the opposite direction to a positive particle moving in the same direction
- charged particle moving in a magnetic field is always at right angles to its velocity, so it experiences a centripetal force and will move in a circular path without changing speed.

Since a charged particle moving in a magnetic field experiences a centripetal force:

$$\frac{mv^2}{r} = BQv$$

Rearranging this gives the radius of the path of a charged particle moving at right angles to a magnetic field:

$$r = \frac{mv}{BQ}$$

This shows how the radius of the path of the charged particle depends on its mass, velocity, charge, and the strength of the magnetic field.

The cyclotron

A cyclotron is a circular particle accelerator which can be used to produce high energy beams for uses such as radiation therapy.

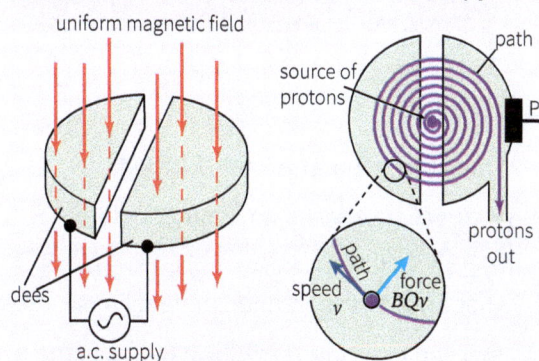

1 charged particles fired into one of the 'dees' near centre of cyclotron

2 magnetic field forces charged particle to follow a circular path out of first dee

3 a p.d. between the dees accelerates charged particle across gap

4 particle is moving faster so its circular path has a bigger radius in the second dee

5 p.d. across gaps reverses so the particle accelerates across the gap agin

6 p article leaves the cyclotron.

A fixed frequency alternating p.d. can be used to accelerate the particle between the dees because the time taken to travel round each dee is independent of speed:

$$r = \frac{mv}{BQ}$$

So, time taken to travel round a semi-circle:

$$r = \frac{\pi r}{v} = \frac{m\pi}{BQ}$$

For a cyclotron of radius R, the speed of particles, of charge Q and mass m, upon leaving is:

$$v = \frac{BQR}{m}$$

The period of the alternating p.d. must be equal to the time taken by a particle to complete a circle:

$$T = \frac{2m\pi}{BQ}$$

So, frequency of alternating p.d. $= \dfrac{1}{T} = \dfrac{BQ}{2\pi m}$

Retrieval

Learn the answers to the questions below, then cover the answers column with a piece of paper and write as many as you can. Check and repeat.

	Questions	Answers
1	What is a magnetic field?	region of space in which objects with magnetic properties experience a force
2	Where are magnetic fields formed?	around all moving charged objects
3	What do the arrows on magnetic field lines represent?	direction of the field, always from north to south
4	What does the separation of field lines indicate?	strength of the field
5	How can the direction of the field around a current-carrying straight wire be found?	using the corkscrew rule
6	How can the direction of the field around a solenoid be found?	using the right-hand grip rule
7	Define magnetic flux density.	force per unit length per unit current on a current-carrying conductor at right angles to the direction of the magnetic field, in tesla ($\phi = BA$, in weber)
8	Give the magnetic flux linkage for a coil of wire with cross-sectional area A and N turns.	$N\phi = BAN$
9	How is magnetic flux linkage denoted?	$N\phi$
10	State the unit of magnetic flux linkage.	weber-turn, Wb turn
11	Give the equation for the flux through a single loop of wire if the magnetic field is at an angle θ to the normal at the face of the loop.	$\phi = BA\cos\theta$
12	Give the equation for the size of the force on a current-carrying wire at 90° to a magnetic field.	$F = BIL$
13	Give the equation for the magnitude of the force on a charged particle moving at right angles to a magnetic field.	$F = BQv$
14	Give the equation relating the centripetal force on a charged particle moving in a magnetic field to the magnetic force it experiences.	$\dfrac{mv^2}{r} = BQv$
15	Give the equation for the radius of the path of a charged particle moving at right angles to a magnetic field.	$r = \dfrac{mv}{BQ}$
16	Give the equation for the speed of particles, of charge Q and mass m, upon leaving a cyclotron of radius R.	$v = \dfrac{BQR}{m}$
17	Give the equation for the time taken by a particle to complete a full circle in a cyclotron.	$T = \dfrac{2\pi m}{BQ}$
18	Give the equation for the frequency of the alternating p.d. in a cyclotron.	$\dfrac{1}{T} = \dfrac{BQ}{2\pi m}$

Put paper here

Practical skills

Practise your practical skills using the worked example and practice questions below.

Measuring force on a wire

When a current passes through a wire that is in a magnetic field, the wire experiences a force. We can use this apparatus to measure the magnitude of this force.

The lower loop of the wire is bent so it is the same length as the magnet, and clamped perpendicular to the magnetic field lines. When a current flows, a downward force acts on the magnets. This is measured as an increase of mass on the balance. The force on the wire is $F = BIL = mg$

Plotting a graph of m against I will give a straight line with a gradient equal to $B\dfrac{L}{g}$.

Worked example

Question

A student measures the force on the top pan balance when a current flows through the loop of wire 49.5 mm long.

Use the graph to determine the magnetic field of the magnets.

Answer

$$\text{gradient} = B\frac{L}{g}.$$

$$\text{gradient} = \frac{14\times10^{-4} - 2\times10^{-4}}{3.4 - 0.6} = 4.3\times10^{-4}$$

$$B = \frac{\text{gradient}\times g}{L} = \frac{4.3\times10^{-4}\times9.8}{49.5\times10^{-3}} = 0.85\,\text{T}$$

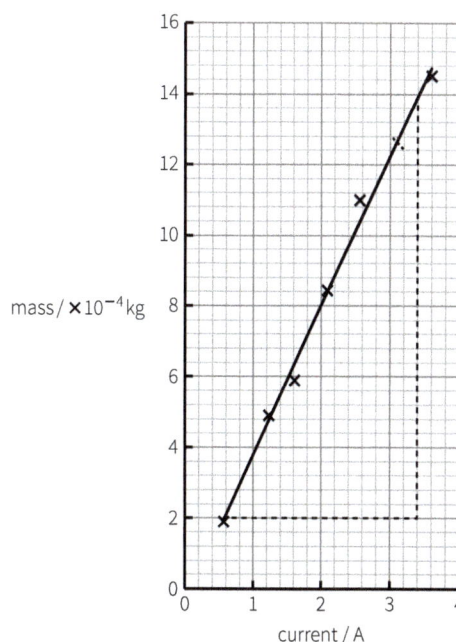

Practice

A student obtains the following data for this experiment.

I / A	m / g
0.56	0.19
1.23	0.49
1.61	0.59
2.09	0.84
2.56	1.10
3.11	1.26
3.59	1.45

1 Use the data to determine the magnetic field of the magnets used.

2 The circuit used in the experiment contains a variable resistor.

Suggest what the purpose of the resistor is.

3 Another student repeats the experiment. They fail to notice that the copper wire has moved and is no longer parallel to the length of the magnets, but is instead at an angle between them.

State and explain what will happen to the measurements on the top pan balance.

Exam-style questions

01 **Figure 1** shows a stiff wire clamped in place and suspended in a uniform magnetic field.

Figure 1

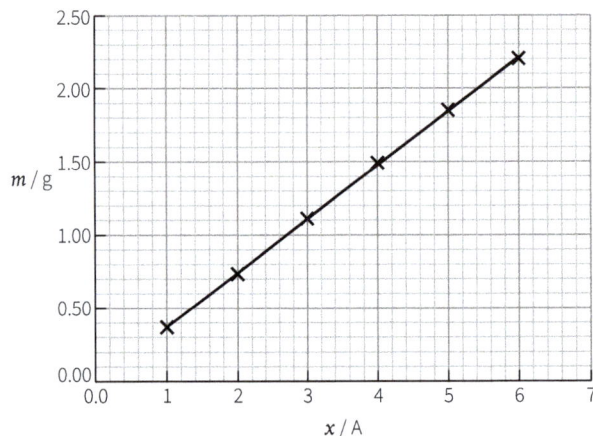

01.1 When current flows through the wire, the wire experiences an upward force. Explain whether the reading on the balance will increase or decrease. **[2 marks]**

_____

_____

01.2 Describe the procedure necessary to accurately and safely investigate the relationship between the force on the wire and the current flowing through it. **[3 marks]**

_____

_____

_____

Synoptic links

3.4.1.5   MS 3.3   MS 3.4

! Exam tip

Remember Newton's third law – if the wire experiences a force, then the magnets must experience an equal and opposite force.

01.3 The results for the experiment are shown in **Figure 2**.

Figure 2

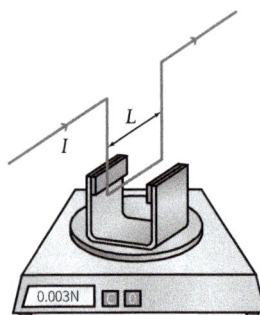

Calculate the gradient of the graph. State the units. **[3 marks]**

gradient = _____ unit = _____

01.4 The length of wire in the magnetic field is 5.0 cm. Determine the magnetic flux density of the magnets using the gradient from **01.3**. **[3 marks]**

magnetic flux density = _____ T

02 **Figure 3** shows part of the structure of a cyclotron.

Figure 3

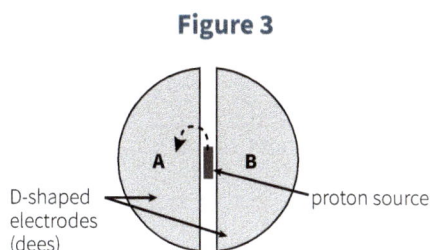

D-shaped electrodes (dees) ← A ⟶ B ← proton source

Synoptic links

3.5.1.4 3.7.3.2
3.7.3.3 3.6.1.1

02.1 Each time a proton passes through the gap between the D-shaped electrodes, it is accelerated by a potential difference V.

Show that the energy gained by the proton is given by the expression: energy gained = neV, where n is the number of times the proton crosses the gap. **[1 mark]**

02.2 State the direction of the magnetic field in **Figure 3**. **[1 mark]**

_____

02.3 Sketch on **Figure 3** the path of the proton as it is accelerated in the cyclotron. **[1 mark]**

02.4 The D-shaped electrodes are hollow.

Explain whether the protons are accelerated inside the D-shaped electrodes and by which fields. **[4 marks]**

_____

_____

_____

_____

02.5 Small cyclotrons can use an alternating potential difference at constant frequency across the plates. This is because the frequency is independent of the radius. This is shown by the following expression, where m is the mass of the proton.

$$f = \frac{Be}{2\pi m}$$

Derive the expression given. **[4 marks]**

Exam tip

If you are unsure how to begin, consider related equations, such as centripetal force and the equation for the force on charged particles moving in a magnetic field.

02.6 When the protons reach relativistic speeds, their mass increases. Suggest what happens to the frequency of the accelerator at that point. **[2 marks]**

_____

_____

03 **Figure 4** shows the structure of a mass spectrometer.

Figure 4

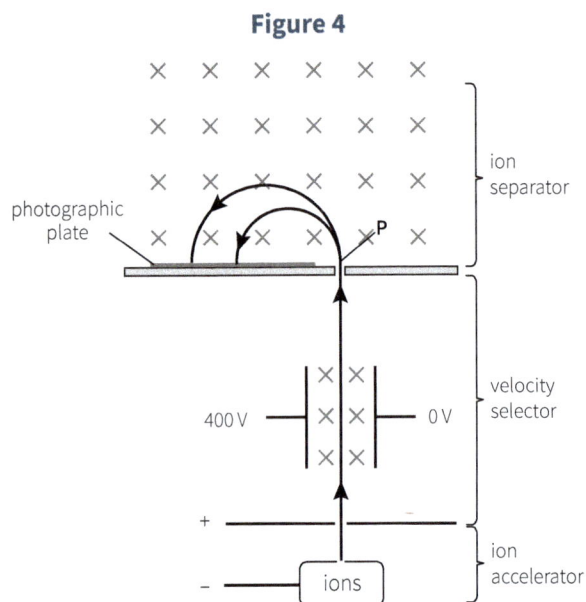

03.1 The velocity selector allows only positive ions of a specific velocity to enter the ion selector stage of the mass spectrometer. The expression for the velocity, v, is given by:

$$v = \frac{E}{B}$$

Explain how this ratio is derived. **[4 marks]**

03.2 The magnetic flux density of the velocity selector is 0.10 T. The velocity of ions leaving the velocity selector is $4.2 \times 10^5 \text{ m s}^{-1}$. Calculate the distance in m between the parallel plates in the velocity selector. The p.d. between the plates is 400 V. **[2 marks]**

03.3 When the positive ions enter the ion selector, they are separated according to their mass and charge.

Derive an expression for the radius of the path of the ions in the ion selector in terms of their mass and charge. **[2 marks]**

03.4 Two singly-charged calcium ions of mass 39.9 u and 43.9 u enter the ion selector. The magnetic field strength is 1.1 T.

Calculate the separation in m between the positions of the two ions when they strike the photographic plate. **[2 marks]**

04 **Figure 5** shows a search coil connected to an oscilloscope and a coil connected to an a.c. supply.

Figure 5

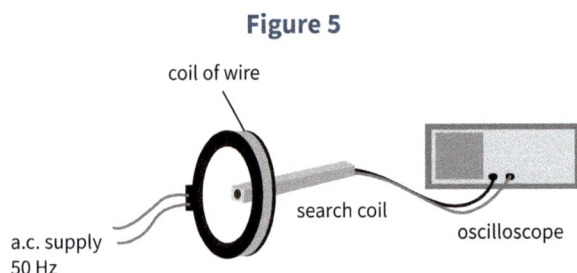

Synoptic links

3.6.1.1   3.7.3.2   3.8.1.6

Exam tip

There are two forces acting on the ion: one due to the electric field and one due to the magnetic field. Do they act in the same direction?

Synoptic links

3.7.5.4   PS1.2
MS3.3   MS3.4

04.1 Explain why the coil must connect to an a.c. supply for this investigation to work. **[2 marks]**

04.2 Describe how you would make accurate measurements of:

- the angle between search coil and magnetic field
- the e.m.f. induced from the oscilloscope screen. **[4 marks]**

04.3 A student collects the results shown in **Table 1**.

Table 1

$\theta/°$	$\cos\theta$	ε/mV
0	1.00	159
10	0.98	157
20	0.94	150
30	0.87	142
40	0.77	124

The student suggests the e.m.f. induced in the search coil is proportional to $\cos\theta$, where θ is the angle between the plane of the coil and the plane of the search coil.

Draw a suitable graph to test this hypothesis. **[3 marks]**

04.4 Explain whether your graph proves that e.m.f. is proportional to $\cos\theta$. Include calculations in your answer. **[3 marks]**

05 A coil of diameter 1.8 cm and 5000 turns is placed perpendicular to a magnetic field. The magnetic flux density is 5.0 mT.

05.1 Calculate the magnetic flux. State the units. **[3 marks]**

05.2 The coil is rotated so that the magnetic field is now at an angle of 40° to the normal of the coil, as shown in **Figure 6**.

Figure 6

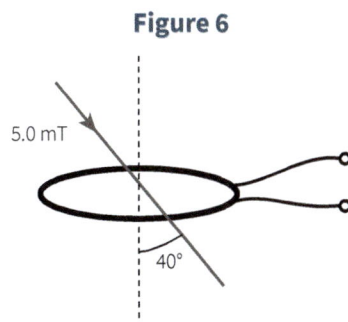

Calculate the new magnetic flux. **[1 mark]**

05.3 The coil is rotated through 40° in a time of 0.2 s. Determine the mean e.m.f. in V generated in the coil. **[2 marks]**

05.4 The magnetic field is produced by an electromagnet. Describe and explain what would happen the coil when the electromagnet is switched off. The coil remains stationary. **[2 marks]**

! Exam tip

Remember to describe how you will ensure the reading is accurate – what will be the difficulty of measuring the angle between the coil and the field? How can you read a voltage from an oscilloscope?

Synoptic link

3.5.7.4

! Exam tip

When the electromagnet is switched off, the magnetic flux linkage must change as it is now zero. Will that be a quick change?

06   A Hall probe is used to measure magnetic flux density directly. It consists of a slice of semiconductor of thickness t and width d carrying a current.

Figure 7 shows the basic structure.

Figure 7

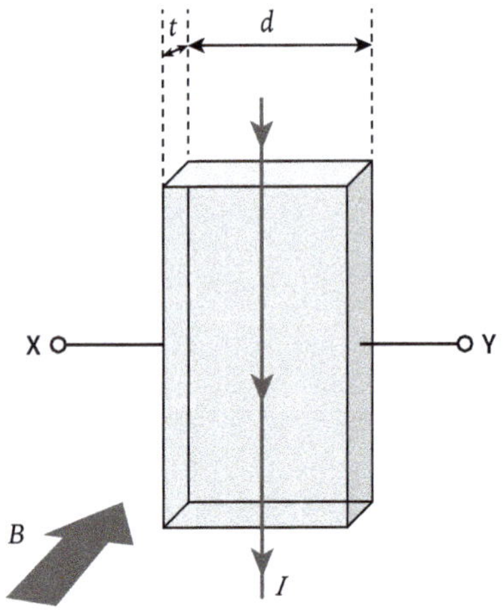

06.1 A magnetic field is applied perpendicular to the flat surface of the semiconductor. The electrons in the current experience a force. Write an equation that gives the magnitude of the force on the electrons.

State and explain the direction of the force.   **[3 marks]**

06.2 Charge builds up at the sides of the semiconductor.

Annotate **Figure 7** to show which side becomes positive and which becomes negative.   **[1 mark]**

06.3 Equilibrium occurs when the accumulation of charge on one side of the semiconductor repels the arrival of more electrons. A potential difference is set up across the semiconductor, called the Hall voltage V_H.

Show that $V_H = Bvd$, where v is the velocity of the free electrons.   **[3 marks]**

06.4 In practice, it is easier to measure the current in the semiconductor rather than the velocity of the free electrons. The relationship between the current and velocity is given by the expression $I = nAvq$, where n is the number of free charge carriers per unit volume and A is the cross-sectional area of the slice.

In **Figure 7** the cross-sectional area $A = d \times t$.

Show that $V_H = \dfrac{BI}{nte}$ **[2 marks]**

06.5 Copper has $n = 1 \times 10^{28}\,\mathrm{m^{-3}}$ and the semiconductor gallium arsenide has $n = 2 \times 10^{12}\,\mathrm{m^{-3}}$.

Explain why, for the same current and magnetic flux density, it would be better to use a Hall probe that was made from gallium arsenide.   **[2 marks]**

Synoptic link

3.7.3.2

Exam tip

This is an unfamiliar equation, but you want to substitute for v – simply follow the algebraic rules you know.

07 A student investigates the forces on a strip of aluminium foil in a magnetic field.

The foil measures 15 mm wide, 50 mm long, and 0.02 mm thick.

07.1 Describe how the thickness of the foil is measured. **[2 marks]**

07.2 The density of aluminium is 2700 kg m^{-3}.

Calculate the mass in kg of the strip of foil. **[2 marks]**

07.3 The 50 mm length of foil is placed in a magnetic field of flux density 0.03 T. A current is passed through the foil such that the foil just begins to lift up.

Calculate the current in A in the foil. **[2 marks]**

07.4 Draw a sketch to show the direction of the magnetic field and current for this experiment to work. **[2 marks]**

Synoptic links

PS4.1   Ate   3.4.2.1

Exam tip

This is a uniform magnetic field – how will you represent that?

08 **Figure 8** shows part of a mass spectrometer. After the ions have passed through **P**, they enter a region over which a uniform magnetic field is applied. The ions are then detected at points such as **R**.

Synoptic link

3.6.1.1

Figure 8

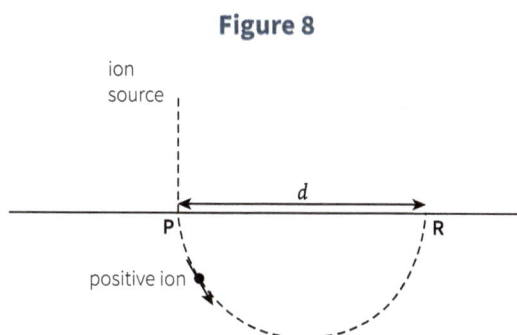

08.1 Annotate **Figure 8** to show the direction of the magnetic field. **[1 mark]**

08.2 Explain why the path of the ion is circular when it enters the uniform magnetic field. **[2 marks]**

08.3 Show that the radius of the path r is related to the mass m, velocity v, charge q, and magnetic flux density B by the expression $r = \dfrac{mv}{Bq}$. **[2 marks]**

08.4 Draw on **Figure 8** the path taken by a heavier isotope arriving at **P** with the same velocity and charge as the path already shown. **[1 mark]**

08.5 Two isotopes of boron have masses 10.012937 u and 11.009305 u. The separation between their detection points is 0.2 mm.

Determine the minimum value of d in m that is necessary for this to happen, assuming that they have the same charge and move at the same velocity at **P**. **[4 marks]**

Exam tip

Use ratios to help you solve **08.5**.

⚙ Knowledge

18 Electromagnetic induction

Experimental phenomena

Electromagnetic induction is the process by which an e.m.f. (electromotive force: a potential difference) is induced in a conductor that experiences a change in magnetic flux. If the conductor is part of a complete circuit, the induced e.m.f. will lead to an induced current.

The change in magnetic flux can be due to relative movement between the conductor and the field, or due to a change in the strength of the field. Examples include:

- moving a wire so that it cuts through the lines of flux of a permanent magnet
- moving a permanent magnet into and out of a coil of wire
- moving a conductor into and out of the field of a magnet
- dropping a magnet through a coil of wire
- spinning a coil of wire in a magnetic field, or spinning a magnet in a coil of wire
- placing a conductor in the field of an electromagnet and changing the current in the electromagnet.

induced current · galvanometer (centre zero)

insulated wire

greater induced current

Lenz's law

Lenz's law states that the direction of an induced current or e.m.f. is always such that it will oppose the change in flux producing it. Lenz's law is due to the law of conservation of energy and can be used to deduce the direction of the e.m.f. or current induced in a conductor.

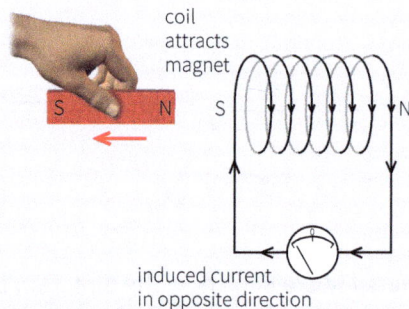

coil repels magnet

induced current

coil attracts magnet

induced current in opposite direction

Lenz's law can also be used to explain why a magnet can cause a force on a non-magnetic conductor if there is relative movement between the two. For example:

- a magnet will fall more slowly than a non magnetic object through a copper tube
- a spinning conductor will spin more slowly in the presence of a magnetic field
- a non-magnetic conductor can be made to move (accelerate or decelerate) by moving a magnet over it.

In these cases, the conductor experiences a change in flux so an e.m.f. is induced, which opposes the change producing it.

Faraday's law

Faraday's law states that the magnitude of an induced e.m.f. in a conductor is equal to the rate of change of flux it experiences or the rate of change of flux linkage in a coil:

$$\text{induced e.m.f. } \varepsilon = -N\frac{\Delta\phi}{\Delta t}$$

$$\text{induced e.m.f. } \varepsilon = \frac{\text{change in flux (linkage)}}{\text{time taken}}$$

The minus sign indicates Lenz's law, showing that the induced e.m.f. will be in the opposite direction to the change producing it.

For a straight conductor (such as a wire) of length l moving at right angles through a magnetic field of strength B at a speed v, the e.m.f. induced is:

$$\varepsilon = Blv$$

Spinning coils and alternating currents

A simple a.c. generator consists of a rectangular coil that spins in a uniform magnetic field. The flux linkage changes continuously, as does the rate of change of flux linkage, so the induced e.m.f. changes sinusoidally.

The angular frequency of the coil is:

$$\omega = 2\pi f$$

where f is the frequency of rotation.

Flux linkage in a coil with N turns, a cross-sectional area of A, rotating with an angular speed ω, in a magnetic field of strength B at time t is:

$$N\phi = BAN\cos\omega t$$

The e.m.f. induced is:

$$\varepsilon = BAN\omega \sin\omega t$$

The maximum induced e.m.f. $= BAN\omega$

The peak-to-peak value of the e.m.f. $= 2BAN\omega$

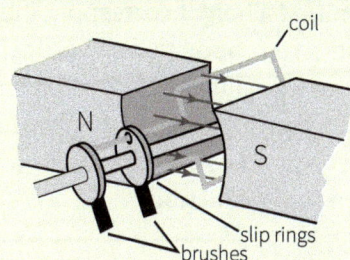

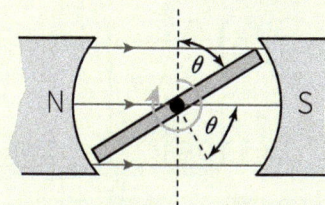

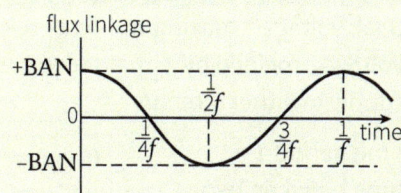

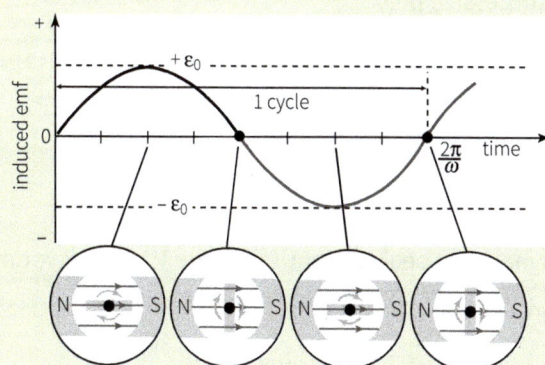

Alternating currents and oscilloscopes

An **oscilloscope** can be used as a voltmeter to show how voltage varies with time – it plots a graph of voltage against time on the screen in real time.

- The vertical axis represents voltage – its scale can be adjusted using the y-gain control so that the height of each square on the screen represents a particular voltage.
- The horizontal axis represents time – its scale can be adjusted using the time-base control so that the width of each square on the screen represents a particular quantity of time.

If the time-base is off, a d.c. voltage appears as a dot, at a height corresponding to the voltage, and an a.c. appears as a vertical line, with length twice the value of peak voltage. If the time-base is on, a d.c. appears as a horizontal line, at a height corresponding to the voltage, and an a.c. appears as a sinusoidal trace.

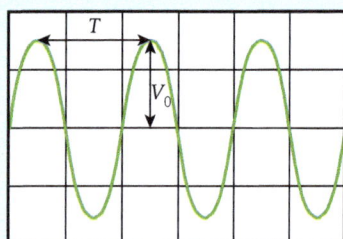

For an alternating current	
peak voltage = maximum voltage provided by supply in either direction	peak-to-peak voltage = twice the peak voltage
time period T = time between two successive peaks	frequency = $\dfrac{1}{\text{time period}}$ $= f = \dfrac{1}{T}$
root mean square (rms) value of the current is the value of the direct current that would have the same heating effect	
$I_{\text{rms}} = \dfrac{I_0}{\sqrt{2}}$ where I_0 = peak current	$V_{\text{rms}} = \dfrac{V_0}{\sqrt{2}}$ where V_0 = peak voltage
in a circuit of resistance R: $P = (I_{\text{rms}})^2 R$ and $P = \dfrac{(V_{\text{rms}})^2}{R}$	

Transformers

A transformer is a device that allows the size of an a.c. voltage to be changed. A transformer consists of two coils of insulated wire wound around a laminated iron core.

- A **step-up transformer** has more turns on the secondary coil than on the primary coil; it steps up the voltage and steps down the current.
- A **step-down transformer** has fewer turns on the secondary coil than on the primary coil; it steps down the voltage and steps up the current.

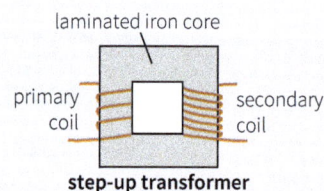

step-up transformer

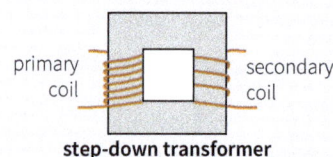

step-down transformer

The ratio of the voltages across the coils is equal to the ratio of the number of turns on the coils:

$$\frac{N_s}{N_p} = \frac{V_s}{V_p}$$

where:

N_s = number of turns on secondary coil

N_p = number of turns on primary coil

V_s = voltage across secondary coil

V_p = voltage across primary coil.

When an a.c. is present in the primary coil, an alternating magnetic field is produced around it. The alternating magnetic field causes a changing flux through the secondary coil and, therefore, induces an alternating e.m.f. in it. The iron core maximises the flux linkage through the secondary coil by channeling the magnetic field produced by the primary coil through the secondary coil.

The iron core is a conductor and the changing magnetic field induces small currents in it, called **eddy currents**.

Transformer efficiency

The efficiency of the transformer is reduced by eddy currents causing the iron core to heat up, and heating of the wires in the coils. Efficiency is improved by using a laminated iron core because the insulated sheets of iron glued together have a higher resistance than a lump of solid iron, so the size of the eddy currents is reduced.

The efficiency of a transformer is:

$$\text{efficiency} = \frac{\text{useful power output}}{\text{total power input}} = \frac{I_s V_s}{I_p V_p}$$

Percentage efficiency is given by:

$$\% \text{ efficiency} = \frac{\text{useful power output}}{\text{total power input}} \times 100\%$$

$$= \frac{I_s V_s}{I_p V_p} \times 100\%$$

Transformers are never 100% efficient but their efficiency is improved by several design features.

Source of power loss	Design feature to improve efficiency
heating of wires in coils	low resistance wires
reduction of flux and heating of iron core by eddy currents induced in iron core	laminated iron core – insulated sheets of iron glued together have a higher resistance than a lump of solid iron, so size of eddy currents is reduced
heating effect of repeated magnetisation and demagnetisation of iron core	soft iron – easily magnetised and demagnetised

Transmission of electrical power

Transformers are useful because the transmission of electrical power over long distances is more efficient at high voltage than at low voltage.

To deliver a power P at a voltage V, the current required is:

$$I = \frac{P}{V} \qquad \textbf{①}$$

The power loss P_{loss} through heating of transmission cables of resistance R with a current I is:

$$P_{loss} = I^2 R \qquad \textbf{②}$$

Combining ① and ② shows the power loss for delivering power P at voltage V is:

$$P_{loss} = \frac{P^2 R}{V^2} \qquad \textbf{③}$$

② and ③ show that transmitting electricity at low currents and high voltages reduces the power lost due to heating the transmission cables.

Retrieval

Learn the answers to the questions below, then cover the answers column with a piece of paper and write as many as you can. Check and repeat.

	Questions	Answers
1	What is electromagnetic induction?	an e.m.f. is induced in a conductor that experiences a change in magnetic flux
2	What can cause a change in magnetic flux through a conductor?	relative movement between conductor and field; change in strength of field
3	State Lenz's law.	direction of induced current or e.m.f. will always oppose the change in flux producing it
4	State Faraday's law.	magnitude of induced e.m.f. in a conductor is equal to the rate of change of flux it experiences or the rate of change of flux linkage in a coil
5	Give Faraday's law in symbols.	$\varepsilon = -N\dfrac{\Delta\phi}{\Delta t}$
6	Give the formula for the e.m.f. induced in a straight conductor moving at right angles through a magnetic field.	$\varepsilon = Blv$
7	How does the induced e.m.f. in a simple alternating current generator vary with time?	sinusoidally
8	Give the equation for the angular frequency of a coil spinning with a frequency f.	$\omega = 2\pi f$
9	Give the equation for the flux linkage in a coil rotating in a magnetic field.	$N\phi = BAN\cos\omega t$
10	Give the equation for the e.m.f. induced in a coil rotating in a magnetic field.	$\varepsilon = BAN\omega\sin\omega t$
11	Give the equation for the maximum induced e.m.f.	maximum induced e.m.f. $= BAN\omega$
12	Give the equation for the peak-to-peak value of the e.m.f.	peak-to-peak value of the e.m.f. $= 2BAN\omega$
13	What is an oscilloscope?	voltmeter showing how voltage varies with time
14	What is the root mean square (rms) value of the current equivalent to?	value of the direct current that would have the same heating effect
15	What is the average power for an alternating current?	$P = I_{\text{rms}} \times V_{\text{rms}}$
16	State the properties of a step-up transformer.	more turns on the secondary coil than on the primary coil; steps up the voltage and steps down the current
17	State the properties of a step-down transformer.	fewer turns on the secondary coil than on the primary coil; steps down the voltage and steps up the current

Put paper here

18	Give the transformer equation relating the voltages across the coils to the number of turns on the coils.	$\dfrac{N_s}{N_P} = \dfrac{V_s}{V_P}$
19	Give the formula for the efficiency of a transformer.	$\dfrac{I_S V_S}{I_P V_P}$

Put paper here

Practical skills

Practise your practical skills using the worked example and practice questions below.

Using a search coil

A search coil is a small flat coil that has 500 to 2000 turns of insulated wire. When placed in a changing magnetic field, a small e.m.f. is generated in the coil that can be viewed on an oscilloscope.

By altering the angle of the search coil, changes in magnetic flux density can be investigated:

flux linkage = $BAN\cos\theta$

where B is the flux density, A is the area of the coil, and N is the number of turns on the coil.

Worked example

Question

This oscilloscope shows the e.m.f. from the search coil. The time-base has been switched off. The y-sensitivity is 5 mV cm^{-1}.

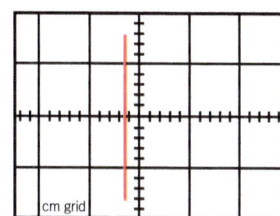
cm grid

1 Determine the maximum e.m.f. generated in the search coil.

2 Suggest how the trace would change if the time-base was on.

Answer

1 max e.m.f. = number of divisions the trace covers $\times$ y-sensitivity
= 3.2 cm $\times$ 5 mV cm^{-1} = 16 mV

2 The trace would move across the display in an approximately sinusoidal pattern.

Practice

A student uses a search coil and oscilloscope to measure the e.m.f. generated when the coil is turned through different angles, θ, in the magnetic flux.

θ/°	0	10	20	30	40	50	60	70	80
e.m.f. / mV	30.0	29.4	28.4	26.1	23.8	19.0	14.9	10.4	4.9

1 Use the data to plot a graph showing the variation of e.m.f. with $\cos\theta$.

2 Explain what the graph shows about the relationship between the angle of the search coil and the magnetic field direction.

3 Calculate the flux linkage for a coil of diameter 12 mm with 25 turns perpendicular to a field of flux density 4.0×10^{-6} T.

Exam-style questions

01 **Figure 1** shows a wire moved upwards through a uniform magnetic field. The wire is connected to a sensitive galvanometer.

Figure 1

motion of wire

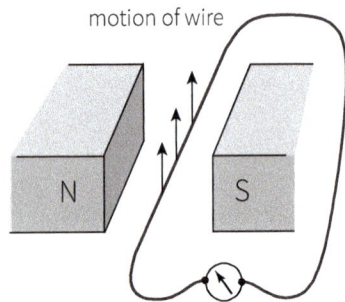

01.2 Annotate **Figure 1** to show the direction of the current in the wire.

[1 mark]

01.3 Explain why a current is induced in the wire. **[2 marks]**

_____

_____

_____

_____

_____

_____

01.4 The length of wire l in the magnetic field B is moving with speed v. Show that the e.m.f. induced ε is given by the expression $\varepsilon = Blv$.

[2 marks]

01.5 The e.m.f. induced is 45 mV when the wire is moved at $15\,\text{ms}^{-1}$. Calculate the magnetic flux density if the length of conductor in the magnetic field is 5.0 cm. State the unit. **[3 marks]**

magnetic flux density = _____ unit = _____

02 **Figure 2** shows the change in flux linkage with time in the coils of a simple generator.

Figure 2

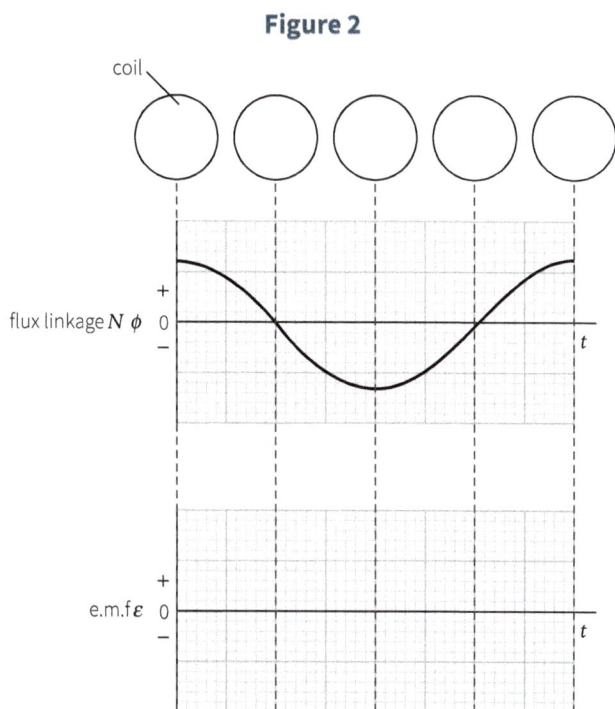

Synoptic link

3.7.5.3

02.1 State the meaning of magnetic flux linkage, and give its unit.

[2 marks]

_____

_____

_____

_____

_____

_____

02.2 Draw the position of the coil of the generator in the circles in **Figure 2**. The positions should match with the flux linkage at that point.

[1 mark]

02.3 Sketch the graph of e.m.f. against time on **Figure 2**. [2 marks]

02.4 The frequency of the generator is 50 Hz and the number of coils is 20. The area of the coils is 1.2×10^{-3} m^2 and the peak e.m.f. generated is 0.45 V.

Calculate the magnetic flux density. [3 marks]

> **Exam tip**
>
> In **02.3**, what is the relationship between flux linkage and e.m.f.? Anchor your graph with key turning points: when gradient of flux linkage graph = zero, what is e.m.f.?

magnetic flux density = _____ T

03 A digital clamp meter can be used to measure the current in a cable without breaking the circuit. The jaws of the clamp meter are placed around the wire under test. **Figure 3** shows a clamp meter being used and a diagram of the insides. This design can only be used to measure a.c.

Synoptic link

3.1.2

Figure 3

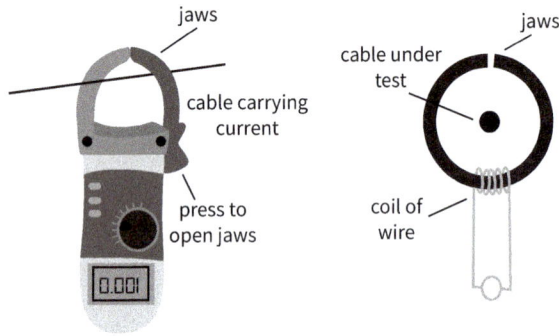

Exam tip

This is an unfamiliar situation, but there are clues to solving it: the jaws are made of iron, there is a coil, and it only works with a.c. In any unfamiliar situation, see if you can spot ideas you have met before – this could be a transformer, where the primary has only one coil.

03.1 The iron jaws of the clamp have a coil of wire wrapped around them. Explain how the clamp meter can measure current without breaking the circuit. **[3 marks]**

_____

_____

_____

_____

_____

03.2 Digital clamp meters can measure currents as high as 600 A a.c. Explain how the meter can differentiate between different size currents. **[2 marks]**

_____

_____

_____

03.3 A clamp meter reads zero when tested on the cable leading to a lamp that is illuminated. The clamp meter is working. Suggest why the clamp meter reads zero. **[2 marks]**

_____

_____

_____

03.4 A different design of clamp meter can measure both a.c. and d.c. It is described as having a resolution of 0.1 mA and 2% accuracy. Explain what this description means. **[2 marks]**

_____

_____

_____

03.5 The d.c. clamp meter works by detecting the presence of a magnetic field. Before use, it must be zeroed, and this must be redone if the orientation of the clamp changes.

Suggest why the clamp meter must be zeroed. **[2 marks]**

_____

_____

_____

04 **Figure 4** shows the trace from an oscilloscope connected across a 20 Ω resistor. The oscilloscope grid is marked in cm. The time-base is set to 0.2 ms cm^{-1}. The y-amplifier sensitivity is set to 0.5 V cm^{-1}.

Figure 4

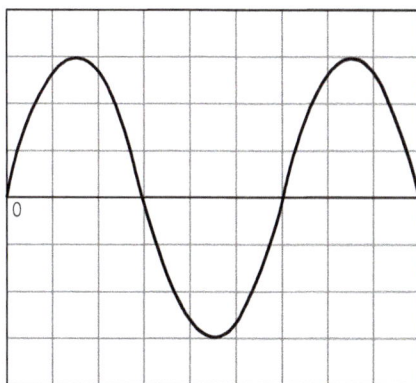

Synoptic link

3.5.1.4

04.1 When the oscilloscope was first connected, the whole wave could not be seen on the screen.
Describe how you adjust an oscilloscope to achieve a screen that is easy to take readings from. **[3 marks]**

04.2 Determine the frequency in Hz of the signal in **Figure 4**. **[2 marks]**

04.3 Determine the rms voltage in V in **Figure 4**. **[3 marks]**

04.4 Calculate the mean power in W transferred to the 20 Ω resistor. **[1 mark]**

Exam tip

You are not being asked how to set up the oscilloscope, simply the steps you would take before trying to take readings from the screen, to get a trace as clear as the one in **Figure 4**.

05 A copper pipe and magnet of smaller diameter can be used to demonstrate Lenz's law. The magnet is dropped into the vertical pipe and takes three seconds to fall through.

05.1 The pipe is 320 mm long.

Show that the expected time for a non-magnetic object to fall through the pipe is 0.3 s. **[2 marks]**

05.2 Explain why the magnet falls so slowly and how this demonstrates Lenz's law. **[3 marks]**

05.3 Explain why the magnet falls at a constant speed in the pipe, by considering the forces acting on the magnet. **[4 marks]**

Synoptic links

3.4.1.3 3.4.1.5

05.4 A second demonstration of Lenz's law uses a coil of wire, an iron clamp stand, and aluminium discs **A** and **B**. These are shown in **Figure 5**.

Figure 5

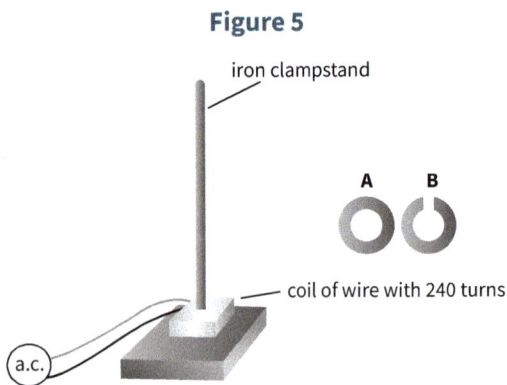

iron clampstand

A   B

coil of wire with 240 turns

a.c.

The transformer is connected to an a.c. supply. When disc **A** is put on, it hovers above the transformer.

Explain why disc **A** hovers. **[4 marks]**

05.5 Explain why this demonstration would not work if the transformer was connected to a d.c. supply. **[1 mark]**

05.6 Disc **B** is not complete. Describe and explain what happens when disc **B** is used instead. **[1 mark]**

06 **Figure 6** shows how the magnetic flux linkage varies with time for a generator.

Figure 6

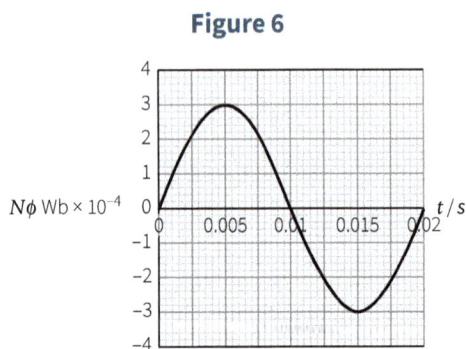

$N\phi$ Wb $\times 10^{-4}$

$t\,/\,s$

The generator is rotated at a steady frequency in a uniform magnetic field. The generator has 330 turns and the magnetic flux density is 0.06 T.

06.1 Calculate the frequency in Hz of the generator shown in **Figure 6**. **[1 mark]**

06.2 Show using **Figure 6** that the area of the generator is $1.5 \times 10^{-3}\,m^2$. **[2 marks]**

06.3 Determine the peak e.m.f. in V induced in the coil of the generator. **[3 marks]**

06.4 A 75 Ω resistor is placed across the output of the generator. Determine the mean power in W dissipated in the resistor. **[2 marks]**

> **(!) Exam tip**
>
> Two repelling magnets hover over each other – how could two repelling magnetic fields have been created?

> **Synoptic link**
>
> 3.5.1.4

> **(!) Exam tip**
>
> The graph is one of flux linkage against time. How are flux density and flux linkage related, and how can you use this to determine the area of the generator?

07 The National Grid is a network of cables and transformers connecting power stations to consumers. A 1500 MW power station has a series of transformers that step up the output pd from 25 kV to 400 kV.

07.1 Calculate the ratio of turns from the first primary coil to the last secondary coil in the power station. Assume that the transformers are ideal. **[1 mark]**

07.2 The resistance in the transmission cables is 30 Ω. Determine the power loss in W in the transmission cable. **[2 marks]**

07.3 Calculate the efficiency of the transmission system. **[2 marks]**

07.4 In a lesson about the National Grid, a student asks why the electricity is stepped up for electrical transmission, since power loss is equal to $\frac{V^2}{R}$. This question demonstrates a misconception of the student. Explain why high p.d. is used for electrical transmission, and correct the misconception. **[3 marks]**

> **! Exam tip**
>
> The 400 kV is the p.d. between the cables and earth; the p.d. across the cables is not the same thing. Think of the cables as the leads that you use to connect components to the battery in your circuit.

08 One method of charging an electric vehicle uses a transformer arrangement. **Figure 7** shows how this achieved.

Figure 7

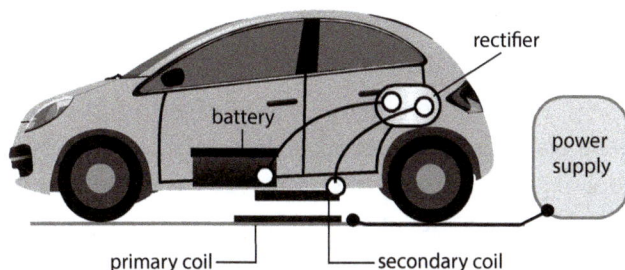

> **🔗 Synoptic links**
>
> 3.5.1.2   3.6.1.4

A charging station on the ground contains a primary coil, which is connected to the mains power supply. The secondary coil is in the car.

08.1 Describe how this transformer arrangement can charge the battery. **[3 marks]**

08.2 The method of charging, called inductive charging or coupling, is used for many household items, such as mobile phones and electric toothbrushes. When charging a vehicle battery, the system used is called resonant inductive coupling.

Suggest why electric vehicles need resonant inductive coupling. Use your understanding of the word resonance. **[2 marks]**

> **! Exam tip**
>
> When in doubt, state the characteristic of a diode – that will be worth a marking point – and then consider why that might be necessary.

08.3 Suggest how transformers could be used in induction charging of cars as they drive along a road.

Discuss the advantages and disadvantages of using this system. **[4 marks]**

⚙ Knowledge

19 Radioactivity

Atomic and nuclear model development

Atoms were originally thought to be tiny spheres that could not be split. Our knowledge and understanding of the structure of the atom and its nucleus has changed and developed over time. Future experiments may lead us to change or replace our ideas about the atom again.

The discovery of electrons

In 1897, J.J. Thomson discovered that atoms contained tiny negatively-charged particles, given the name electrons. As a result, some scientists thought electrons were embedded inside a sphere of positive charge to make a neutral atom – called the **plum pudding model**.

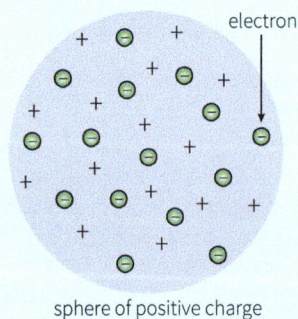

sphere of positive charge

The energy level model

The modern accepted model of the atom consists of electrons occupying **energy levels** at specific distances from the nucleus. The nucleus consists of neutrons and protons, which themselves are composed of **quarks**.

The nuclear model

In 1911, the plum pudding model was replaced by the **nuclear model** of the atom, where the nucleus contains the positive charge of the atom and the negative electrons orbit the nucleus. Most of the mass is concentrated in the nucleus, and most of the atom is empty space.

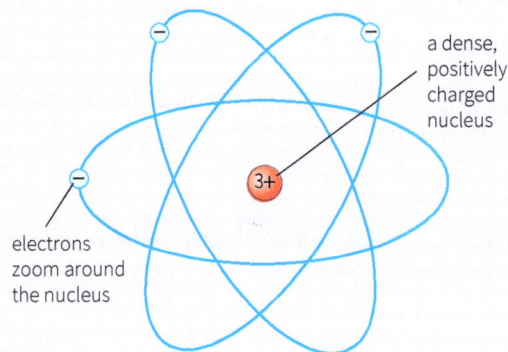

a dense, positively charged nucleus

electrons zoom around the nucleus

Alpha particle scattering

The nuclear model was based on the results of an experiment done in a laboratory run by Ernest Rutherford. A beam of positively-charged alpha particles was fired at a very thin sheet of gold. Experimental observations could only be explained by the nuclear model:

- Most of the particles were straight through the foil – most of an atom is empty space.

- A few particles were deflected by large angles – the nucleus is positively charged so will repel alpha particles.

- Very few particles were deflected backwards – the mass of the atom is concentrated in a tiny space in the atom, so direct collisions between alpha particles and the nucleus are rare.

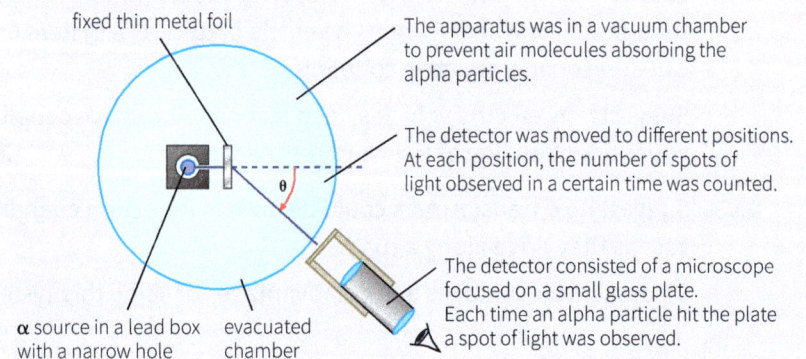

fixed thin metal foil

The apparatus was in a vacuum chamber to prevent air molecules absorbing the alpha particles.

The detector was moved to different positions. At each position, the number of spots of light observed in a certain time was counted.

The detector consisted of a microscope focused on a small glass plate. Each time an alpha particle hit the plate a spot of light was observed.

α source in a lead box with a narrow hole

evacuated chamber

Nuclear radiation

Nuclear radiation is emitted by unstable atomic nuclei to become more stable. Nuclear radiation is known as **ionising radiation** because it can produce ions by knocking electrons out of atoms it collides with.

Types of nuclear radiation

An unstable nucleus can decay by emitting an alpha particle (α), a beta particle (β), a gamma ray (γ), or sometimes a neutron (see *Chapter 2* Particles).

Identifying nuclear radiation

The different types of radiation can be identified experimentally using their properties. Penetrating power or absorption by different materials can be measured using the equipment shown. Penetrating power is used in industry to control the thickness of sheets of aluminium foil, paper, and steel.

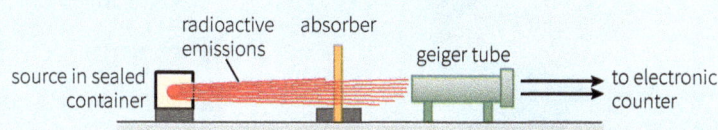

Hazards of exposure

Ionising radiation is dangerous because ionisation can damage the atoms or molecules in living cells. High doses of ionising radiation can kill living cells, and even low doses can cause cell mutation and cancerous growth.

The ionising effects of alpha, beta, and gamma radiations are different, so the health risks from them vary:

- Alpha radiation is dangerous inside the body as it affects all surrounding tissue, but outside the body it only affects cells in the skin and eyes.

- Beta and gamma radiation are dangerous inside and outside the body as they can reach cells throughout the body.

Uses in medicine

Nuclear radiation from radioactive isotopes is used in medicine where the risks posed by ionisation are balanced against the potential benefits.

Medical use	Method	Type of radiation	Half-life of source
Tracers – trace flow of a substance through an organ. Cameras – take images of internal body organs.	Radioactive isotope is injected into, or swallowed by, the patient – radiation detected outside the body is converted to an image, video, or other information by a computer.	Gamma – can pass through the body without causing much damage to cells.	Long enough for tracing or imaging; short enough so it does not remain inside the body for long.
Radiotherapy – external beams to treat cancer.	Beam of gamma radiation from outside the body is aimed at cancerous cells.	Gamma – can penetrate deep into the body.	Up to a few years as patient is only exposed when treated.
Radiotherapy – implants to treat cancer.	Small 'seed' or rod of radioactive material is implanted inside cancerous tumour.	Beta or gamma – depending on how far into tissue radiation needs to penetrate.	Long enough to kill cancerous cells; short enough that it does not remain radioactive inside the body once treatment finished.

Safe handling of radioactive sources

People who work with ionising radiations must take precautions to protect against irradiation and contamination, including:

- keeping as far away as possible from the radiation source
- spending as little time as possible in the presence of the radiation source
- shielding themselves from the radiation using concrete barriers or lead plates.

Background radiation

The process by which an unstable nucleus emits radiation to become more stable is radioactive decay. This is a random process – it is not possible to predict when a particular nucleus will decay.

Measuring radioactive decay

The **activity** A (unit becquerel; 1 Bq = 1 decay per second), of a radioactive isotope is the number of nuclei that decay every second. Activity is equal to the rate of change of number of nuclei of the radioactive isotope, and is proportional to the number of nuclei N remaining at time t:

$$A = \frac{\Delta N}{\Delta t} = -\lambda N$$

Background radiation

Background radiation is radiation that is around us all the time. The level of background radiation varies depending on location. It comes from:

- naturally occurring radioactive isotopes present in the air, ground, rocks, food, buildings, and so on
- cosmic rays from space (mainly the Sun)
- human activity, such as medical use of radioactive isotopes, fallout from nuclear weapons testing, or nuclear power station accidents.

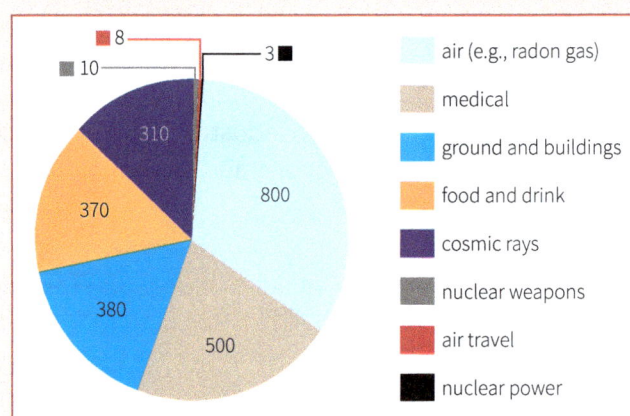

Background radiation levels should be measured and subtracted from data in experiments involving other radioactive sources.

Inverse square law

The intensity I of radiation is the radiation energy passing through an area of $1\,m^2$ normal to the radiation per second. Gamma radiation spreads out in all directions from a source and its intensity I varies with the inverse square of distance r from the source:

$$I = \frac{k}{r^2}$$

where k is a constant determined by the activity of the source.

The decay constant

The **decay constant** λ (s^{-1}) is the probability of an individual nucleus decaying per second, and the minus sign indicates ΔN is decreasing.

This leads to:

$$N = N_0\,e^{-\lambda t}$$

where N_0 = number of radioactive nuclei in sample at time $t = 0$. This shows that the number of radioactive nuclei decreases exponentially with time.

Similarly, for the activity A and mass m of the radioactive isotope:

$$A = A_0\,e^{-\lambda t}$$

$$m = m_0\,e^{-\lambda t}$$

Radioactive decay can be modelled by other events that have a constant decay probability, such as tossing a coin.

Half-life

The half-life $T_{1/2}$ of a radioactive isotope is the time it takes for its activity to halve.

Calculating half-life

The half-life is related to the decay constant of a radioactive sample by:

$$T_{1/2} = \frac{\ln 2}{\lambda}$$

Linking half-life to uses

Radioactive isotopes with short half-lives are useful as radioactive tracers in medicine because they only remain in the body for a short time.

The long half-life of carbon-14 makes it useful for radioactive dating of objects. Long half-lives of isotopes in radioactive waste make it difficult to store safely.

Using graphs

Half-life can be found from graphs of decay data such as activity against time or number of radioactive nuclei remaining against time.

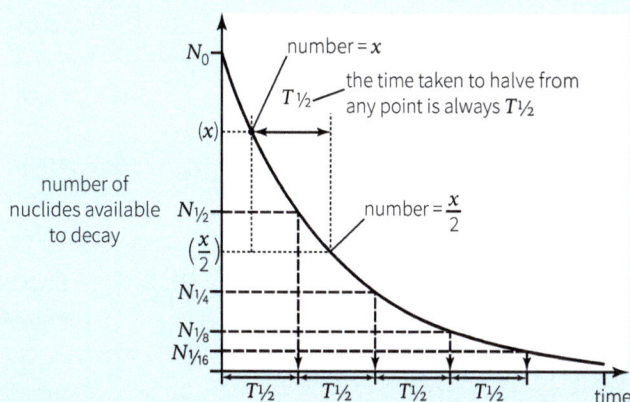

$N = N_0\,e^{-\lambda t}$ can be written as:

$$\ln N = \ln N_0 - \lambda t$$

So, a graph of $\ln N$ against t is a straight line with a gradient of $-\lambda$, from which the half-life can be calculated using

$$T_{1/2} = \frac{\ln 2}{\lambda}.$$

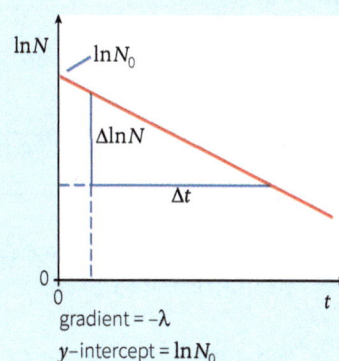

Retrieval

Learn the answers to the questions below, then cover the answers column with a piece of paper and write as many as you can. Check and repeat.

	Questions	Answers
1	What did J.J. Thompson discover in 1897?	electrons
2	What experiment led to the development of the nuclear model of the atom?	Rutherford alpha scattering experiment
3	List the key features of the current accepted model of the atom.	electrons occupy energy levels at specific distances from the nucleus, which consists of neutrons and protons
4	What property of nuclear radiation makes it ionising radiation?	can produce ions by knocking electrons out of atoms with which it collides
5	What types of radiation can an unstable nucleus emit as it decays?	alpha particles α, beta particles β, gamma radiation γ, neutrons
6	What is the detection of radiation through materials used for in industry?	measure and control the thickness of sheets made from aluminium foil, paper, and steel
7	What are radioactive tracers used for in medicine?	trace the flow of a substance through an organ and take images of internal body organs
8	What are radiotherapy beams and implants used for in medicine?	treat cancers
9	What precautions must be taken to protect against ionising radiation?	keep far away from the source of radiation; spend only a short time in the presence of the radiation source; shielding using concrete barriers or lead plates
10	Where does background radiation come from?	naturally-occurring radioactive isotopes present in the air, ground, rocks, food, buildings, and so on
11	What should be done to account for background radiation in experimental data?	measure levels and subtract from data in experiments
12	Define the intensity of radiation.	radiation energy passing through an area of $1\,\text{m}^2$ normal to the radiation per second
13	State the inverse square law for the intensity of gamma radiation.	$I = \dfrac{k}{r^2}$
14	Why is it not possible to predict when a particular nucleus will decay?	radioactive decay is a random process
15	What is the activity of a radioactive isotope proportional to?	number of nuclei remaining at a given time: $A = \dfrac{\Delta N}{\Delta t} = -\lambda N$
16	What is the decay constant?	probability of an individual nucleus decaying per second
17	Give the equation showing that the mass of a radioactive isotope decreases exponentially with time.	$m = m_0\, e^{-\lambda t}$
18	What is the half-life of a radioactive isotope?	time it takes for its activity to halve

19	How is the half-life related to the decay constant?	$t_{\frac{1}{2}} = \dfrac{\ln 2}{\lambda}$
20	How can half-life be found from graphs of decay data?	directly from a graph of activity against time or number of radioactive nuclei remaining against time, or from gradient of a graph of $\ln N$ against t

Put paper here

Practical skills

Practise your practical skills using the worked example and practice questions below.

Inverse square law for gamma radiation

The equipment shown is used to demonstrate the relationship between intensity of gamma radiation (count rate) I and the distance between source and detector x.

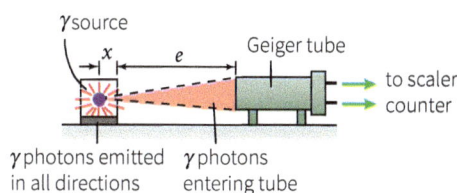

γ source, x, e, Geiger tube, to scaler counter, γ photons emitted in all directions, γ photons entering tube

Gamma radiation obeys the inverse square law, where k is the constant of proportionality. For this equipment the equation must include the distance e between the source and its exit from the holder:

$$I = \frac{k}{(x+e)^2}$$

Rearranging, including taking the square root of both sides:

$$x = \sqrt{\frac{k}{I}}\, e$$

Plotting a graph of $I^{-0.5}$ against x will give a straight line with an intercept equal to the unknown e.

Worked example

Question

A student used the equipment shown to obtain data for distance and count rate. Before bringing the radioactive source into the laboratory, they measured the background count rate.

1 Describe how the student should use the background count rate to correct the measured count rate. Explain why this is necessary.

2 How could the student use a graph of (corrected count rate)$^{-0.5}$ against x to determine the value of e?

Answer

1 The student should subtract the background count rate from the measured count rate to give a corrected count rate that takes account of the presence of background radiation at all times.

2 e is the intercept on the x-axis.

Practice

1 State **two** precautions that should be taken when working with radioactive sources.

2 A student measured the background count for 20 minutes and obtained a count of 360, then measured the count rate of a radioactive source at different distances from the source.

x / cm	Count rate / s^{-1}
1.0	43
2.0	26
3.0	20
4.0	14
5.0	11
6.0	8

a Calculate the background count rate.

b Use the background count rate to calculate the corrected count rates.

c Plot a graph of (corrected count rate)$^{-0.5}$ against x.

d Estimate a value for e, the distance between the source and the point at which the radiation is emitted.

Practice

Exam-style questions

01 Some articles suggest Rutherford did not suspect there was a nucleus at the centre of the atom before he asked his students, Geiger and Marsden, to conduct the gold foil experiment.

This is more of a history question than a science one.

01.1 Explain how **one** finding from the gold foil experiment led to a change in the model of the atom. **[2 marks]**

_____

_____

_____

> **! Exam tip**
>
> In answers about changing models, be clear about the difference between the evidence and the impact of the evidence on the prevailing model of the time.

01.2 The nuclear model as described by Rutherford is not the atomic model that we use today.

Describe **one** piece of evidence that led to the atomic model being changed to the one that we use today. **[1 mark]**

_____

_____

01.3 Rutherford used a variety of sources for the radiation needed in the gold foil experiment, but the main source was 30.0 mg of pure radium bromide, which he bought in 1903. The half-life of radium is 1600 years.

Show that the decay constant of radium in the sample is about $1.4 \times 10^{-11}\,\text{s}^{-1}$. **[2 marks]**

01.4 Calculate the mass of radium remaining today, approximately 120 years since Rutherford bought the sample, if 58% of the radium bromide is radium. **[3 marks]**

mass = _____ g

01.5 Calculate the change in sample activity that the change in mass from **01.5** would cause. The relative atomic mass of radium in the sample is 226. **[2 marks]**

activity = _____ Bq

02 A student models radioactive decay using tokens that have two sides, with a symbol on one side. When the student throws a cupful of tokens on the table, some land with the symbol side up and some land with the symbol side down. After each throw, the student deems the tokens that land symbol side up to have 'decayed' and removes them. They write down the number of sweets remaining.

02.1 **Table 1** shows the predicted and actual numbers of tokens remaining.

Table 1

Throw	Predicted number	Actual number
0	240	240
1	120	118
2	60	65
3	30	31
4	15	10

Describe how the student determined the predicted numbers in **Table 1**, and suggest why the actual numbers are different.

[2 marks]

_____

_____

_____

02.2 Explain why the rate of 'decay' is proportional to the number of 'undecayed' tokens. **[2 marks]**

_____

_____

_____

02.3 Another student argues that each of the 'decayed' tokens should not be removed but should be replaced by another item.
Suggest why the student made this suggestion, and describe the physical situation modelled by replacing the 'decayed' token with:

1 a spherical token

2 a token with four sides, and the same symbol on one of them.

[3 marks]

_____

_____

_____

_____

_____

02.4 The second student records the data in **Table 2** using a box of dice that do not have six sides. Each time a dice lands on a side with only one dot they are deemed to have decayed.

Use **Table 2** to work out the number of sides on each dice.

Explain your method. [2 marks]

Table 2

Throw	Number of dice remaining
0	250
1	219
2	191
3	167
4	147

> **! Exam tip**
>
> You should be clear about the meaning of 'λ' in the context of the decay of an unstable nucleus.

_____

_____

03 Everyday objects, including all foods, emit radiation. This radiation is measured as background radiation.

03.1 State **two** other sources of background radiation. [2 marks]

03.2 The average banana contains roughly 0.05 g of potassium, 0.0117% of which is the radioactive isotope potassium-40. In total, there are about 8.7×10^{16} atoms of potassium-40 in an average banana, which decay by the emission of beta particles. The half-life of potassium-40 is 1.3 million years.

Calculate the decay constant. [1 mark]

03.3 Explain, using your answer to **03.2**, whether the activity of a banana would be noticeable in an area where the background count is 0.15 Bq. [2 marks]

03.4 The effect of radiation on the human body depends on the type of radiation and the part of the human body that the radiation hits. The average banana emits about 4.0×10^{-3} Bq of alpha radiation owing to the decay of radium-226.

Suggest why it is difficult to judge which radioactive isotope presents the most risk to a consumer of a banana. [1 mark]

> **⊗ Synoptic links**
>
> 3.1.1.2   3.1.1.3
> 3.1.2.1

> **! Exam tip**
>
> You need to remember the link between relative atomic mass in grams and the Avogadro constant.

04 A radioactive source emits only gamma rays.

04.1 Describe how to show that the source does not emit alpha or beta radiation. [2 marks]

04.2 Describe a method to obtain data to show that the radiation emitted by the source obeys an inverse-square law. [3 marks]

04.3 A student obtains the data shown in **Table 3**:

Table 3

Distance / cm	Activity / counts min^{-1}	$\frac{1}{d^2}$ / m^{-2}
5	67	40.0
10	22	10.0
15	13	4.4
20	10	2.5

Draw a graph using data in **Table 3** to show that the gamma radiation obeys an inverse-square law. **[3 marks]**

04.4 A teacher looks at the data and suggests that the student did not subtract the background count from the activity measurements.

Explain how the graph drawn in **04.3** confirms this suggestion, and calculate the background count in Bq. **[3 marks]**

05 A sample of rock contains potassium-40 (half-life 1.3 billion years). The activity of the sample is 0.48 Bq.

05.1 Calculate the number of potassium-40 nuclei in the rock. molar mass of potassium-40 = 40 g mol^{-1} **[3 marks]**

05.2 The rock was found in an area where other rocks had been dated to 3.2 billion years old. The potassium decays to a stable isotope of argon.

Calculate the mass in g of argon that the scientist would expect to find in the sample, stating any assumptions that you make. The atomic mass of argon is 40 g mol^{-1}. **[3 marks]**

05.3 Potassium-40 also decays into an unstable isotope of calcium. Suggest how this would affect the conclusion the scientist might make about the age of the rock. **[1 mark]**

05.4 A collector wants to date artefacts using potassium-40 in the glaze. However, limitations to detection techniques means that only a change of at least 0.0053% of the potassium in a rock can be detected. Show that this means that objects younger than 100 000 years old cannot be dated using potassium-40. **[3 marks]**

06 Technicians follow several safety precautions when handling radiation sources for use in nuclear medicine.

06.1 Suggest **one** reason for each of the following precautions:
- handling all sources with thick rubber gloves
- handling the sources from behind a screen made of glass impregnated with lead. **[2 marks]**

> **! Exam tip**
> Remember that to show the relationship between variables it is useful to manipulate data to produce a straight line.

> **! Exam tip**
> Remember that the number of atoms is always conserved.

> **Synoptic links**
> 3.2.1.2 3.2.1.3 3.2.1.7

06.2 A sample used in a hospital is checked by measuring the count rate over time. **Table 4** gives the data collected.

Table 4

t / min	Activity / Bq	ln(activity)
0	1682	
30	1382	
60	1136	
90	933	
120	767	

Calculate the values for ln(activity), and draw a graph of ln(activity) against time.

Use your graph to calculate the half-life of the material.   **[3 marks]**

Exam tip

Remember that the unit of time used for λ will then be the unit of time when you calculate half-life.

06.3 Some isotopes used in nuclear medicine are listed in **Table 5**.

Table 5

Radioisotope	Half-life / min	Type of radiation emitted
carbon-11	14.0	positron
oxygen-15	2.0	positron
fluorine-18	109.7	positron

The isotope in the sample in **06.2** is used in hospitals as part of positron emission tomography. Patients are injected with a small amount of the radioisotope and then sit in a scanner for between 30 and 60 minutes.

Identify the isotope in the sample using **Table 5**, and suggest why the other two isotopes are not used for the same purpose.   **[2 marks]**

06.4 The emitted positrons annihilate when they interact with electrons to produce two gamma rays.

Explain why two gamma rays are produced.   **[1 mark]**

06.5 Calculate the frequency in Hz of the gamma rays emitted.
$1 \, MeV = 1.6 \times 10^{-13} \, J$ **[3 marks]**

06.6 In beta decay, a neutrino is produced.

Explain whether a neutrino is produced in annihilation.   **[1 mark]**

07 Houses with cellars built on granite rock need to have extraction fans fitted to circulate fresh air. The granite contains isotopes that decay to produce radon, which is an alpha-emitting gas.

07.1 Explain why alpha-emitting isotopes are not used as tracers in nuclear medicine.   **[1 mark]**

07.2 An alpha particle of energy 6.4 MeV is emitted from a nucleus of radon-210 (symbol Ra). The daughter nucleus, polonium (symbol Po), contains 84 protons.

Write an equation to show the decay of the nucleus of radon-210.   **[2 marks]**

Synoptic links

3.2.1.2   3.4.1.6
3.4.1.8   3.8.1.5

Exam tip

Check that all equations obey the laws of conservation of mass and charge.

07.3 Compare the radius of the alpha particle with the radius of radon. Use the following equation, where $R_0 = 1.25 \times 10^{-15}$ m. **[3 marks]**

$$R = R_0 A^{\frac{1}{3}}$$

07.4 Calculate the speed in m s^{-1} of the alpha particle using a non-relativistic approximation.
1 MeV = 1.6×10^{-13} J **[2 marks]**

07.5 Radon is a monatomic gas. If the radon nucleus emits an alpha particle, the nucleus will recoil.

Calculate the velocity in m s^{-1} of the polonium nucleus after the decay. **[2 marks]**

07.6 The true value for the velocity is lower than that calculated in **07.5**. Suggest **one** reason why. **[1 mark]**

08 Water pipes that are buried underground are difficult to inspect, so radioactive isotopes, such as those shown in **Table 6**, are used as tracers.

Table 6

Radioisotope	Half-life	Type of radiation
bromine-82	1.47 days	gamma
iodine-131	8.05 days	gamma
sodium-24	15 hours	beta and gamma

Synoptic link

3.2.1.2

08.1 Explain **two** reasons why sodium-24 is used to trace leaks. **[2 marks]**

08.2 Compare the mechanisms that produce the emission of beta and gamma radiation from the nucleus of a sodium-24 isotope. **[2 marks]**

08.3 Complete the equation for the beta decay of sodium-24. Explain why an antineutrino is produced. **[4 marks]**

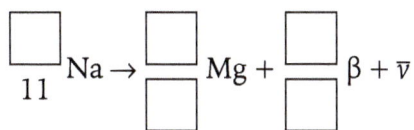

$$\boxed{}_{11}\text{Na} \rightarrow \frac{\boxed{}}{\boxed{}}\text{Mg} + \frac{\boxed{}}{\boxed{}}\beta + \bar{v}$$

Exam tip

Think about what else needs to be conserved besides mass, momentum, and charge in particle physics.

08.4 Show that the decay constant of sodium-24 is about 1.3×10^{-5} s^{-1}. **[1 mark]**

08.5 An engineer adds a sample containing 1.3×10^{-2} µg of sodium-24 to 100 cm^3 of water.

Show that the activity of the sodium before it is added to the water is about 4 GBq. **[2 marks]**

08.6 The radiation dose deemed safe for people who work in the nuclear industry is 50 times that of the average annual dose of background radiation. The background count in the area is 0.24 Bq.
Calculate how many days it would take for the sample to decay to a level deemed safe in the nuclear industry.

Explain why the sample would not pose a substantial risk. **[3 marks]**

20 Nuclear energy

Nuclear instability

A graph of neutron number N against proton number Z can be plotted for all known isotopes.

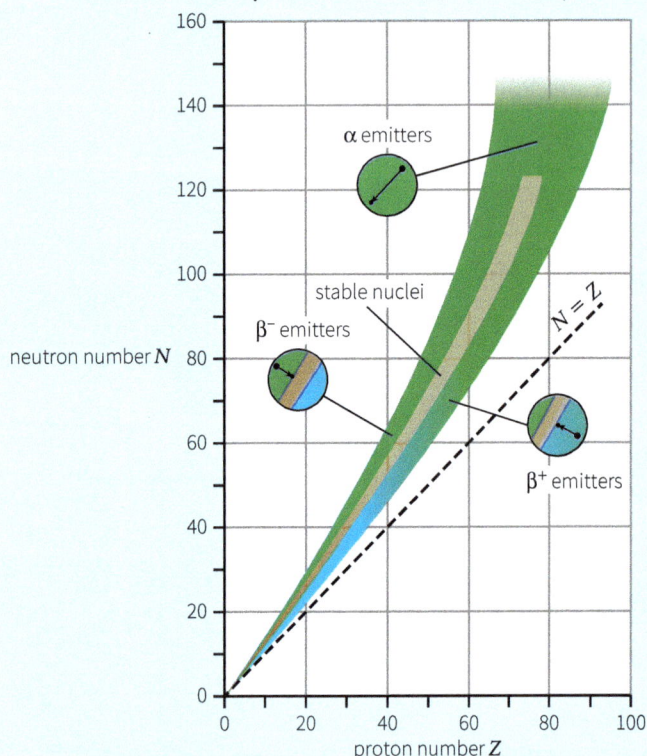

Stable nuclei

Stable nuclei lie along the central orange belt that curves upwards.

- Light stable nuclei (Z up to 20) have approximately the same number of protons and neutrons $N \approx Z$.
- Heavy stable nuclei ($Z > 20$) have more neutrons than protons (ratio $\frac{N}{Z} > 1$) to help bind the neutron together.

Emitters

- Alpha α emitters tend to be large nuclei $Z \geq 60$) because the strong nuclear force is unable to overcome the electrostatic force of repulsion between protons.
- Beta minus β^- emitters occur to the left of the stability belt where isotopes are neutron-rich.
- Beta plus β^+ emitters occur to the right of the stability belt where isotopes are proton-rich.

Electron capture

This occurs when a nucleus captures an orbiting electron – this is to the right of the stability belt.

Decay equations

Changes in N and Z caused by radioactive decay can be represented by the following general equations (remember, $Z + N = A$). There is no change to the composition of a nucleus due to gamma radiation.

alpha decay	$^A_Z X \rightarrow {}^4_2\alpha + {}^{A-4}_{Z-2}Y$
beta minus decay	$^A_Z X \rightarrow {}^{\ 0}_{-1}\beta + {}^{\ A}_{Z+1}Y + \bar{\nu}_e$
beta plus decay	$^A_Z X \rightarrow {}^{\ 0}_{+1}\beta + {}^{\ A}_{Z-1}Y + \nu_e$
electron capture	$^A_Z X + {}^{\ 0}_{-1}e \rightarrow {}^{\ A}_{Z-1}Y + \nu_e$

$N-Z$ graphs

Changes in N and Z can also be represented on the $N-Z$ graph.

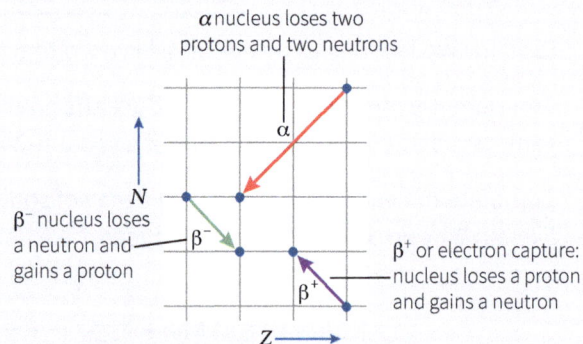

A **radioactive decay series**, when a radioactive isotope decays into an isotope which itself is unstable, can also be represented on the $N-Z$ graph.

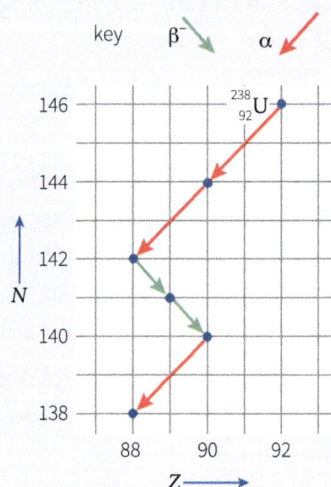

Nuclear energy level diagrams

Excited states

The emission of alpha or beta radiation, or electron capture, can leave a nucleus in an excited state. The nucleus loses energy and returns to its ground state by emitting gamma radiation.

The energy changes in a nucleus can be shown using a nuclear energy level diagram.

Metastable states

Excited states that last for a significant period (> 1 ns) are called metastable states. Isotopes, such as technetium-99m, in a metastable state can be useful in medicine because they can be separated and used as a source that emits only gamma radiation.

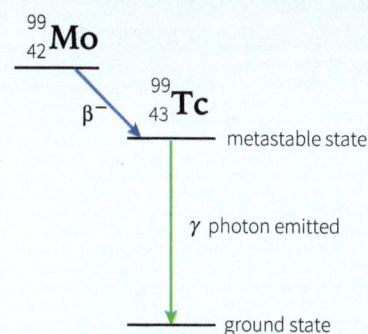

Nuclear radius

The radius of a typical nucleus is of the order of 10^{-15} m (1 fm).

Estimating the nuclear radius

The maximum size of a nuclear radius can be estimated from the distance of closest approach of an alpha particle. At the distance of closest approach to a nucleus of charge Q_n, the electric potential energy of the alpha particle (charge Q_a) will be equal to its initial kinetic energy E_k:

$$E_k = \left(\frac{1}{4\pi \varepsilon_0}\right)\left(\frac{Q_n Q_a}{r}\right)$$

$$r = \frac{Q_n Q_a}{4\pi \varepsilon_0 \times E_k}$$

Measuring nuclear radius

Electron diffraction can be used to determine the radius of nuclei more accurately. High energy electrons have a de Broglie wavelength small enough that they are diffracted by the nuclei of atoms. Graphs can show how the intensity varies with the angle for electron diffraction by a nucleus.

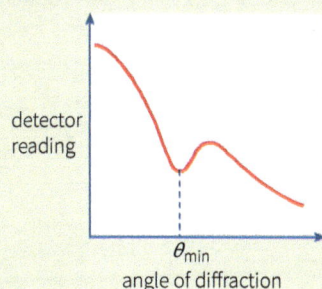

The angle of the first minimum θ_{min} is given by $\sin\theta = \frac{1.22\lambda}{d}$ where λ is the de Broglie wavelength of the electron and d is the diameter of the nucleus.

Nuclear radius and nucleon number

Data from electron diffraction experiments show that the radius of a nucleus R is related to its nucleon number A by:

$$R = R_0 A^{\frac{1}{3}}$$

where $R_0 = 1.05\times10^{-15}$ m.

This equation can be used to show that all nuclear material has the same density:

- assuming nucleus is spherical, its volume is:

$$V = \frac{4}{3}\pi R^3 = \frac{4}{3}\pi (r_0 A^{\frac{1}{3}})^3 = \frac{4}{3}\pi r_0^3 A$$

- substituting in equation for density:

$$\rho = \frac{m}{V}$$

density of nucleus $\rho = \frac{m_{nucleus}}{\frac{4}{3}\pi r_0^3 A}$

- protons and neutrons have approximately the same mass, u = 1 atomic mass unit, so mass of nucleus with nucleon number $A = A$ u, giving:

$$\rho = \frac{A\text{u}}{\frac{4}{3}\pi r_0^3 A} = \frac{3\text{u}}{4\pi r_0^3} = \text{constant}$$

This shows density of any nucleus is independent of its nucleon number A and of its radius r. Substituting appropriate values into the above equation gives the nuclear density:

$$\rho = 3.4\times10^{17}\,\text{kg m}^{-3}$$

Mass and energy

The mass m of any object increases or decreases when it gains or loses energy E.

$$E = mc^2$$

where c = speed of light, 3.00×10^8 m s^{-1}.

This applies to all energy changes.

Atomic and nuclear masses are often expressed in atomic mass units (u), where:

$$1\,u = \frac{1}{12} \text{ mass of carbon-12 atom}$$

$$1\,u = 1.661 \times 10^{-27} \text{ kg}$$
$$= 931.5 \text{ MeV}$$

The mass defect

In any change where energy is released, such as radioactive decay, the total mass after the change is always less than the total mass before the change because some of the mass is converted to released energy. The mass of a nucleus is less than the total mass of the individual nucleons of which it is formed.

The **mass defect** Δm of a nucleus is the difference between the mass of the separated nucleons and the mass of the nucleus. This is equal to the energy released when a nucleus forms from separate protons and neutrons.

Nuclear fission and fusion

Nuclear **fission** is the process by which a large unstable nucleus splits into two smaller nuclei.

Nuclear **fusion** is the process by which small nuclei join to form a larger nucleus.

Energy is released when nuclear fission and fusion take place because the resulting nuclei have a higher binding energy per nucleon.

Induced nuclear fission is the splitting of heavy nuclei such as uranium-235 and plutonium-239 by firing slow-moving **thermal neutrons** at them. Thermal neutrons are those made in thermal equilibrium with the surroundings.

A **chain reaction** can occur when the neutrons released by the fission of a nucleus go on to induce fission in other nuclei.

Critical mass is the minimum mass a **fissile** material (able to undergo fission) must have for a self-sustaining chain reaction to occur.

Binding energy

The binding energy (B.E.) of a nucleus is the work that would need to be done to separate it into its constituent nucleons, and is equal to the mass defect:

$$\text{binding energy of nucleus} = \text{mass defect} \times c^2$$
$$= \Delta m c^2$$

The binding energy per nucleon is a measure of the stability of a nucleus – the greater the binding energy, the more stable a nucleus is.

The change in binding energy per nucleon is equal to the energy released from **fission** or **fusion**.

- Iron-56 has the highest binding energy per nucleon, so is the most stable nucleus.

- Binding energy per nucleon is increased, and so energy is released, by fusion of small nuclei.

- Binding energy per nucleon is increased, and so energy is released, by fission of large nuclei.

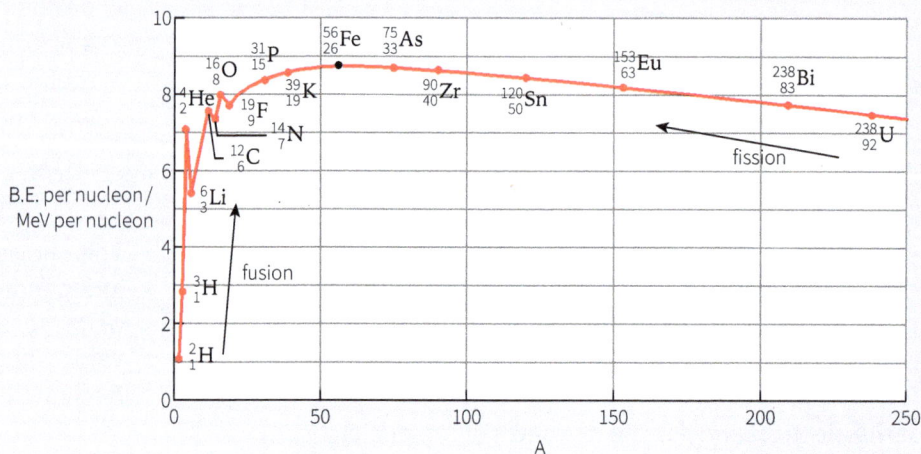

Nuclear reactors

Induced nuclear fission is used to create a controlled chain reaction in nuclear reactors in power stations.

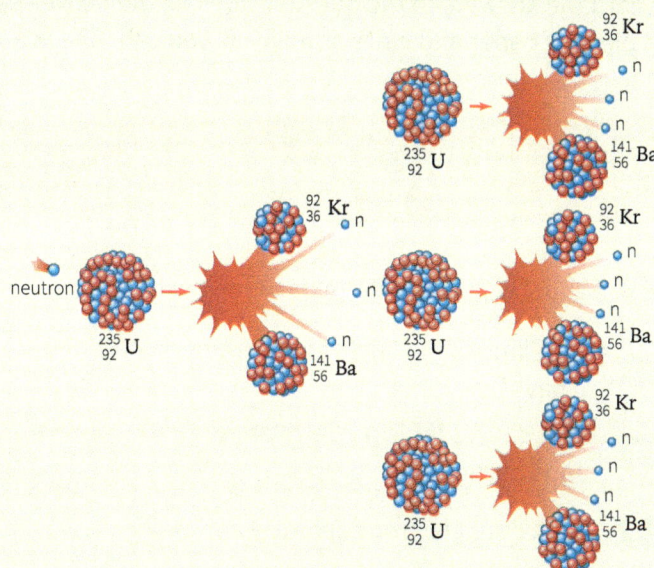

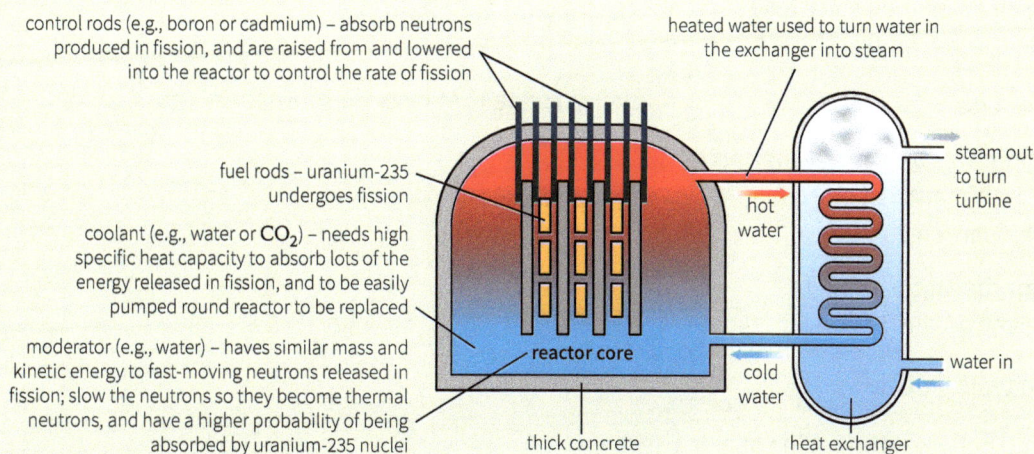

control rods (e.g., boron or cadmium) – absorb neutrons produced in fission, and are raised from and lowered into the reactor to control the rate of fission

heated water used to turn water in the exchanger into steam

fuel rods – uranium-235 undergoes fission

coolant (e.g., water or CO_2) – needs high specific heat capacity to absorb lots of the energy released in fission, and to be easily pumped round reactor to be replaced

moderator (e.g., water) – haves similar mass and kinetic energy to fast-moving neutrons released in fission; slow the neutrons so they become thermal neutrons, and have a higher probability of being absorbed by uranium-235 nuclei

reactor core

thick concrete

hot water

cold water

heat exchanger

steam out to turn turbine

water in

Nuclear reactor safety

Safety features of nuclear reactors

The reactor is surrounded by thick concrete walls (shielding) to absorb the radiation emitted.

Fuel rods are radioactive, so they are inserted and removed from the reactor using remote handling devices.

In emergencies, the control rods are automatically inserted fully into the core to bring fission to a complete stop.

Disposing of fuel rods

Spent fuel rods are hot and radioactive because they contain the products from the fission of uranium, which are themselves unstable and emit radiation. Remote handling devices are used to remove the

spent fuel rods from the reactor and place them in cooling ponds. After cooling, the spent fuel rods are reprocessed so that some products can be used for practical applications.

Disposing of radioactive waste

The remaining radioactive waste is stored, usually in sealed steel containers in concrete vaults to minimise the possibility of the waste entering the environment.

Advantages and disadvantages of nuclear power

Nuclear power can generate a lot of electricity to meet high power needs without producing greenhouse gases. However, this benefit must be weighed carefully against the risks posed by nuclear waste and the possibility of reactor accidents.

Retrieval

Learn the answers to the questions below, then cover the answers column with a piece of paper and write as many as you can. Check and repeat.

	Questions	Answers
1	What property do large stable nuclei have?	more neutrons than protons to help bind the nucleus together
2	What kind of nuclei do alpha emitters tend to have?	large nuclei ($Z \geq 60$)
3	What causes large nuclei to emit alpha particles?	strong nuclear force is unable to overcome the electrostatic force of repulsion between protons
4	When does electron capture occur?	when a nucleus captures an orbiting electron
5	How does a nucleus in an excited state lose energy and return to its ground state?	emitting gamma radiation
6	Define a metastable state.	excited state that lasts for more than 1 ns
7	How can the maximum size of a nuclear radius be estimated from the distance of closest approach of an alpha particle?	$r = \dfrac{Q_n Q_a}{4\pi \varepsilon_0 \times E_k}$
8	Give the equation for the angle of the first minimum θ_{min} for electron diffraction.	$\sin\theta = 1.22\dfrac{\lambda}{d}$
9	State the radius of a typical nucleus.	$\sim 10^{-15}$ m (1 fm)
10	State the relationship between the radius of a nucleus and nucleon number.	$R = R_0 A^{\frac{1}{3}}$
11	Give the equation for calculating the increase in mass when any object gains energy.	$E = m c^2$
12	What is the mass defect is equal to?	energy released when a nucleus forms from separate protons and neutrons
13	Why can small nuclei undergo fusion?	binding energy per nucleon is increased so energy is released when they fuse
14	What is nuclear fission?	large unstable nucleus splits into two smaller nuclei
15	What is nuclear fusion?	small nuclei join to form a larger nucleus
16	Why is energy released when nuclear fission and fusion take place?	resulting nuclei have a higher binding energy per nucleon
17	What is induced fission?	splitting of heavy nuclei by firing slow-moving thermal neutrons at them
18	When can a chain reaction occur?	when neutrons released by the fission of a nucleus go on to induce fission in other nuclei
19	Define critical mass.	minimum mass a fissile material must have for a self-sustaining chain reaction to occur
20	What does the coolant in a nuclear reactor do?	absorbs the energy released in fission

Put paper here

21 What does the moderator in a nuclear reactor do? ⋮ slows down the fast-moving neutrons released in fission so that they become thermal neutrons

22 Define a thermal neutron. ⋮ neutron that is in thermal equilibrium with its surroundings and has a similar kinetic energy to the moderator molecules

23 What do controls rods do in a nuclear reactor? ⋮ absorb the neutrons produced in fission and control the rate at which fission occurs in it

24 What makes the spent fuel rods hot and radioactive? ⋮ contain the products from the fission of uranium, which are themselves unstable and emitting radiation

25 What happens to the radioactive waste from nuclear reactors? ⋮ stored in sealed steel containers in concrete vaults

Put paper here

🖩 Maths skills

Practise your maths skills using the worked example and practice questions below.

Orders of magnitude estimations

Calculations based on order of magnitude allow us to make an estimate of a value without needing to know the exact value.

It can be helpful to use an order of magnitude calculation to sense-check values that are obtained from a calculator.

For an order of magnitude calculation, the values are written with only one significant figure, or sometimes to the nearest power of 10.

Worked example

Question

In Rutherford, Geiger, and Marsden's alpha scattering experiment, helium nuclei with an initial kinetic energy of 8.0×10^{-13} J were deflected backwards from a sheet of gold foil.

Using the formula for electric potential energy, estimate a maximum size for the radius of the gold atoms.

$$Z\,Au = 79$$
$$\varepsilon_0 = 8.85 \times 10^{-12}\,\text{F m}^{-1}$$
$$e = 1.6 \times 10^{-19}\,\text{C}$$

Answer

The He atom is deflected backwards in a direct collision. Just before collision, the He atom starts to move backwards – the initial kinetic energy of the He atom will be equal to the potential energy of the nucleus.

Rearranging electric potential energy equation gives $r = \dfrac{qQ}{4\pi\varepsilon_0 E_k}$.

Order of magnitudes of the different quantities: $E_k \approx 10 \times 10^{-13}\,\text{J} \approx 10^{-14}\,\text{J}$; elementary charge $q \approx 10^{-19}\,\text{C}$; $Q \approx 10^{-18}\,\text{C}$; $4\pi\varepsilon_0 \approx 10^{-10}\,\text{F m}^{-1}$

$$r \approx \frac{10^{-19} \times 10^{-18}}{10^{-10} \times 10^{-14}} \approx 10^{-13}\,\text{m}$$

The maximum size of the radius is $\approx 10^{-13}\,\text{m}$.

Practice

1 Using an initial energy of 8.0×10^{-13} J for the helium nucleus, estimate the maximum size for an aluminium nucleus that could be obtained from the alpha scattering experiment. $Z\,Al = 13$

2 The gold foil used in the scattering experiment was approximately $0.5\,\mu\text{m}$ thick.

Use an order-of-magnitude calculation to estimate how many atoms thick the gold foil is.

3 Suggest why Rutherford and his team used gold foil in their experiment.

Practice

Exam-style questions

01 Alpha particles are scattered by the nuclei in a piece of very thin gold foil, as shown in **Figure 1**.

Figure 1

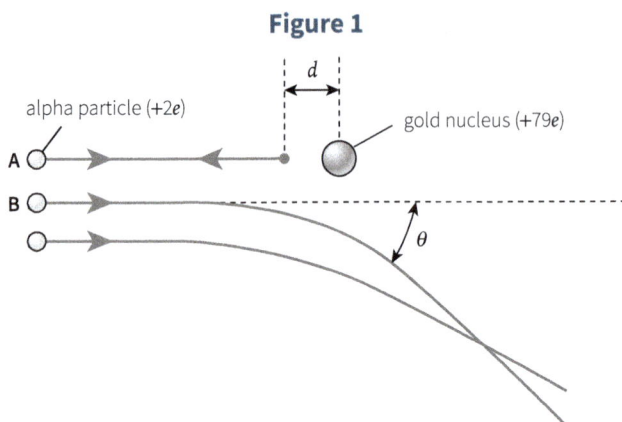

alpha particle (+2e)

gold nucleus (+79e)

A

B

θ

01.1 Explain the differences between the paths of the alpha particles that start at points **A** and **B**. **[2 marks]**

_____

_____

01.2 An alpha particle of energy 5 MeV is incident on the nucleus along path **A**.

Show that the distance of closest approach d is about 10^{-13} m.

$1\,\text{MeV} = 1.6 \times 10^{-13}\,\text{J}$ **[2 marks]**

> **! Exam tip**
> Remember that the charges on the alpha particle and nucleus determine its path.

01.3 Using the distance of closest approach as an estimate for the nuclear radius, estimate the volume of a person who would have the same density as a gold nucleus. The relative atomic mass of gold is 197.

Calculate the length of the side of a cube with the same volume. **[3 marks]**

length = _____ m

01.4 The experiment is repeated with zirconium foil. The atomic mass and atomic number of zirconium are about half that of gold.

Suggest the effect on your answer to **01.3**. Assume that the energy of the alpha particle is the same. **[3 marks]**

_____

_____

_____

_____

02 An isotope of radium, radium-226, decays to radon by emitting alpha radiation.

02.1 Describe the changes to the neutron number and proton number when a nucleus of radium decays. **[1 mark]**

_____

_____

02.2 **Table 1** shows data for the decay of the radium nucleus.

Table 1

| Transition | | Energy of alpha particle emitted |
From	To	
radium-226	radon-222*	4.59
radium-226	radon-222 (ground state)	4.78

Calculate the wavelength of the gamma ray emitted by the radon when the nucleus decays to the ground state.
$1\,MeV = 1.6 \times 10^{-13}\,J$ **[2 marks]**

wavelength = _____ m

02.3 Radium-225 decays by beta emission to an isotope of actinium. The atomic number of radium is 88.

Figure 2

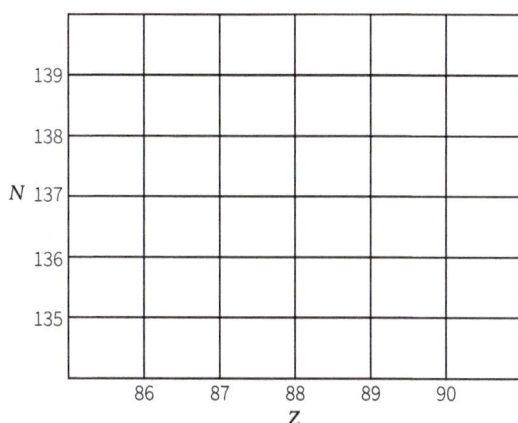

Draw an arrow on **Figure 2** to show the decay of radium-225. Explain the position and direction of the arrow. **[4 marks]**

_____

_____

_____

_____

> **! Exam tip**
>
> You can treat nuclear energy levels like electron energy levels in the Bohr model of the atom.

02.4 Describe **two** differences between the arrow for the beta decay of radium-225 in **Figure 2** to the arrow that would show the alpha decay of radium-226. **[2 marks]**

_____

_____

_____

03 An experiment to investigate nuclear diameters uses electrons. Electrons are accelerated through a high potential difference and are transmitted through a film that is only a few atoms thick.

A detector on the other side of the film records an intensity shown in **Figure 3**.

Figure 3

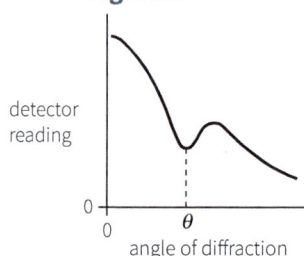

detector reading

0

0 θ

angle of diffraction

03.1 The wavelength of the electrons used is about 10^{-15} m. Explain why. **[1 mark]**

03.2 In a diffraction pattern produced by light transmitted through a diffraction grating, the minima have zero amplitude. In an electron diffraction pattern, the minima do not have zero amplitude. Suggest why. **[1 mark]**

03.3 Electron diffraction techniques produce a more accurate measurement of the radius of a nucleus than measurements of the distance of closest approach in alpha particle scattering experiments. Suggest **two** reasons why. **[2 marks]**

03.4 Despite producing more accurate measurements than using alpha particles, the results for the nuclear radius produced by electron diffraction still have a range of values.

Suggest why, using information from **Figure 3**. **[1 mark]**

> **! Exam tip**
>
> Remember that electrons and alpha particles interact differently with the constituent parts of the nucleus.

04 Heavy water, which is used in some nuclear power stations, is made with oxygen and heavy hydrogen, also known as deuterium $_{1}^{2}$H.

04.1 Explain why there is a difference between the mass of the particles in a deuterium nucleus and the atomic mass of deuterium. Calculate the mass difference using the data sheet and information given.

Give your final answer to 3 significant figures.
atomic mass of deuterium = 2.013553 u **[4 marks]**

04.2 Show that the binding energy of a deuterium nucleus in MeV is about 2 MeV. **[2 marks]**

04.3 The binding energy of tritium $_{1}^{3}$H is 8.482 MeV. Explain the difference between the binding energies of deuterium and tritium. **[2 marks]**

> **! Exam tip**
>
> Find the energy in joules first, then convert to MeV.

04.4 **Table 2** shows the relative atomic mass and binding energy for tritium and helium-3.

Table 2

Isotope	Relative atomic mass / u	Binding energy / MeV
tritium	3.01604	8.482
helium-3	3.01603	7.718

Suggest why, despite having approximately the same relative atomic mass, tritium and helium-3 have different binding energies.

[3 marks]

05 The world's first artificial nuclear reactor, Chicago Pile-1, produced the first self-sustaining nuclear chain reaction in 1942.

05.1 State the meaning of self-sustaining nuclear chain reaction. Suggest **one** condition where a nuclear reaction is not self-sustaining.

[2 marks]

05.2 The reactor used 4.9 tonnes (4.9×10^3 kg) of natural uranium and 41 tonnes of uranium oxide. **Table 3** give some information about Chicago Pile-1 and one of the newest nuclear reactors.

Table 3

Reactor (age)	Chicago Pile-1 (1942)	Russian nuclear reactor (2019)
Fuel	4.9 t natural uranium / 46 t uranium oxide	90 t uranium dioxide
Moderator	graphite	pressurised water
Control rods	cadmium / wood	boron
Cooling system	none	pressurised water
Radiation shielding	none	in containment building
Output power	0.5 W	1114 MW

Suggest **one** reason why Chicago Pile-1 did not use a cooling system or radiation shielding, despite the mass of uranium used.   **[1 mark]**

05.3 Suggest why the power output of Chicago Pile-1 was less than 0.05% of the Russian reactor, even though the mass of fuel used was 50% of that of the Russian reactor.

Explain your answer in terms of what happens when a uranium nucleus undergoes fission.   **[3 marks]**

05.4 Describe the purpose of the cadmium and boron in terms of the nuclear reactions occurring in the reactors.   **[1 mark]**

06 Sodium-24 is used as a tracer to follow the path that sodium takes in the human body. It is used to check that a person's uptake levels of sodium are within the normal range. The half-life of sodium-24 is 15 hours. It decays by beta decay to an excited state of magnesium.

> **! Exam tip**
>
> Try to make logical connections between the data you are given and the question that you are asked.

> **Synoptic links**
>
> 3.2.1.2   3.2.1.4   3.8.1.3

06.1 State why a neutrino was suggested to be emitted alongside a beta particle in beta decay. **[1 mark]**

06.2 State the exchange particle for the type of interaction responsible for this decay. **[1 mark]**

06.3 Suggest a situation where the use of sodium-24 isotopes would represent an acceptable risk. **[1 mark]**

Exam tip

Remember that alpha, beta, and gamma radiation affect the body in different ways.

06.4 A nucleus of sodium-24 is produced when a stable nucleus, sodium-23, absorbs a neutron.
Write the changes to the atomic number and atomic mass when the neutron is absorbed, and when the beta particle is emitted from sodium-24. **[2 marks]**

06.5 The excited states of magnesium are shown in **Table 4**.

Table 4

Energy level	Energy / MeV	Percentage of decays to this level / %
ground state	0.00	0.000
1st	1.38	0.003
2nd	4.14	99.905
3rd	4.23	0.002
4th	5.22	0.090

The decay of the sodium nucleus produces two gamma rays.
Suggest the energy of the two gamma rays. Justify your answer. **[2 marks]**

07 Two billion-year-old natural nuclear reactors can be found in Gabon. The power output of each reactor is about 100 kW.

The reactors are a result of uranium deposits underground.

Synoptic link

3.4.1.7

07.1 Natural deposits of uranium are not sufficient to start a nuclear reaction.
Describe what is meant by a chain reaction, and suggest why it does not start until groundwater enters the uranium deposits. **[3 marks]**

07.2 Around two billion years ago, the deposits were composed of about 3% the uranium isotope $^{235}_{92}$U, which is fissile, and 97% of $^{238}_{92}$U, which is non-fissile. This is comparable to the composition of uranium fuel rods in a modern reactor.

Describe the difference between fissile and non-fissile. **[1 mark]**

07.3 One of the products of the fission reaction is neodymium $^{142}_{60}$Nd.
Deduce the atomic mass and atomic number of the other fission product if the reaction is induced by one neutron and produces three further neutrons. **[2 marks]**

07.4 The energy released per reaction is 151.9 MeV. Calculate an estimate for the number of decays per second required to produce the power output of the reactor.
1 MeV = 1.6×10^{-13} J **[3 marks]**

Exam tip

You need a clear understanding of nuclear binding energy.

07.5 Calculate the volume of steam in cubic metres produced per second. Assume the water enters the uranium deposit at a temperature of 20 °C.

specific heat capacity of water = 4200 J kg^{-1} °C^{-1}; specific latent heat of vaporisation of water = 2260 kJ kg^{-1}; density of steam = 0.6 kg m^{-3}

[3 marks]

08 All nuclear power stations use nuclear fuel rods made of uranium or plutonium. Both nuclei can undergo induced fission, and they can also undergo spontaneous fission.

08.1 Describe the difference between induced and spontaneous fission.

[1 mark]

Synoptic links

3.8.1.3   3.5.1.1
3.8.1.2   3.5.1.4

08.2 Uranium is mined and uranium-235 and uranium-238 are extracted. Uranium-238 is non-fissile and makes up about 99% of natural uranium. However, plutonium is made from the $^{238}_{92}$U. A uranium-238 nucleus absorbs a neutron, then undergoes two β decays.

Deduce the atomic mass and atomic number of the plutonium nucleus produced. **[2 marks]**

08.3 Uranium-235, which makes up 0.72% of natural uranium, is fissile.

Table 5

Fuel	Probability of undergoing induced fission	Energy released in one fission / MeV	Half-life / years	Number of spontaneous fissions / kg
uranium-235	medium	202	703 million	5000
plutonium	high	207	24 110	10

Suggest **one** reason, using the information in **Table 5**, why plutonium appears to be a better fuel for use in a nuclear power station.

[1 mark]

08.4 An RTG is a type of nuclear battery that uses thermocouples. In a thermocouple, one end of a piece of wire is heated while the other is kept at a lower temperature.

Suggest why this produces a current in the wire. **[1 mark]**

08.5 A pacemaker is a device implanted inside a person that produces electrical pulses to maintain a regular heartbeat.

Suggest why it was thought that plutonium-238 could be used in a pacemaker's RTG without significant risk to the patient.

Plutonium-238 decays by emitting an alpha particle of energy 5.6 MeV (8.9610^{-13} J). It has a half-life of 87.7 years. **[1 mark]**

08.6 A pacemaker produces a p.d. of ~ 1.5 V and a current of ~ 2 mA.

Calculate an estimate for the mass in g of plutonium in an RTG required for these outputs, and suggest why the mass of plutonium used in a pacemaker would be much less. **[3 marks]**

21 Telescopes

Reflecting and refracting telescopes

Refracting telescopes (Galilean)	Reflecting telescopes (Cassegrain)
contain a short focal length eyepiece lens and a long focal length objective lens	contain a parabolic concave primary mirror and a convex secondary mirror
suffer from **spherical aberration** and **chromatic aberration**	suffer from spherical aberration if spherical, not from chromatic aberration
have large lenses which are hard to make and are likely to break under their own weight	can be made much larger than refracting telescopes because the mirrors can be supported from behind

Telescopes with two lenses

The **objective lens** collects light from a distant object and forms a **real image** at focal length f_o.

The **eyepiece lens** forms a magnified, **virtual image** at infinity of the image formed by the objective lens.

The **focal plane** of the eyepiece f_e coincides with the focal plane of the objective.

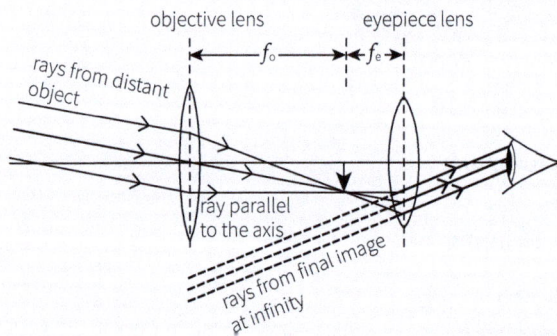

Telescopes with two mirrors

The **primary** (parabolic **concave**) mirror collects light from a distant object.

The **secondary** (**convex**) mirror reflects the light through a hole and into an eyepiece lens:

- the eyepiece forms a magnified image at the eye
- the secondary mirror produces a slight dimming, but no missing section in the image.

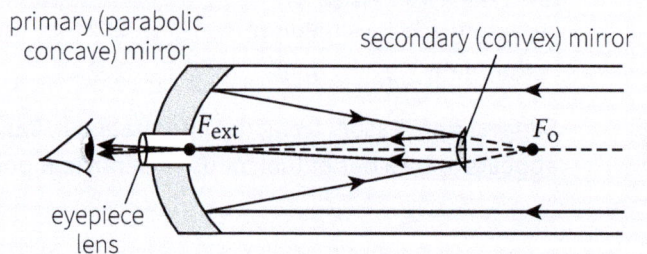

The eye and the charge-coupled device (CCD)

Apparatus	eye	CCD
Quantum efficiency (percentage of incidence photons that release electrons)	1% at 550 nm in low light; colour vision lost, but pupil expands	70–80% number of electrons $\propto \dfrac{\text{number of photons}}{\text{intensity}}$
Resolution (smallest object / difference detectable)	2 μm = size of light sensitive cell, diameter of fovea ≈ 1 mm	depends on number of pixels e.g. 4 μm × 4 μm, typically 6 million pixels in 1 cm^2
Ease of use	very convenient, but needs time to adapt to light conditions	can produce images with long exposure / detect wavelengths the eye can't see

Non-optical telescopes

Radio telescopes:

- have a similar design to optical reflecting telescopes (parabolic mirrors)
- have a receiving aerial instead of a secondary mirror.

Satellite-based telescopes:

- infrared, ultraviolet, x-ray, and gamma telescopes
- all reflecting telescopes
- placed on balloons, in orbit, and on mountains in deserts (such as the Atacama Desert, Chile) because the atmosphere and/or water vapour absorbs radiation.

Radians and magnification

Angles in radians

If an object of height h is distance d away, then the angle θ subtended by the object:

$$\theta \text{ (rad)} = \frac{\text{height (m)}}{\text{distance (m)}}$$

A larger object at the same distance subtends a larger angle at the eye.

Calculating magnification

angular magnification M

$$M = \frac{\text{angle subtended by image at eye (rad)}}{\text{angle subtended by object at eye (rad)}} = \frac{f_o}{f_e}$$

where:

f_o = focal length of objective lens

f_e = focal length of eyepiece lens.

If the image subtends a larger angle, then it appears magnified.

Spherical and chromatic aberration

In **spherical aberration**, spherical lenses or mirrors do not bring parallel beams to the same focus.

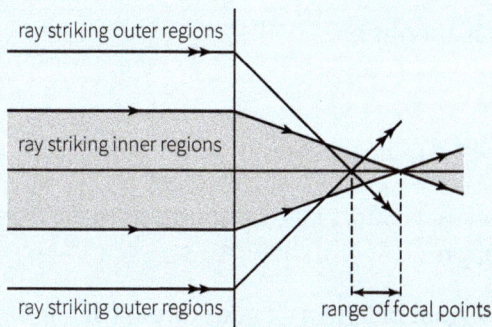

ray striking outer regions

ray striking inner regions

ray striking outer regions

range of focal points

In **chromatic aberration**, the edges of lenses behave like prisms, so different colours focus at different points.

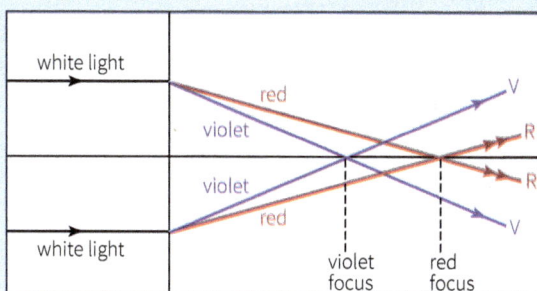

white light

red

violet

violet

red

white light

violet focus

red focus

V

R

R

V

Large diameter telescopes

- light collected is subject to diffraction
- two sources are just resolved when the maximum of one pattern coincides with the minimum of the other.
- The **Rayleigh criterion** says the angle at which this occurs depends on the wavelength λ and the diameter D of the mirror or lens larger apertures have better resolving power:

$$\theta \text{ (rad)} = \frac{\lambda \text{ (m)}}{D \text{ (m)}}$$

- brightness $\propto$ collecting power $\propto D^2$

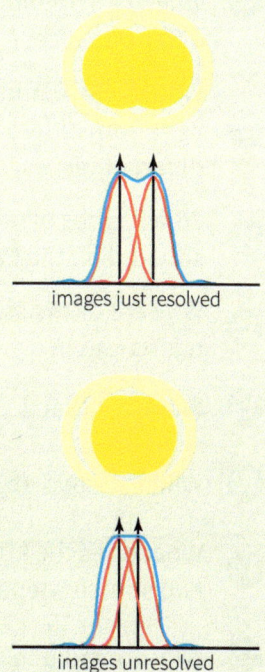

images just resolved

images unresolved

⇄ Retrieval

Learn the answers to the questions below, then cover the answers column with a piece of paper and write as many as you can. Check and repeat.

	Questions	Answers
1	What are the two lenses found in a refracting telescope?	eyepiece and objective
2	Which lens produces a real image in a refracting telescope?	objective lens
3	What type of image is produced by a refracting telescope?	magnified, virtual image
4	What shape is the primary mirror in a reflecting telescope?	parabolic and concave
5	What shape is the secondary mirror in a reflecting telescope	convex
6	What happens to the angle subtended by an object at the eye as the object moves away?	decreases
7	Why do objects appear magnified?	subtend a larger angle at the eye
8	What is the name of the phenomenon when light of different colours focuses on different places?	chromatic aberration
9	What is the name of the phenomenon when light from different parts of a lens focuses on different places?	spherical aberration
10	Which types of radiation do telescopes situated on the Earth's surface detect?	visible light, radio waves
11	Why are ultraviolet, infrared, and X-ray telescopes in orbit around the Earth?	atmosphere absorbs large amounts of these types of radiation
12	What is diffraction?	spreading out of light as it goes through a gap
13	When are two objects just resolved?	maximum of one diffraction pattern coincides with the minimum of the second pattern
14	What does the ability of a telescope to resolve two sources depend on?	relative size of the aperture and the wavelength
15	Why are radio telescopes so large?	wavelength of radio waves is large, and a bigger dish can resolve two sources more easily
16	What is quantum efficiency?	percentage of incidence photons that cause the release of electrons
17	Compare the quantum efficiency of the eye and a charge-coupled device.	eye – 1%; CCD – 70–80%
18	What is resolution?	smallest size object or difference detectable

Put paper here (repeated in central divider)

19 Compare the resolution of the eye and a charge-coupled device.

comparable resolutions in micrometres

20 Give two advantages of a charge-coupled device over the eye.

can take long exposure images and record wavelengths the eye cannot see

Put paper here

🖩 Maths skills

Practise your maths skills using the worked example and practice questions below.

Rayleigh criterion	Worked example	Practice
If two stars are close together when viewed through a telescope, the Rayleigh criterion is used to determine if they can be resolved into two separate objects. The two objects emitting light of wavelength λ have an angular separation θ, and are viewed through a telescope with diameter D. For such small angles, the small-angle approximations $\sin\theta \approx \theta$, $\tan\theta \approx \theta$ and $\cos\theta \approx 1$ hold true. If $\theta \approx \dfrac{\lambda}{D}$ the two objects will be just resolvable.	**Question** Two stars have an angular separation of 7.0×10^{-5} rad. A telescope with an objective lens of 15 cm is used to observe the stars on a wavelength of 600 nm. Determine whether the two stars can be resolved by the telescope. **Answer** First convert values to SI units: $D = 15\,cm = 0.15\,m$ $\lambda = 600\,nm = 6.00 \times 10^{-7}\,m$ The telescope can resolve objects with an angular separation of: $\theta \approx \dfrac{6.00 \times 10^{-7}}{0.15} = 4 \times 10^{-6}\,rad$ This is smaller than the angular separation of the two stars, so they can be seen separately using the telescope.	1 A double star has an angular separation of 1.1×10^{-5} rad. The wavelength of light from the star is 570 nm. Calculate the minimum diameter of objective lens for a telescope to just resolve the star. 2 A radio telescope has a diameter of 76.2 m. Calculate the minimum angular separation of objects that the radio telescope can resolve using the 21 cm wavelength. 3 Suggest why radio telescopes need to be much larger than optical telescopes.

Exam-style questions

01 **Figure 1** shows a simple refracting telescope. Point **F** represents the common foci of the objective and eyepiece lenses.

Figure 1

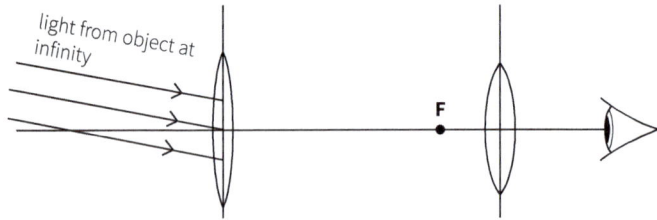

light from object at infinity

F

01.1 Complete the ray diagram for the refracting telescope. **[3 marks]**

01.2 This is the type of telescope Galileo used to observe Jupiter.
Calculate the angle subtended by the image of Jupiter at the eye.
distance to Jupiter = 9.3×10^{11} m; diameter of Jupiter = 1.4×10^8 m
[1 mark]

angle = _____ rad

01.3 The magnification of this telescope is approximately 20×.

Determine an estimate of how far away you would need to place a 1.0 cm coin to represent the angle subtended by Jupiter with the naked eye. **[3 marks]**

distance = _____ m

01.4 The telescope has an objective lens with a diameter of 37 mm and a focal length of 980 mm.

Determine an estimate for the focal length of the eyepiece. **[1 mark]**

focal length = _____ mm

01.5 There is some evidence that Galileo observed the planet Neptune in the vicinity of Jupiter but mistook it for a star.

Explain why. **[2 marks]**

_____

_____

_____

_____

_____

_____

02 The largest refracting telescope built has an aperture of 1.25 m. The largest reflecting telescopes in current use have diameters that exceed 10 m.

02.1 Suggest **two** reasons why it is easier to make reflecting telescopes with larger apertures. **[2 marks]**

_____

_____

_____

_____

_____

_____

02.2 Draw a diagram showing what happens to parallel rays when they enter a spherical mirror.

Describe how your diagram shows why mirrors in reflecting telescopes are not spherical. **[5 marks]**

_____

_____

_____

_____

_____

02.3 Describe the construction on a Cassegrain telescope. **[1 mark]**

_____

_____

_____

_____

> **! Exam tip**
>
> Make sure you include whether the mirrors are convex or concave.

02.4 An amateur astronomer uses a Cassegrain telescope with an aperture of 12 cm to look at the Moon.
Explain why the astronomer can see the Moon through the telescope.
mean Earth–Moon distance = 380 000 km; diameter of the Moon = 3.5×10^6 m; mean wavelength of visible light = 550 nm **[2 marks]**

_____

_____

_____

_____

_____

_____

03 A student uses the resolving power of the eye to measure pupil diameter. They draw two black parallel lines 1 mm apart on a piece of paper and stick the paper to the wall.

The student walks back until they can only just distinguish the two lines.

03.1 Sketch a diagram of the diffraction patterns from the lines that are formed on the back of the student's eye when the lines are just resolved. **[2 marks]**

03.2 The distance at which the lines are just resolved is 10 m. The average wavelength of visible light is 550 nm.

Calculate the diameter in mm of the student's pupil. **[2 marks]**

03.3 Another student suggests that they would obtain a more accurate diameter value if they measured the distance for pairs of lines that are different distances apart.

Explain why this would produce a more accurate value.

Describe how to use the data to plot a graph so that the gradient was a function of the pupil diameter. **[4 marks]**

03.4 A student looks at a star in the night sky with the naked eye and through a telescope with an aperture of 10 cm.

Determine an estimate for how much brighter the star appears through the telescope than with the naked eye. **[2 marks]**

Exam tip

Remember that brightness is related to aperture^2.

04 The search for extraterrestrial intelligence (SETI) uses ground-based and space-based telescopes to search for signals coming from space that may have been produced by other civilisations.

04.1 Suggest why space-based telescopes are used in addition to ground-based telescopes. **[1 mark]**

04.2 Compare the construction of a radio telescope with that of a reflecting optical telescope. **[2 marks]**

04.3 Radio telescopes can be based on the ground.

Two sources in the night sky emit both visible light, with a wavelength of 470 nm, and radio waves, with a wavelength of 52 cm.

An optical telescope with an aperture of 22 cm can resolve the sources.

Calculate the aperture of the radio telescope that would be able to resolve them.

State any assumptions that you make. Comment on your answer. **[4 marks]**

04.4 Suggest **one** source of interference affecting radio astronomy that does not affect optical astronomy, and describe the measures that astronomers take to minimise this interference. **[2 marks]**

Exam tip

For compare questions, check the number of marks and focus on similarities and differences.

05 In a telescope, the resolving power is related to the diameter.

05.1 Explain why the resolving power is not measured in watts. **[2 marks]**

05.2 The three images in **Figure 2** were produced by three different telescopes.

Figure 2

A   B   C

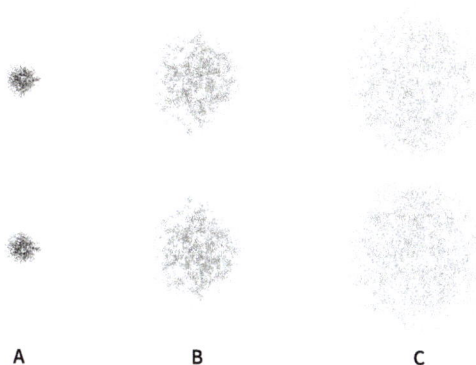

Suggest a physical difference between the three telescopes, and explain how the difference accounts for the appearance of the images. **[2 marks]**

05.3 The telescopes use the same charge-coupled device (CCD) as a detector.

The detector has a comparable resolution to the human eye. The size of a light-sensitive cell of the eye is 2 μm.

Calculate the number of pixels on the 1.2 cm² CCD wafer. **[2 marks]**

05.4 Compare the eye and the CCD in terms of quantum efficiency. **[2 marks]**

> **Exam tip**
>
> You need to know how to convert from cm² to m².

06 **Figure 3** shows the chromatic aberration of a lens.

Figure 3

> **Synoptic link**
>
> 3.3.2.3

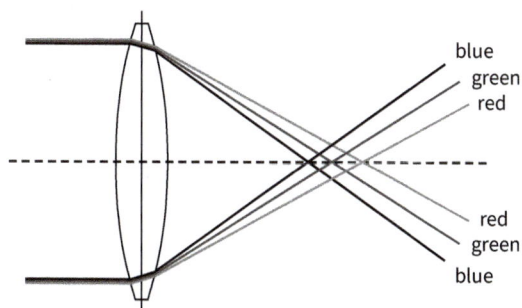

06.1 Explain why chromatic aberration happens, and describe how the image would appear. **[3 marks]**

06.2 The refractive index of the glass used for the lens is 1.45. Calculate the speed of light in m s⁻¹ in the glass. **[1 mark]**

> **Exam tip**
>
> When discussing chromatic aberration, you should talk about the shape of the lens.

06.3 The chromatic aberration can be removed by adding another piece of glass to the lens, making it a doublet, as shown in **Figure 4**.

Figure 4

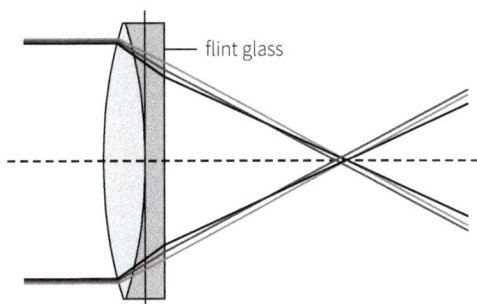

Suggest what the flint glass does to the light to bring the white light to a focus at the same point. **[3 marks]**

06.4 A student uses lenses to make a simple telescope. Some data about the lenses that they used are shown in **Table 1**.

Table 1

Telescope	Focal length of objective / m	Focal length of eyepiece / m
A	0.3	0.2
B	0.5	0.2
C	1.0	0.2

Suggest and explain through which telescope the student would notice the most chromatic aberration. **[2 marks]**

07 In 2016, the world's largest radio telescope, the FAST, started collecting data. The diameter of the telescope is 500 m, and it is built into a natural dent in the landscape in a remote region of China.

07.1 About 10 000 people were relocated from the surrounding area. Suggest why. **[1 mark]**

07.2 The telescope can detect sources that emit radio waves in the range 70 MHz to 3.0 GHz. Two sources emit radio waves of the same frequency.

Explain which frequency the most distant of the pair could emit that would still be resolvable. **[2 marks]**

07.3 The FAST telescope will detect microwaves produced by transitions between two levels in a hydrogen atom.

Describe what happens in the atom to produce the emission of a photon in the microwave region of the electromagnetic spectrum. **[1 mark]**

Synoptic links

3.3.2.1   3.2.2.3

Exam tip

When writing about transitions, you need to be clear whether there is a transition from a higher to a lower level or the other way around.

07.4 The energy of the photon emitted is 5.87 µeV.

Show that this transition is in the range detected by the telescope.

[2 marks]

07.5 Suggest why this telescope will be able to produce more detailed images than existing telescopes. **[1 mark]**

08 The Hubble Space Telescope (HST) has a mirror that is 2.4 m in diameter and orbits the Earth every 95.42 minutes.

08.1 Show, using Newton's law of gravitation, that the radius of orbit of the HST is:

$$\sqrt[3]{\frac{G\,M\,T^2}{4\pi^2}}$$

where G is the gravitational constant, M is the mass of the Earth, and T is the time of orbit. **[3 marks]**

08.2 Calculate the radius in m of the orbit. **[1 mark]**

08.3 Calculate the energy in J required to move the HST from the surface of the Earth to the orbit.

HST launch mass = 11 110 kg; radius of the Earth = 6.37×10^6 m

[2 marks]

08.4 It costs about £20 000 per kg to put an object into orbit. Despite the cost, astronomers put telescopes that detect gamma rays, X-rays, ultraviolet, and infrared radiation into orbit.

Explain why. **[1 mark]**

08.5 Describe **one** disadvantage, apart from the cost, of putting a telescope in orbit. **[1 mark]**

08.6 Gamma ray and X-ray telescopes have a different design to other telescopes.

Explain why you cannot use a refracting or reflecting telescope to focus gamma rays. **[1 mark]**

08.7 Some people think that the HST produces good images of astronomical objects because it is closer to the stars, or because it is very powerful.

Deduce whether the HST could resolve two astronauts standing 25 m apart on the surface of the Moon.

The closest distance that the surface of the Moon gets to the Earth is 355 600 km.

Comment on your answer.

Mean wavelength of visible light = 550 nm **[4 marks]**

Synoptic links

3.7.2.1   3.7.2.3

Exam tip

Remember that you cannot use the difference in distance when calculating a difference in gravitational potential energy.

Knowledge

22 Stars and cosmology

Measuring distance in astronomy

A **light year** (ly) is the distance light travels in a year.

An **astronomical unit** (AU) is the distance from the Earth to the Sun.

A **parsec** (pc, 1 pc = 3.26 ly) is the distance to a star that subtends an angle of one second at a distance of 1 AU.

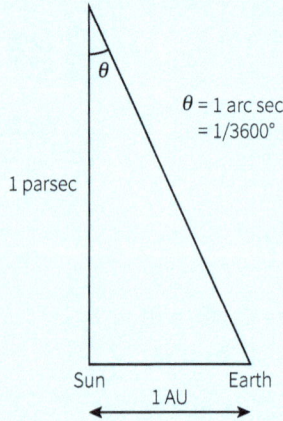

θ = 1 arc sec = 1/3600°

1 parsec

Sun   Earth

1 AU

Classifying by magnitude

Luminosity

The total power radiated by a star.

Intensity I

The power radiated per unit area at the observer.

Brightness

A subjective scale of measurement; how bright a star appears depends on both its distance and the power emitted at visible wavelengths. Brightness is measured on the **Hipparchus scale**.

The faintest star visible with the naked eye from the Earth has a value of 6; the brightest is 1.

Apparent magnitude m

The brightness of an astronomical body as seen from the Earth. An observer receives 100 times more light from a star with $m = 1$ than $m = 6$. A difference of 1 is equal to an intensity ratio of 2.51. Stars can have a negative m.

Absolute magnitude M

The apparent magnitude an astronomical body would have if it was 10 pc from the observer with d in pc:

$$m - M = 5 \log_{10}\left(\frac{d}{10}\right) \qquad d = 10\,\text{pc}, m = M$$

$$d < 10\,\text{pc}, m < M \qquad d > 10\,\text{pc}\, m > M$$

Exoplanet detection

Detection techniques include:

- **transits** – light curves show decrease / periodic variation in apparent magnitude
- **radial velocity** – orbiting planets cause stars to wobble, and light is blue and red shifted.

Hubble's law

The Universe is expanding – all galaxies are moving away from each other, with:

$$v = H d$$

where:

H = Hubble constant in km s^{-1} Mpc^{-1}

More distant galaxies recede faster.

$\frac{1}{H}$ = approximate age of Universe

= gradient of d against v graph

Evidence for the **Big Bang** comes from Hubble's law, supported by **cosmic microwave background radiation** (CMBR) and helium production in the early Universe.

Quasars

- emit all types of EM radiation
- come from supermassive black holes in the centre of galaxies
- have very large powers.

Doppler effect

The frequency f and wavelength λ detected depend on the relative velocity v of the source/observer:

$$\frac{\Delta f}{f} = \frac{v}{c}$$

and red shift:

$$z = \frac{\Delta \lambda}{f} = -\frac{v}{c}$$

Binary stars show ± red shift due to moving towards or away in the plane of orbit.

Classifying by temperature

Everything we know about stars comes from the electromagnetic radiation that they emit.

Stefan's law	Wien's displacement law	Inverse square law
$l_{max} T = 2.9 \times 10^{-3} \text{ m K}$ T in Kelvin	$P = s A T^4$ T in Kelvin, A in m^2	$I = \dfrac{I_0}{d^2}$ I in $W\,m^{-2}$
Hotter stars appear white/blue, cooler stars appear red. 	If Star X and Star Y are at the same temperature, X will appear brighter if it has a larger area than Y: $$\frac{P_X}{P_Y} = \frac{A_X}{A_Y}\left(\frac{T_X}{T_Y}\right)^4$$	Intensity is $\frac{1}{4}$ at double the distance. Assume no light is absorbed or scattered between the source and the observer.

Supernovae, neutron stars, and black holes

Supernovae show a sudden, huge increase in magnitude over days and are billions of times brighter than the Sun. Type 1a supernovae are used as **standard candles** to work out distances, using the inverse square law.

Neutron stars are made of neutrons, are small with a huge density, similar to that of atomic nuclei.

Black holes have escape velocity $v_e > c$, which defines the event horizon; this implies a

(Schwarzschild) radius of $R_s = \dfrac{2GM}{c^2}$ They can produce gamma ray burst during formation and probably exist as supermassive black holes at the centre of galaxies.

Dark energy could be responsible for the accelerated expansion of the Universe.

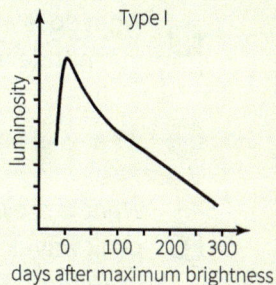

Spectral classes and Hertzsprung–Russell (HR) diagram

Stars are classified by temperature O (hot, ionised), B, A, F, G, K, M (cool).

- above 10 000 K (O-A)– hydrogenis ionised.
- below 10 000 K (A-M)– electronsmove up, then down by different steps, producing dimming = absorption lines

The HR diagram - absolute magnitude (15 to −10) vs. temperature / spectral class (from 50 000 K to 2500 K).

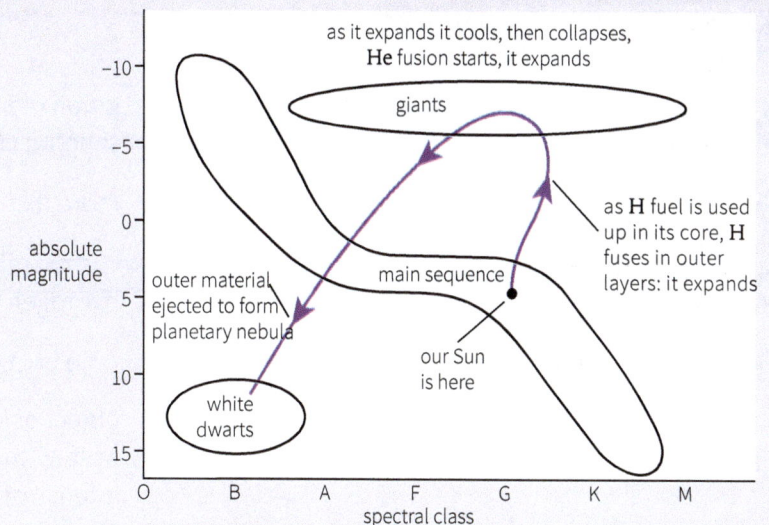

Retrieval

Learn the answers to the questions below, then cover the answers column with a piece of paper and write as many as you can. Check and repeat.

	Questions	Answers
1	What is the definition of a light year?	distance light travels in a year
2	What is an astronomical unit (AU)?	distance between the Earth and the Sun
3	What is a parsec (pc)?	distance to a star that subtends an angle of 1 second $\left(\frac{1}{3600^\circ}\right)$ at a distance of 1 AU
4	What is luminosity?	total power radiated by a star
5	What is brightness?	how bright a star appears; a subjective measurement
6	What is values are given to the faintest and brightest stars on the Hipparchus scale?	faintest = 6; brightest = 1
7	What is intensity?	power radiated per unit area at the observer
8	What does a difference in apparent magnitude of 1 imply?	apparent magnitude is the brightness seen from Earth; a difference of 1 implies an intensity ratio of 2.51.
9	What is absolute magnitude?	apparent magnitude a star would have if it was 10 pc from the observer
10	What is the relationship between the peak wavelength emitted by a star and its temperature?	peak wavelength × temperature = constant (Stefan's law)
11	How can a star at a lower temperature appear brighter?	it is closer or has a larger area
12	What is the order of spectral classes?	O, B, A, F, G, K, M
13	To which state are the Balmer transitions in the hydrogen atom?	2
14	What is the Hertzsprung–Russell (HR) diagram, and what are the scales?	graph of absolute magnitude (15 to −10) against spectral class / temperature in Kelvin (50 K to 2.5 K)
15	What did Hubble discover?	more distant galaxies were receding faster
16	What is the evidence for an expanding Universe?	cosmic microwave background; red shift; the ratio of hydrogen to helium
17	What is red shift a measure of?	ratio of change in wavelength to wavelength
18	What is a standard candle?	object of known brightness / distance that can be used with the inverse square law to work out the intensity of stars

Put paper here

19 What is a supernova, and how long does it last?

Put paper here

sudden huge increase in magnitude of a star over days

20 Why do binary stars show blue and red shift?

they move towards and away from an observer as they move around each other

Maths skills

Practise your maths skills using the worked example and practice questions below.

Wein's law and Stefan's law

Ensure you are confident using your calculator to find and use powers, exponential functions, and logarithms. This topic will give you some more practice.

We can use Wien's displacement law to estimate the surface temperature of a star.

$$\lambda_{max} T = \text{Wien's constant}$$

Wien's constant $= 2.90 \times 10^{-3}$ m K

If we know the temperature and luminosity of a star, we can estimate its area and thus its radius using Stefan's law:

$$P = A \sigma T^4$$

where P is the luminosity in W, A is the surface of the star in m², T is the surface absolute temperature in K, and σ is the Stefan constant.

$$\sigma = 5.67 \times 10^{-8} \text{ W m}^{-2} \text{ K}^{-4}$$

Worked example

Question

1 The radius of the Sun is 7.0×10^5 km and its surface temperature is 5800 K.

Calculate the luminosity of the Sun.

2 The peak wavelength of a nearby star is 450 nm.

Calculate the temperature of the star.

Answer

1 First the known quantities should be converted to SI units:

radius $= 7.0 \times 10^8$ m; area $= 4\pi r^2$
$= 6.2 \times 10^{18}$ m²

Substituting into the equation using Stefan's law:

$P = 6.2 \times 10^{18}$ m² $\times 5.67 \times 10^{-8}$ W m⁻² K⁻⁴ $\times (5800$ K$)^4$
$P = 3.98 \times 10^{26}$ W

2 Rearrange for equation for Wien's constant:

$$T = \frac{\text{Wien's constant}}{\lambda_{max}}$$

Convert the known values to SI units:

$\lambda = 450 \times 10^{-9}$ m

Substitute into the equation:

$$T = \frac{2.90 \times 10^{-3} \text{ m K}}{450 \times 10^{-9} \text{ m}}$$
$$= 6440 \text{ K}$$

Practice

1 Betelgeuse, a large reddish star in the constellation of Orion, has a peak wavelength of 970 nm.

Calculate the surface temperature of Betelgeuse.

2 A star has a surface temperature of 6000 K and a radius of 7.9×10^7 m.

Calculate the luminosity of the star.

3 A large star has a surface temperature of 25 000 K and a luminosity of 5.5×10^{32} W.

Calculate the radius of the star.

Practice

Exam-style questions

01 The constellation of Orion contains many bright stars. **Table 1** summarises the properties of some of them.

Table 1

Star	Distance / pc	Apparent magnitude	Spectral class
Betelgeuse	220	0.50	M
Rigel	260	0.13	B
Bellatrix	77	1.64	B

01.1 Describe the difference between apparent magnitude and absolute magnitude. **[1 mark]**

01.2 Deduce which star in **Table 1** is the dimmest. **[1 mark]**

01.3 Show that the absolute magnitude of Bellatrix is about −3. **[2 marks]**

01.4 The absolute magnitude of Rigel is −7.84.

Deduce the relative size of Rigel compared to Bellatrix. **[4 marks]**

Exam tip

Remember that spectral class is an indication of temperature.

02 Sirius is the brightest star as seen from Earth and has a surface temperature of 9940 K.

02.1 Calculate the peak wavelength emitted by Sirius.

Use this value to sketch the black-body radiation curve for Sirius.

[5 marks]

wavelength = _____ m

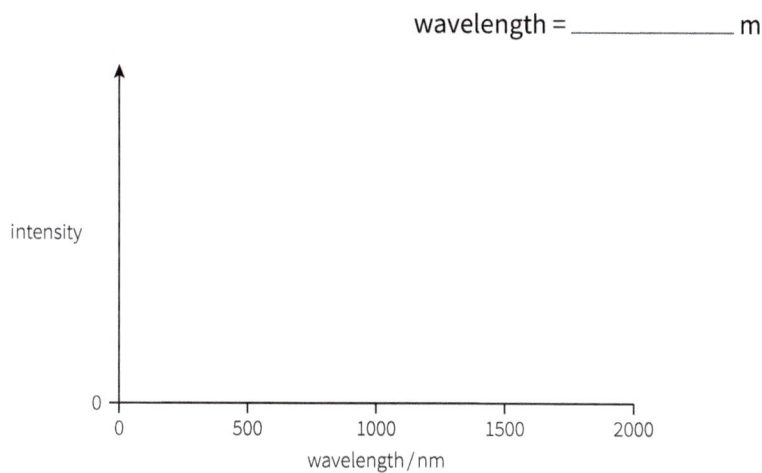

02.2 State the spectral class of Sirius. **[1 mark]**

_____

02.3 Arcturus is the fourth brightest star as seen from Earth and has a surface temperature of 4290 K.

Describe **two** ways in which the black-body radiation graph would be different for Arcturus. **[2 marks]**

_____

_____

_____

_____

_____

02.4 Use the data in **Table 2** to compare the intensity of light received on Earth from Sirius and Arcturus.

State **one** assumption that you make. [5 marks]

Table 2

Star	Sirius	Arcturus
Diameter / million km	2	35
Distance / ly	8.6	34

> **! Exam tip**
>
> You need to be aware of the assumptions behind each of the relationships you learn.

_____

_____

_____

_____

02.5 The magnitude of Sirius is −1.5, and the magnitude of the star Adhara is +1.5.

Compare the intensities of the two stars. [3 marks]

_____

_____

_____

_____

02.6 Barnard's Star has an apparent magnitude of +9.5 in visible light. Suggest whether it is visible to the naked eye. [1 mark]

_____

_____

02.7 Before astronomers used the link between intensity and brightness to produce the magnitude scale, the relative brightness was given using the Hipparchus scale.

State **one** reason why the Hipparchus scale of relative brightness is no longer used. [1 mark]

_____

_____

03 An astronomer analyses the light arriving at Earth from a star. The spectrum is continuous, but contains black lines.

03.1 The black lines form part of the Balmer series of the hydrogen spectrum. Explain how they are produced. [4 marks]

03.2 Explain which class of star produces the most prominent Balmer lines. [2 marks]

> **! Exam tip**
>
> Always use the number of marks as an indication of the level of detail that you need in an answer.

03.3 Describe the effect of the black lines on the black-body spectrum of the star. **[1 mark]**

03.4 Epsilon Eridani is a star with a spectrum that shows lines that indicate the presence of mostly neutral metals and some titanium oxide.

Suggest the spectral class of the star. **[1 mark]**

03.5 The distance to Epsilon Eridani is calculated using the parallax method. The parallax angle is 3.22 seconds. Calculate the distance to the star in light years. **[2 marks]**

! Exam tip

Remember to familiarise yourself with the data on the Data and Formula Sheet.

04 The Hertzsprung–Russell diagram shows the relationship between absolute magnitude and temperature or spectral class for stars.

Figure 1

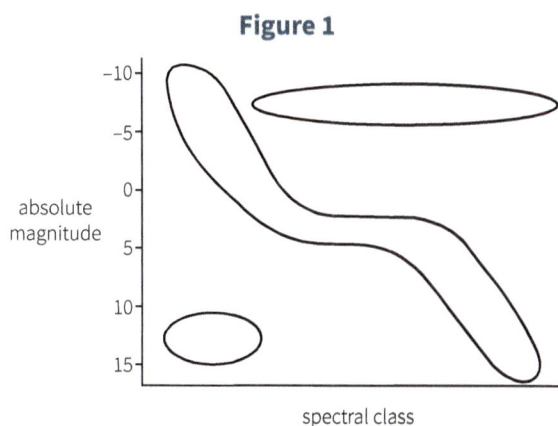

04.1 Annotate **Figure 1** with the three main types of star, and add values to the x-axis. **[3 marks]**

04.2 Draw a line with an arrow on **Figure 1** to show the approximate evolution of the Sun. **[3 marks]**

! Exam tip

You should practise drawing the Hertzsprung–Russell diagram, including labels and axes.

04.3 Describe the factors that affect how long the Sun will take to evolve in this way. **[2 marks]**

04.4 Sirius B is a white dwarf that is about the same size as the Earth, but that has about the same mass as the Sun.

Explain why the mass of Sirius B is less than it was before it became a white dwarf. **[3 marks]**

04.5 Sirius B has not emitted gamma ray bursts. Explain why. **[2 marks]**

05 The recessional velocities of four stars are shown in **Table 3**.

Table 3

Galaxy	Recessional velocity / km s⁻¹	Distance / millions of parsecs
A	2172	30.99
B	488	6.96
C	2804	40.00
D		9.57

05.1 A line in a spectrum emitted by galaxy **D** is shifted from 402.5 nm to 403.4 nm.

Calculate the recessional velocity of galaxy **D** in km s^{-1}. **[2 marks]**

05.2 Deduce a value for the age of the Universe using a graph of these data.

Explain your method. **[5 marks]**

05.3 Describe **two** other pieces of evidence that the Universe is expanding. **[2 marks]**

05.4 Before the launch of the Hubble Space Telescope (HST), the estimates of the Hubble constant were subject to errors of 50%. Measurements made by the HST mean that the accuracy is estimated as ±10%.

Suggest **one** reason why. **[3 marks]**

> **! Exam tip**
>
> Distances in cosmology are measured in parsecs, light years, or astronomical units.

05.5 The HST has imaged the most distant quasar, which has a red-shift of 7.54. This red-shift predicts an age of the Universe of 13.8 billion years.

Deduce the distance to the quasar. State the unit. **[3 marks]**

06 A supermassive black hole has a mass equal to 5 million solar masses.

06.1 Calculate the Schwarzschild radius in m of the black hole. **[2 marks]**

> **Synoptic links**
>
> 3.7.5.1   3.7.5.2

06.2 Compare the density of the black hole with the density of the Sun. Explain why the Sun will not become a black hole. **[4 marks]**

06.3 A supermassive black hole can form a quasar. A quasar in the constellation Lynx has a power of 10^{42} W. The Sun has a power of about 4×10^{26} W.

Calculate how close to the Earth in m the Sun would need to be to appear to have the same power as the quasar. **[2 marks]**

06.4 Some quasars are too distant to be imaged by single radio telescopes.

Suggest why. **[1 mark]**

> **! Exam tip**
>
> The ability of a telescope to resolve an object is related to both the wavelength of the radiation and the size of the aperture of the telescope.

06.5 Radio telescopes detect varying radio signals from spinning neutron stars, which have strong magnetic fields. The field of a neutron star is 109 T. An alpha particle is moving at $\frac{1}{2}c$ in this field.

Calculate the force in N on the alpha particle. **[2 marks]**

06.6 Explain why the force will not affect the velocity of the alpha particle. **[1 mark]**

07 A binary star system contains two stars that are a distance of 23 AU apart. The system is 4.3 ly distant and can be resolved with an optical telescope.

> **Synoptic links**
>
> 3.9.1.4   3.7.2.1
> 3.7.2.4   3.6.1.1

07.1 Calculate an estimate for the aperture of the telescope in m. Comment on your estimate, and discuss whether it is a maximum or minimum. **[5 marks]**

07.2 The stars rotate about a common centre of mass with a period of 80 years.

Calculate the mass in kg of each star. Assume that the stars have the same mass and orbit the same distance from the centre of mass. **[4 marks]**

07.3 In an eclipsing binary system, the two stars cannot be resolved, but the light curve of the stars show changes, as in **Figure 2**.

Figure 2

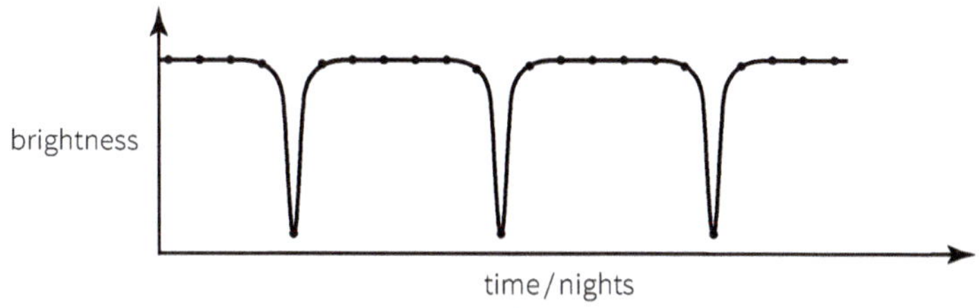

Describe the orientation of the binary system that produces this light curve. **[1 mark]**

07.4 Explain what the light curve tells you about the two stars in the binary system. **[2 marks]**

07.5 A spectral line of an element is detected from a binary star system; the line has wavelength in the laboratory of 438 nm. The line splits so that there is a difference of 0.032 nm and 0.018 nm.

Calculate the orbital velocities in m s^{-1} of each star. **[2 marks]**

07.6 Suggest and explain which star is closer to the centre of mass. **[2 marks]**

08 The Wide Angle Search for Planets (WASP), uses the transit method to search for exoplanets. The program uses two robotic observatories, each having a 2048×2048 charge-coupled device (CCD).

08.1 Sketch the light curve for a star with an orbiting exoplanet. **[2 marks]**

08.2 The CCD sits on a wafer that is 1 cm×1 cm. Compare the resolution of the CCD and the eye. **[2 marks]**

08.3 Each pixel in a CCD array is attached to a capacitor that produces a p.d. depending on the charge accumulated by the pixel. Suggest how the light can cause charge to be released, and how that charge is converted to a p.d. **[2 marks]**

08.4 Describe an alternative method for detecting exoplanets. **[2 marks]**

> **! Exam tip**
>
> You should practise drawing light curves for different types of eclipsing binary stars.

> **Synoptic links**
>
> 3.9.1.4   3.2.2.2   3.7.4.1

Knowledge

Image formation by the eye

Light is mainly refracted at the **cornea**, then adjusted by the lens to produce an image on the **retina**.

Ciliary muscles adjust the thickness of the lens. Focusing on a nearby object (<5 m), the ciliary muscles contract, giving the lens more curvature – this is called **accommodation**.

Fluids either side of the eye reduce refraction at the surface of the lens, allowing for accommodation.

Light entering through the pupil is controlled by the coloured extension of the ciliary muscle, which is the **iris**.

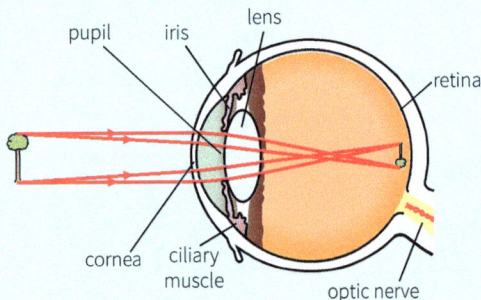

Sensitivity of the eye

Rods and cones

The retina is made up of light-sensitive (**photoreceptor**) cells called rods and cones.

Rods detect light at low intensity. There are about 120 million rods around the boundary of the retina. They give greyscale vision with little detail. Several rods connect to a single nerve fibre.

Cones detect light of high intensity. There are about 5 million cones (mostly packed densely in the **fovea**). There are three types – detecting red, blue, or green – providing detailed colour vision. A single rod connects to a single nerve fibre.

Resolution and colour

Spatial resolution is good when you look directly at an object because the fovea is densely packed with cone cells. It is less good with peripheral vision as rod cells are used.

The eye is sensitive to wavelengths between 380 nm and 760 nm.

Lenses and defects of vision

The **principal focus** is the point on the principal axis where rays cross (or where they appear to come from). When incident rays are parallel to principal axis, distance to focus = **focal length** f in metres.

Lenses have a power measured in dioptres (D):

$$\text{power} = \frac{1}{f}$$

- for converging/convex lenses, power is positive
- for diverging/concave lenses, power is negative.

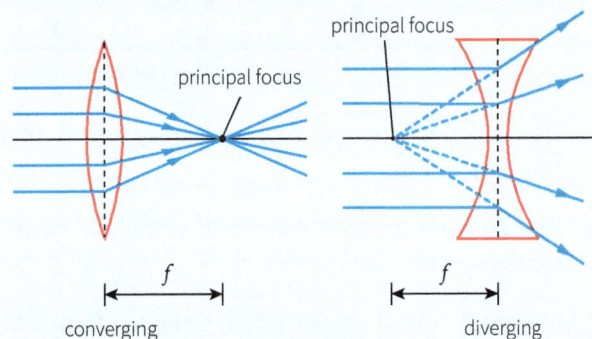

When rays are not parallel to the principal axis, v is positive, u is negative, and:

$$\frac{1}{v} = \frac{1}{u} = \frac{1}{f}$$

The diameter of the eye is 2 cm, which is the image distance when the image is in focus.

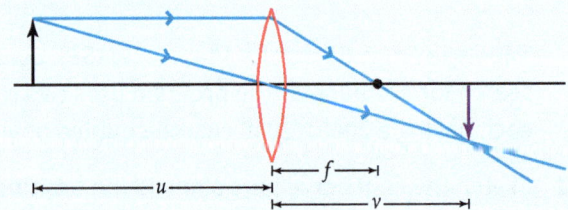

Defects in eyesight

Myopia (short sightedness): distant objects are seen out of focus because the image is formed in front of the retina; corrected with a diverging lens.

Hypermetropia (long sightedness): close objects are seen out of focus because the image is formed behind the retina; corrected with a converging lens.

Astigmatism: the cornea or lens is an oval shape, distorting light entering the eye so all objects are seen out of focus; corrected with a cylindrical lens.

Structure of the ear

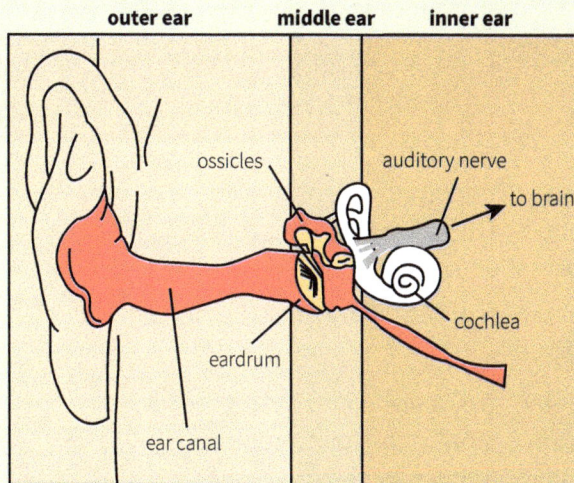

Sound waves pass through the **outer ear** into the **ear canal** causing the **eardrum** to vibrate; the sound waves transfer through the **ossicles** (three tiny bones: **malleus**, **incus**, **stapes**) to fluid in the **cochlea**. The movement of fluid makes **hair cells** in the cochlea bend, changing movement into electrical pulses which are transmitted to the **auditory nerve** which sends signals to the brain to translate into meaningful sounds.

Intensity and loudness

Intensity is the energy per second per unit area (unit $W\,m^{-2}$):

$$\text{minimum threshold intensity } I_0 = 1\,pW\,m^{-2} \text{ (at 1 kHz)}$$

The maximum intensity tolerable is $1\,W\,m^{-2}$.

Loudness scales

Perceived loudness is *not* proportional to intensity, so the scale for sound intensity is logarithmic, with $I_0 = 1\,pW\,m^{-2}$.

The dBA (decibel A) scale is adjusted to account for the sensitivity of the ear to different frequencies compared with the dB (decibel) scale, and can underestimate the perceived loudness of low frequency sounds.

$$\text{relative intensity (bels)} = \log_{10}\left(\frac{I_1}{I_0}\right)$$

for an incident intensity of I_1

$$\text{relative intensity (decibels, dB)} = 10\log_{10}\left(\frac{I_1}{I_0}\right)$$

1 dB means an increase in intensity $\times$ 1.26. Intensity $\times$ 100 means an increase of 20 dB on the scale.

Sensitivity, frequency response, and damage

Loudness curves

A loudness curve is a graph of the relative loudnesses needed for a person to hear particular frequencies. A graph of intensity against frequency is an **equal loudness curve**. Each curved line shows equal loudness in phons.

The curves show that:

- the ear is not sensitive to low frequencies, so we are not aware of physiological noises such as blood flow

- the ear is very sensitive between 2 kHz and 5 kHz

- the ear is most sensitive at a frequency of 3 kHz.

Damage to hearing

If a person suffers from hearing loss due to injury resulting from exposure to excessive noise, the equal loudness curve may be generally higher, showing that all sounds need to be louder to be perceived. This is significantly higher at 4 kHz.

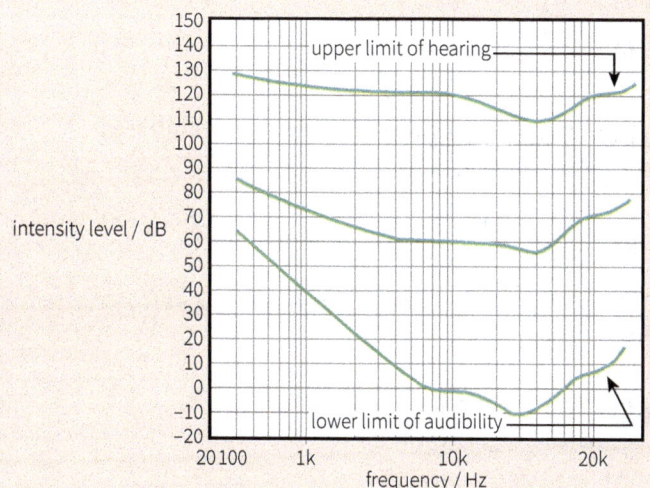

⇄ Retrieval

Learn the answers to the questions below, then cover the answers column with a piece of paper and write as many as you can. Check and repeat.

Questions	Answers
1 What is accommodation?	changing of curvature of lens due to the tension / relaxation of the ciliary muscles to focus on near and distant objects
2 What is the iris, and what does it do?	coloured part of the ciliary muscle; controls size of pupil
3 What is the difference between rod and cone cells in terms of function, location, and density?	rods detect greyscale, cones detect colour; cones are mainly in the fovea; rods are generally ~20 times as dense as cones
4 What is the power of a lens, and what are the units?	$\text{power (in dioptres, D)} = \dfrac{1}{\text{focal length (m)}}$
5 When is the distance from the lens to the focus equal to the focal length?	when the incident rays are parallel to the principal axis / the focus is on the principal axis
6 Roughly how big is the focal length of the focusing system of the eye?	2 cm
7 In terms of sign, which is negative: object distance or image distance?	object distance
8 Name one cause of myopia, and the lens that corrects it.	eyeball is too long; concave
9 Name one cause of hypermetropia, and the lens that corrects it.	eyeball is too short; convex
10 What is astigmatism, and why is there an axis in the prescription?	cornea is barrel shaped; axis shows the axis of the barrel
11 What is the minimum power detectable by the human eye?	$1\,\text{pW}\,\text{m}^{-2}$
12 Which part of the ear that is necessary for hearing is filled with fluid?	cochlea
13 What are the sound sensitive cells of the cochlea attached to?	cilia
14 Why is the perceived loudness scale logarithmic?	perceived loudness is not proportional to intensity
15 If a sound is 20 dB louder, by what factor is it more intense?	100
16 Which parts of the ear are responsible for amplifying sound?	ossicles in the middle ear

Put paper here

17	What is a bel related to?	logarithm of the ratio of the intensity of the sound to the minimum detectable power
18	What is plotted on a loudness curve?	minimum intensity detectable versus frequency
19	What is the effect of prolonged exposure to loud noise on the loudness curve?	curve is higher
20	What change happens to the loudness curve as people age?	fewer sounds are detectable at high frequencies

Put paper here

Maths skills

Practise your maths skills using the worked example and practice questions below.

Properties of converging lenses

The formula used to calculate the position of an image obtained using a lens is:

$$\frac{1}{f} = \frac{1}{u} + \frac{1}{v}$$

The formula for the power of a lens is:

$$P = \frac{1}{f} \text{ dioptres}$$

These two formulae will give you further practice in working with algebraic equations.

Worked example

Question

Some long-sighted people use reading glasses. For a pair of reading glasses, when the object is 25 cm away from the lens it forms a virtual image appearing at 0.75 m from the lens.

Calculate the power of the lens.

Answer

Convert the values to SI units:

$$u = 25\,\text{cm} = 0.25\,\text{m}$$

We use the 'real is positive' convention, so $v = -0.75\,\text{m}$

Substituting values into the equation:

$$\frac{1}{f} = \frac{1}{0.25} + -\left(\frac{1}{0.75}\right)$$

$$\frac{1}{f} = 4 - 1.33 = 2.7\,\text{D}$$

Practice

1. A short-sighted person uses a diverging lens that produces a virtual image 0.25 m from the lens when the object is 26 cm from the lens.

   Calculate the power of the lens.

2. A convex lens is placed 0.25 m from an object. The power of the lens is 2.3 D.

   Calculate the position and nature of the image formed by the lens.

3. An eye has an unaided near point of 0.65 m. A correcting lens is used that allows the eye to see clearly an object 0.25 m away from the eye.

   Calculate the power of the lens.

Exam-style questions

01 In an experiment to determine the power of a diverging lens, the lens was placed next to a converging lens of known power, as in **Figure 1**.

Figure 1

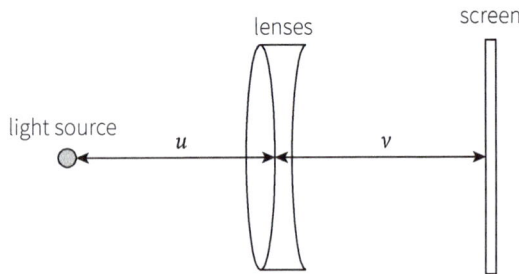

01.1 Explain which lens, converging or diverging, has the greater power.

[2 marks]

_____

_____

_____

_____

01.2 Describe how the distances u and v could be measured accurately.

[2 marks]

_____

_____

_____

_____

01.3 The results for the experiment are shown in **Figure 2**.

Determine the power of the lens combination in dioptres using **Figure 2**.

[2 marks]

_____

_____

Figure 2

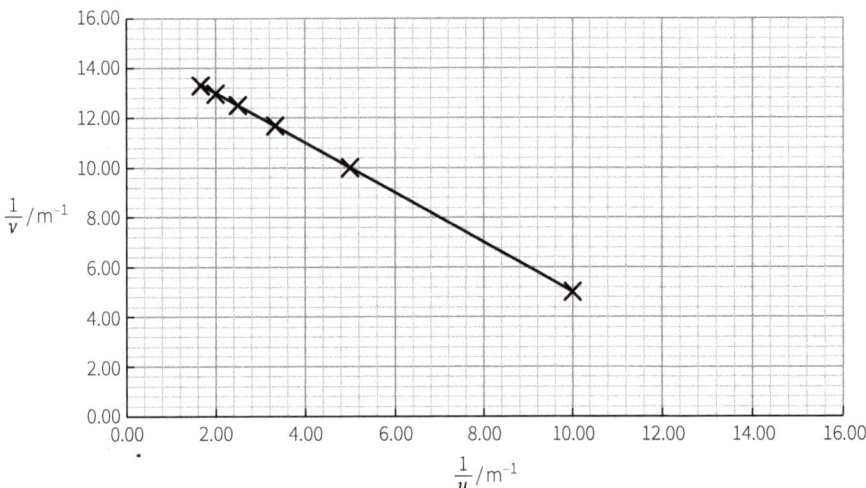

> **!** **Exam tip**
>
> Look carefully at the axes of the graph and use the equation for a straight line ($y = mx + c$) to find $\frac{1}{f}$, or the power of the lens combination.

01.4 The power of the converging lens is 20 dioptres.

Determine the power of the diverging lens in dioptres and explain whether you are confident in your answer. **[2 marks]**

_____

_____

_____

_____

02 A person suffers from myopia. They have a far point of 26 cm.

02.1 Explain why the person cannot clearly see objects placed further than 26 cm from the eye. **[2 marks]**

_____

_____

_____

_____

02.2 Calculate the lens power needed for this person to see distant objects clearly.

Assume each lens is 2.0 cm from the eye. **[3 marks]**

power = _____ doptres

02.3 Unaided, the person has a near point of 12 cm.

Calculate the new near point when they are wearing correcting glasses. **[2 marks]**

near point = _____ cm

> **! Exam tip**
>
> The near point is where the image has to form for the person to be able to see, so the near point will be v in the equation. Take care that the near point is in reference to the eye – not the glasses.

02.4 The person also has astigmatism.

Describe what causes astigmatism, the effect it has on vision, and how it is corrected. **[3 marks]**

_____

_____

_____

_____

_____

_____

03 **Figure 3** shows the distribution of cones in the retina.

Figure 3

03.1 Identify points **X** and **Y** in **Figure 3**. [2 marks]

X _____

Y _____

03.2 Sketch the distribution of the rods in the retina on **Figure 3**.
[3 marks]

03.3 Sketch on **Figure 4** the response curves for each of the three colour cones.

Annotate the curves with their corresponding colour. [3 marks]

Figure 4

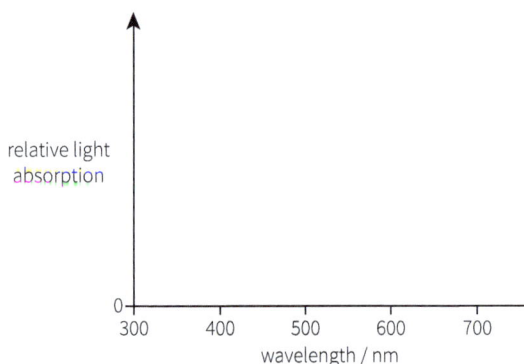

> **! Exam tip**
>
> Questions about response curves occur frequently in different formats. Make sure you know this graph and the ranges for the cones.

03.4 Explain how the resolution and appearance of an image changes when you move from bright light to dim light. [3 marks]

_____

_____

_____

_____

_____

_____

04 A student reads the board from the back of a classroom, then looks down at their book.

04.1 Describe the changes that occur in the eye as the student does this. Name this process. **[3 marks]**

_____

_____

_____

_____

_____

_____

04.2 **Figure 5** shows the cross-section of an eye of a student with hypermetropia. They are not wearing their corrective glasses.

Complete the paths of the two rays. **[3 marks]**

Figure 5

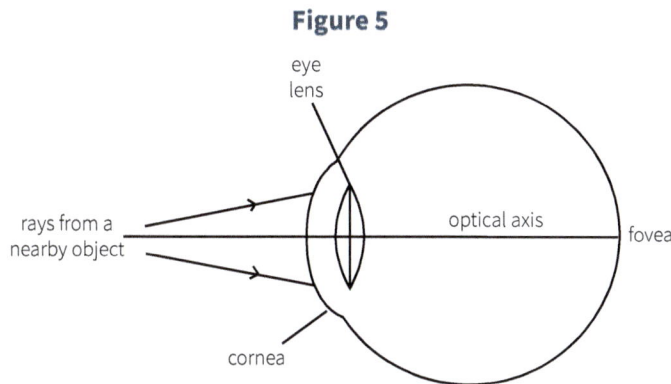

04.3 The student has a near point of 80 cm.

Determine the power in dioptres of the lens needed for them to comfortably read at a distance of 20 cm.

Assume each lens is 2 cm from the eye. **[2 marks]**

> **!** **Exam tip**
>
> Remember that signs are very important in lens questions – follow the 'real is positive' convention.

power = _____

04.4 Lenses are made of modern polymers that have a much higher refractive index than glass.

Suggest an advantage of using these polymers. **[1 mark]**

_____

_____

_____

05 A person compares the frequency response of two loudspeakers.
Speaker **A**: 230 Hz to 13.3 kHz ± 10 dB.

Speaker **B**: 55 Hz to 32 kHz ± 4 dB.

05.1 Compare the frequency range of speaker **A** and speaker **B** to the range of human hearing. **[2 marks]**

05.2 Comment on how your answer to **05.1** would change for a 50-year-old person. **[1 mark]**

05.3 For speaker **B**, the intensity level does not vary by more than 4 dB from the average.

Show that for speaker **B**, the maximum intensity changes by a factor of 2.5 from the average intensity. **[3 marks]**

05.4 Another person suggests that the maximum intensity change of speaker **A** is approximately twice that of speaker **B**.
Comment on this suggestion and explain your answer. **[2 marks]**

> **! Exam tip**
>
> If you are unsure how to proceed with a 'Show' question, you could leave it until last, then spend time playing with the equation and work backwards from the solution.

06 **Figure 6** shows a cross-section of the ear.

Figure 6

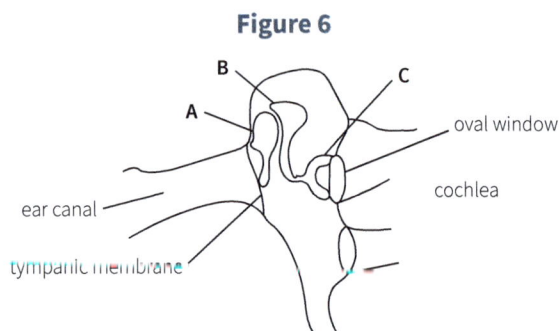

06.1 Annotate parts **A**, **B**, and **C** on **Figure 6**. **[2 marks]**

06.2 The pressure on the oval window is 20 times the pressure on the tympanic membrane.

Explain how this increase in pressure is achieved. **[3 marks]**

06.3 Describe the role of the cochlea in hearing. **[3 marks]**

06.4 An elderly person has a threshold of hearing at an intensity level of 60 dB at a particular frequency.

Calculate the intensity of the sound incident on the ear.
State the units. **[3 marks]**

> **! Exam tip**
>
> Learn these basic labels – these are easy marks to get in an exam.

07 **Figure 7** shows the equal loudness curves for different phons. The phon is a unit of loudness perception. 1 phon is equal to 1 dB at 1000 Hz.

Figure 7

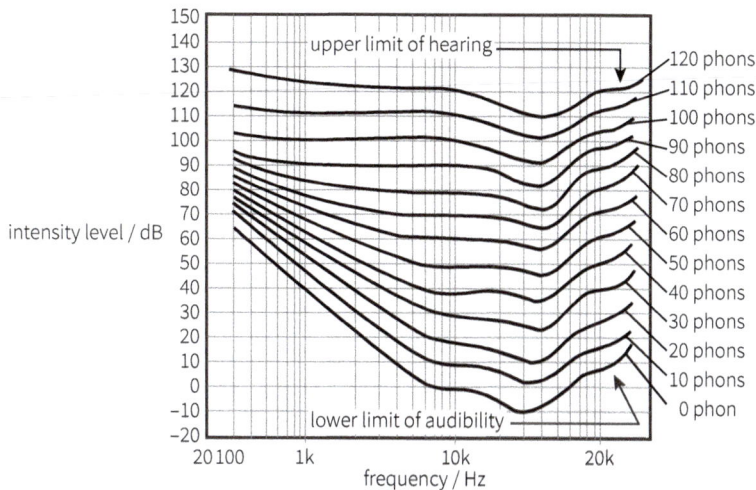

07.1 State what the 0 phon curve represents. **[1 mark]**

07.2 Determine, using information from **Figure 7**, the loudness in phons of a 500 kHz sound of 20 dB. **[1 mark]**

07.3 For human conversations, the range of frequency is usually between 250 and 4000 Hz and the intensity level is between 40 and 60 dB. Distinguish between the difficulties in hearing loss, at all frequencies, of a conversation at 40 phons with one of 60 phons. **[4 marks]**

07.4 **Figure 7** is based on normal hearing. State the feature of a graph for a person with noise-related hearing loss. **[1 mark]**

> ⚠ **Exam tip**
>
> It may help to sketch on the graph the area of the graph relating to normal conversation.

08 The Health and Safety Executive issues guidelines on noise levels at work:

'The lower exposure action values are 80 dB for daily exposure and 135 dB for peak noise.
The upper exposure action levels are 85 dB for daily exposure and 137 dB for peak noise.
The limits, which must not be exceeded, are 87 dB for daily exposure and 140 dB for peak noise.'

08.1 Suggest why the guidance has values for both daily exposure and peak noise. **[2 marks]**

08.2 An employer reading the guidance on peak noise may feel that the difference between 135 dB and 140 dB is very little. Calculate the increase in intensity level between these two values. **[2 marks]**

> ⚠ **Exam tip**
>
> Where there are multiple values mentioned, make sure you suggest reasons for both.

08.3 A chainsaw has a sound intensity is 0.1 W m^{-2} at a distance of 1 m. Calculate the sound intensity level in dB at that distance. **[2 marks]**

08.4 Tree surgeons always work in pairs. The person operating the chainsaw wears ear defenders for hearing protection.

Their colleague is working 5 m from the chainsaw.
Determine whether the colleague should also wear ear defenders. **[4 marks]**

Knowledge

24 Medical imaging

Using electrical signals

Electrodes attached to the body (using gel to reduce contact resistance) detect a changing potential difference, showing the electric activity of the heart (atria and ventricles), called an electrocardiogram. The signal is amplified, and is subject to noise.

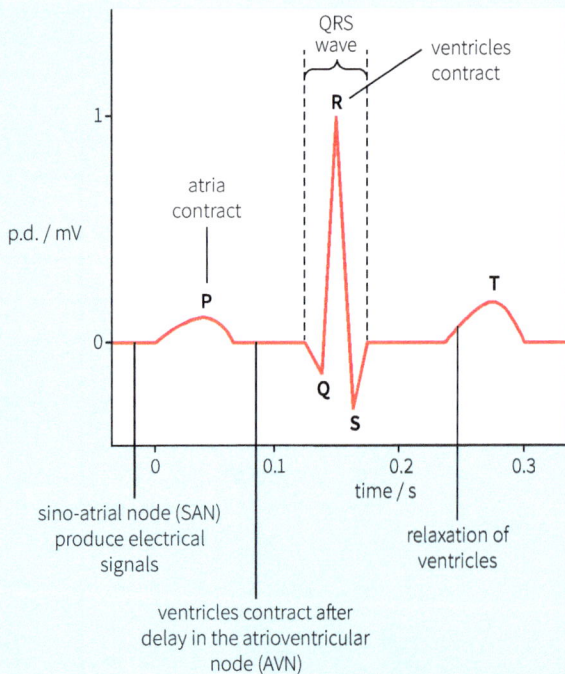

Magnetic resonance (MR) imaging

The patient is placed in a strong magnetic field of a superconducting magnet:

- spinning hydrogen nuclei (protons) precess about the magnetic field lines
- 'gradient' field coils are used to scan a cross-section
- short radio frequency (RF) pulses cause excitation and a change of spin state in successive small regions
- protons excited during the scan emit RF signals as they de-excite
- RF signals are detected and the resulting signals are processed by a computer to produce a visual image.

Ultrasound imaging

Piezoelectric effect

The piezoelectric effect is the deformation of a piezo crystal when a p.d. is applied across two metal plates, producing ultrasound.

Conversely, a p.d. is produced when ultrasound hits the crystal, so the same crystal can be used to send and receive ultrasound waves.

Scanning uses ultrasound pulses reflected at tissue boundaries:

$$\text{speed} \times \frac{\text{time for echo}}{2} = \text{distance}$$

Acoustic impedance

The acoustic impedance Z is a measure of the resistance to the transmission of sound:

$$Z = \text{density} \times \text{velocity}$$

Intensity of reflection

The intensity of reflection I_r at a boundary between media 1 and 2 depends on the impedances of media 1 (Z_1) and media 2 (Z_2):

$$\frac{I_r}{I_i} = \left(\frac{Z_2 - Z_1}{Z_2 + Z_1}\right)^2$$

where I_i = incident intensity.

A-scans and B-scans

A-scans

A-scans are amplitude scans, used when surfaces of interest lie along a particular line.

They are commonly used in ophthalmology and optometry to measure the axial length of the eye for cataract surgery.

B-scans

B-scans are brightness scans.

They are used for complex structures; for example, to produce an image of the inside of the eye when a patient has cataracts.

Fibre optics and endoscopy

An **optical fibre** is a long, thin, transparent fibre clad in material of lower refractive index. Light is **totally internally reflected** when the angle of incidence is bigger than the critical angle.

An **endoscope** comprises one bundle of fibres carrying light from a source, and another bundle carrying reflected light. Coherent fibre optic bundles are needed for imaging, and non-coherent bundles are used for illumination/surgery.

Radionuclide imaging and therapy

Most **radionuclide** imaging produces gamma rays, which are detected using a gamma camera containing photomultiplier tubes.

Gamma rays produce scintillation in crystals at the end of tubes (so their direction can be determined). Light photons eject electrons from photocathodes, which are accelerated, and further collisions amplify the signals.

Molybdenum-technetium generator

Molybdenum ($t_{1/2}$ = 66 hours) decays to technetium-99$_m$.

This can be easily transported, is used in a wide range of radiopharmaceuticals, and is cheap.

PET scans

A labelled radionuclide is injected and absorbed by an organ; this produces positrons, which are annihilated by electrons to produce two gamma rays travelling in opposite directions. The gamma rays are detected and used to produce an image using time differences in detection.

Injected radionuclides decay in the body with an effective half-life T_E:

$$\frac{1}{T_E} = \frac{1}{T_B} + \frac{1}{T_P}$$

The physical (radioactive) half-life T_P is due to radioactive decay, and the biological half-life T_B is due to excretion.

Beta-emitting implants

These can be placed close to cancer cells, limiting the dose to healthy cells.

Use of high-energy X-rays in therapy

Tumours deep in the body can be treated with a fine collimated beam of high-energy X-rays after scans pinpoint their location. The correct energy and multiple beams are needed to limit exposure to healthy cells.

Radioisotope	Radiation	$t_{1/2}$	Energy / keV	Uses
technetium-99$_m$	gamma	6 hours	140	gamma camera, labelling / targeting organs
iodine-131	beta and gamma	8 days	364	taken up by thyroid, diagnosis and therapy
indium-111	gamma	2.8 days	245	labelling antibodies and blood cells

Producing diagnostic X-rays

Electrons emitted from a heated cathode are accelerated / decelerated when they hit the target anode – p.d.s of 20 kV to 120 kV produce:

- a spectrum that shows intensity against photon energy
- a smooth spectrum (from deceleration)
- specific X-rays due to transitions between inner energy levels (from knocking out inner electrons).

A rotating anode X-ray tube spins at 3000 revolutions per minute (rpm). Rotation reduces the chance of the anode melting while the p.d. can still be varied.

The patient dose is controlled with:

- filters that reduce beam intensity, increase penetrating power, and eliminate radiation at low energies; they need to be made of material that allows the **photoelectric effect** to dominate, as effect proportional to Z^3, such as copper
- a diaphragm, to limit the extent of the X-ray beam.

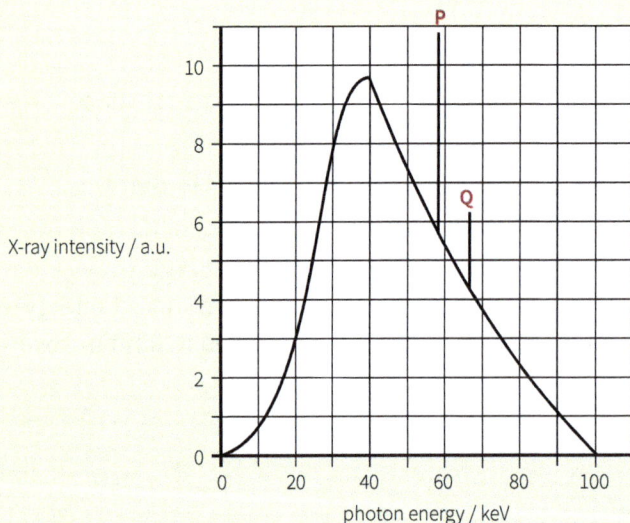

X-ray attenuation

X-rays are attenuated depending on the density, thickness, and atomic number of the medium.

The effect is exponential: $I = I_0\,e^{-\mu x}$

The linear coefficient μ is like λ for radioactive decay:

$$\text{half-thickness } x_{1/2} = \frac{\ln 2}{\mu}$$

Attenuation depends on density, which is useful in imaging, so the mass linear coefficient $\mu_m = \dfrac{\mu}{\rho}$.

Two materials with similar atomic numbers and thicknesses but different densities will attenuate differently, so they can be distinguished.

Computerised tomography (CT) scanners

CT scans are produced by a collimated (narrow), monochromatic X-ray beam being rotated around an object.

An array of detectors is located in the scanner gantry and a computer is used to process the signals and produce a visual image.

X-ray image detection and enhancement

Interaction of X-rays is by chance – material sensitive to X-rays needs to be used to detect them (originally photographic plates).

Contrast enhancement uses a contrast medium (for example, a barium meal) to increase the difference between the proton number of the tissue being imaged and the surrounding tissue.

In scattered radiation, a grid of lead strips is used to reduce the amount of scattered radiation reaching the film.

Flat-panel detectors:

These are more sensitive, faster, and more easily transportable than photographic film. They contain:

* X-ray scintillator (converts X-rays to visible photons)
* photodiode pixels that convert visible photons to p.d. or charge, which is read by electronic scanning.

Photographic detection:

* uses intensifying screen
* is made of fluorescent material that absorbs X-rays and emits visible light
* reduces the intensity of X-rays needed (the dose) to patient.

Advantages and disadvantages of imaging methods

Method	Advantages	Disadvantages
ultrasound	• non-invasive • no side effects	• resolution limited to 1 mm ($f \approx 3\,\text{mHz}$)
endoscopes	• can be non-invasive • can be combined with tools • live visible image	• only a small area can be imaged
MR imaging	• non-invasive • no ionising radiation • no harmful effects • can show 3D and cross-sectional images • gives a better image than CT scan with soft tissue	• cannot use if the patient has metallic implants such as pacemakers • patient has to be still for a long time • bone and calcium not visible • more expensive than any other scan
X-ray imaging and CT scans	• good images of bone fractures and organ calcification, brain / abdominal organs • better resolution than ultrasound • gives full cross-sectional image • non-invasive	• highly ionising • limited contrast between tissues of similar density; images of brain can be distorted by nearby bone • more expensive than a conventional X-ray picture • often requires patients to hold their breath
radionuclide	• good images of organs, abnormalities • non-invasive	• highly ionising

Retrieval

Learn the answers to the questions below, then cover the answers column with a piece of paper and write as many as you can. Check and repeat.

	Questions		Answers
1	What does an electrocardiogram detect?		changing potential difference
2	What is acoustic impedance?		measure of the resistance to transmission of a sound wave through a medium
3	What is total internal reflection?		optical phenomenon occurring when waves are totally reflected because the angle of incidence at a boundary is larger than the critical angle
4	What is the difference between an ultrasound A-scan and an ultrasound B-scan?		A-scan shows surfaces along a single line; B-scan produces an image by measuring reflections in many directions
5	For what purpose would you need the optical fibres in an endoscope to be coherent?		produce images of internal structures of the body
6	When are radio frequency signals produced in a magnetic resonance scan?		when protons that have been excited due to short RF bursts de-excite
7	In a gamma camera, how is the signal from a single gamma ray detected and amplified?		a photon is produced by a scintillator, which is amplified by a photomultiplier tube using photocathodes
8	How is technetium-99 produced in a hospital?		using a molybdenum–technetium generator
9	How do gamma ray photons come to be emitted in a positron emission tomography scan?		a radionuclide that emits positrons is taken up by tissue; when the positrons annihilate with electrons gamma rays are produced
10	What is effective half-life?		the half-life that results from excretion (biological) half-life, and physical (radioactive) half-life
11	How is the damage to healthy tissue in the treatment of cancer with X-rays limited?		X-ray beams are accurately targeted to cancerous tissue, to reduce exposure to healthy tissue
12	How are X-rays produced?		electrons emitted from a heated cathode are accelerated and hit a target; deceleration produces a spectrum of X-rays
13	Why do some X-ray tubes use a rotating anode?		reduce the chance of the anode melting
14	What do filters in an X-ray tube do?		reduce the intensity of the beam and eliminate low energy radiation
15	In terms of X-rays, what is attenuation?		decrease in intensity as the X-rays move through a material, like tissues and organs of the human body
16	What is the linear coefficient in the attenuation of X-rays m analogous to in radioactivity?		probability of decay per second λ

Put paper here

17	What does a contrast medium do in the context of X-ray imaging?	increases difference in proton number of tissue being imaged compared to surrounding tissue
18	What does a flat panel detector contain?	X-ray scintillator (produce visible photons), and photodiodes (produce a p.d. from the photons)
19	Name methods of medical imaging that involve non-ionising radiation.	ultrasound, endoscopes, magnetic resonance imaging
20	Which is the most expensive medical imaging process?	magnetic resonance (MR) imaging

Put paper here (×3)

🖩 Maths skills

Practise your maths skills using the worked example and practice questions below.

Choosing suitable radioactive materials	Worked example	Practice

Choosing suitable radioactive materials

Radioactive materials are used as tracers in medical imaging. The choice of radioactive material will depend on the medical procedure that is used.

The radioactive material should:

- contain an isotope with a half-life of a few minutes or hours
- be non-toxic to the patient
- be utilised by the organ or body process under observation.

Worked example

Question

Iodine-123 is a radioactive isotope of iodine used to image the thyroid gland. ^{123}I decays by electron capture to form tellurium-123. A gamma camera is used to detect this decay. The half-life of ^{123}I is approximately 13.13 hours.

1 Calculate the decay constant for ^{123}I.

2 A patient is given a tablet containing ^{123}I. Its initial activity is 14.8 MBq. The test lasts for 24 hours from the time the tablet is taken.

Calculate the activity of the radioactive material at the end of the test.

Answer

1 $T_{1/2} = \dfrac{\ln 2}{\lambda}$

so $\lambda = \dfrac{\ln 2}{13.13} = 0.053\,\text{h}^{-1}$

Note: The decay constant is usually calculated in s^{-1}. However, the next part of the question uses hours, so it is easier to keep the decay constant as h^{-1}.

2 Substitute the known quantities:

$A = A_0\,e^{-\lambda t}$

$A = 14.8 \times 10^6\,\text{Bq} \times e^{-0.053\,\text{h} \times 24\text{h}}$

$A = 4.15 \times 10^6\,\text{Bq}$

Practice

The radioactive isotope oxygen-15 is used to observe blood flow in the brain during PET scans. ^{15}O decays to nitrogen-15 by β^+ decay, with a half-life of 2.03 minutes.

1 Suggest why ^{15}O is a suitable isotope to observe blood flow in the brain.

2 Write the nuclear decay equation for ^{15}O.

Calculate the decay constant for ^{15}O.

3 A patient information leaflet for ^{123}I imaging includes the following paragraph.

"After receiving radioactive iodine treatment you should sleep in a separate bed from other people for 11 days. You should also limit the time you spend in public places for three days."

Suggest why these precautions should be taken and why those particular time periods were chosen. Refer to the worked example for data to use in calculations of ^{123}I activity to support your answer.

01   **Figure 1** shows an electrocardiogram (ECG) trace for a healthy person.

Figure 1

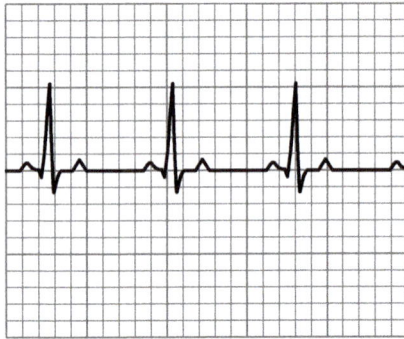

01.1   The width of each small square on the x-axis is equal to 0.1 s.
Measure the period of one heartbeat and determine the pulse rate per minute of this individual.   **[2 marks]**

pulse rate = _____ beats per minute

01.2   Annotate the trace with a suitable scale for the y-axis.   **[2 marks]**

01.3   Explain the characteristic shape of the waveform.
You may sketch a diagram to help your explanation.   **[4 marks]**

> **! Exam tip**
>
> A labelled diagram may help you order your thoughts – it is important that you explain the steps in order for **01.3**.

_____

_____

_____

_____

_____

01.4   Describe **two** changes that would occur in the waveform if the patient began to exercise.   **[2 marks]**

_____

_____

_____

01.5   Suggest **two** ways of ensuring good electrical contact between the ECG electrodes and the skin.   **[2 marks]**

_____

_____

_____

02 **Figure 2** shows a transducer used in an ultrasound A-scan.

Figure 2

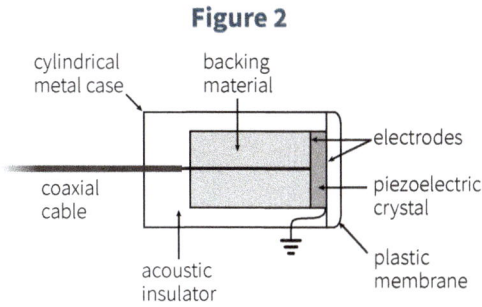

cylindrical metal case
backing material
electrodes
coaxial cable
piezoelectric crystal
acoustic insulator
plastic membrane

Synoptic link

3.3.1.1

02.1 Describe how the transducer produces ultrasound pulses and detects reflected signals in an A-scan. **[4 marks]**

_____

_____

_____

_____

_____

02.2 **Figure 3** shows the A-scan of a leg bone.

Figure 3

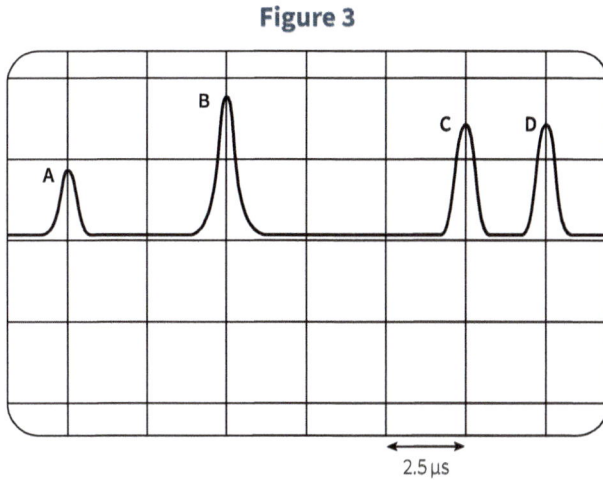

2.5 µs

Explain why the strength of the signal varies over time. **[2 marks]**

_____

_____

_____

02.3 The oscilloscope time-base setting is set at 2.5 µs per division. The velocity of sound in bone is 4080 ms⁻¹.

Determine the thickness of the bone. **[3 marks]**

thickness = _____ m

02.4 An ultrasound wave has a frequency of 8.0 MHz.
Calculate its wavelength when traveling through bone. **[1 mark]**

wavelength = _____ m

02.5 Explain how the resolution of the image is related to the wavelength of the wave. **[2 marks]**

_____

_____

_____

_____

Exam tip
Start by defining resolution and then consider how wavelength could affect it.

03 An X-ray beam is incident on a patient's arm.

03.1 X-rays cannot be focused.
Describe how the sharpness of the image can be improved. **[2 marks]**

_____

_____

_____

_____

03.2 The X-ray beam is passed through an aluminium filter.
Explain why the filter is used. **[2 marks]**

_____

_____

_____

_____

03.3 The beam travels through 3 cm of muscle, 1.5 cm of bone, and then 2 cm of muscle.

Sketch a graph, using the axes in **Figure 4**, to show how the intensity of the radiation changes with distance through the arm.
Annotate your graph. **[2 marks]**

Exam tip
How might the attenuation of bone and muscle compare? How can you show that clearly on a graph?

Figure 4

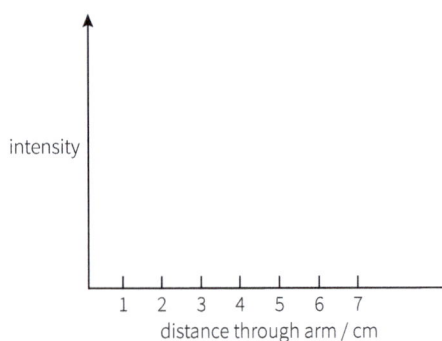

03.4 X-rays of bones show good contrast.

Explain why an X-ray of the heart and surrounding blood vessels is much more difficult to analyse and how this can be overcome.

[3 marks]

_____

_____

_____

_____

_____

_____

04 An optical fibre has a core refractive index of 1.55 and a cladding refractive index of 1.43.

04.1 Calculate the critical angle in degrees for the boundary between the two types of glass in the optical fibre. [2 marks]

Synoptic link

3.3.2.3

boundary = _____°

04.2 The optical fibre can be bent when it is in use.

Describe how this affects the probability of light escaping from the fibre. [2 marks]

_____

_____

_____

_____

04.3 An endoscope contains two types of bundle, coherent and non-coherent.

Compare the two types of bundle. [3 marks]

Exam tip

Point out the differences in each type of bundle; don't just list properties of each.

_____

_____

_____

04.4 Describe how laser light might be used in an endoscope.
Suggest **one** advantage of using laser light in medical procedures.

[2 marks]

_____

_____

05 Different types of scanners are available in hospitals.

05.1 Outline the principles of a magnetic resonance (MR) scanner.

[4 marks]

05.2 Discuss the advantages and disadvantages of computerised tomography (CT) scans and MR scans for use on patients. **[6 marks]**

Exam tip

Learn the basic principles behind each of the different types of medical scanner.

06 **Figure 5** shows the structure of a gamma camera.

An array of gamma cameras is used in a positron emission tomography (PET) scanner to produce three dimensional images.

Synoptic link

3.1.2.3

Figure 5

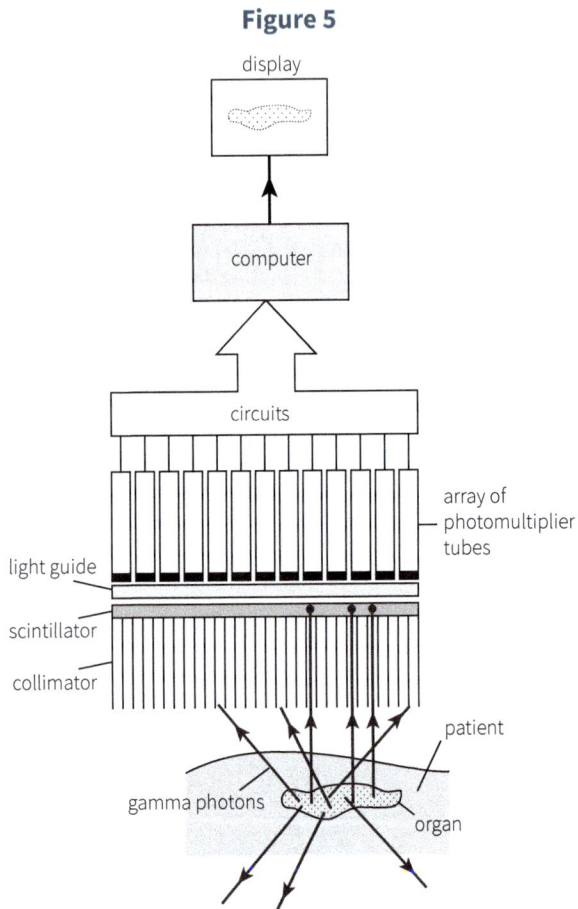

06.1 Explain the functions of the collimator, scintillator, and photomultiplier tubes in the gamma camera. **[3 marks]**

06.2 The radionuclide used in PET scans emits positrons when it decays. Explain how this results in the emission of two photons with equal and opposite momentum from inside the body. **[2 marks]**

06.3 Calculate the minimum energy of each photon in MeV. **[3 marks]**

06.4 Explain how the detection of these gamma photons allows an image to be created. **[2 marks]**

07 This question is about the use of radionuclides in medical physics. One widely used radionuclide is technetium-99m.

It has a biological half-life of 86 400 s and a physical half-life of 21 700 s.

07.1 State the meaning of biological half-life. [1 mark]

07.2 Calculate the effective half-life of technetium-99m in hours. [2 marks]

07.3 Suggest how long after an injection of technetium-99m a patient should have to wait before being scanned. Explain your reasoning. [3 marks]

07.4 Technetium-99m is a gamma source.

Describe and explain the properties of a radionuclide used in a medical implant in a tumour. [2 marks]

Synoptic link

3.1.2.3

Exam tip

Why do you think the word 'effective' is important in **07.2**?

08 **Figure 6** shows the components of an X-ray tube.

Figure 6

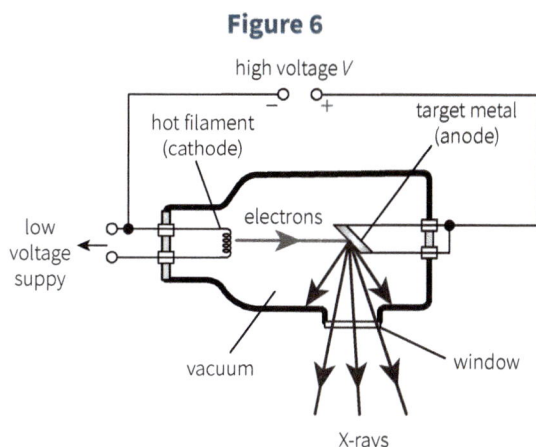

high voltage V

hot filament (cathode)

target metal (anode)

low voltage suppy

electrons

vacuum

window

X-rays

Synoptic links

3.2.1.3

08.1 State the name of the process by which electrons are supplied. [1 mark]

08.2 Explain why the X-ray tube is evacuated, and the target metal continually rotates. [2 marks]

08.3 To produce high energy electrons for radiotherapy, the electrons are accelerated by a voltage of 180 kV.

Calculate the shortest possible wavelength in m of X-rays produced. [2 marks]

08.4 A patient needs 1.2 J of energy to treat a tumour. The machine uses a current of 25 mA to produce the beam.

The efficiency of photon production is 1% and only 4% of the photons produced are used on the patient.

Calculate the time in s that the beam needs to be switched on for. [4 marks]

Exam tip

How can you use the current to determine the number of electrons hitting the surface per second? Then consider how many photons are actually needed to treat the tumour.

⚙ Knowledge

25 Rotational mechanics

Rotational motion

Rotational mechanics is analogous to linear mechanics.

- Moment of inertia I in kg m^2 behaves like mass.
- Torque T in N m behaves like force.
- Angular momentum L in kg m^2 rad s^{-2} behaves like linear momentum.

Properties	Linear mechanics	Rotational mechanics
Inertia	m	$I = \Sigma m r^2$
Force / torque	F	$T = F r$
Momentum	$p = m v$	$L = I \omega$
Newton's second law	$F = m a$ $F \Delta t = \Delta(m v)$	$T = I a$ $T = \Delta t = \Delta(I \omega)$

Moment of inertia

Inertia is the resistance to force or the tendency to continue linear motion; mass is the measure of inertia.

The **moment of inertia** (MoI) is the resistance to **torque** or the tendency to continue rotational motion, which depends on how mass is distributed.

For an isolated mass m at a radius r from the axis of rotation: $I = m r^2$

For an extended body: $I = \Sigma m r^2$

Adding mass m at distance r to an object with existing I means $I_{\text{new}} = I + m r^2$.

Torque and angular acceleration

A torque (also called turning force or moment; unit N m) is produced when a force F acts at a distance r from a pivot:

$$T = F r$$

Torques produce angular accelerations:

$$T = I a$$

In experiments to find the effect of torque on angular acceleration, there is always a frictional torque acting at the bearing. Frictional torque can be calculated by allowing a spinning wheel to decelerate to a stop.

Rotational kinematics

Rotational motion is analogous to linear motion:

- **angular displacement** θ in radians behaves like x, where $360° = 2\pi$ radians, $x = r \theta$

x, s	θ

- **angular velocity** ω in rad s^{-1} behaves like v, and $v = \omega r$, $1 \text{ rpm} = \dfrac{2\pi}{60}$ rad s^{-1}

$v = \dfrac{\Delta x}{\Delta t}$	$\omega = \dfrac{\Delta \theta}{\Delta t}$

- **angular acceleration** α in rad s^{-2} behaves like a, and $a = \alpha r$

$a = \dfrac{\Delta v}{\Delta t}$	$\alpha = \dfrac{\Delta \omega}{\Delta t}$

- **equations of motion** follow the same format.

$v = u + a t$ $s = u t + \frac{1}{2} a t^2$ $v^2 = u^2 + 2 a s$ $s = \frac{1}{2}(u + v) t$	$\omega_2 = \omega_1 + \alpha t$ $\theta = \omega_1 t + \frac{1}{2} \alpha t^2$ $\omega_2^2 = \omega_1^2 + 2 \alpha \theta$ $\theta = \frac{1}{2}(\omega_1 + \omega_2) t$

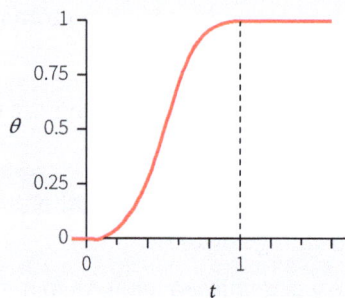

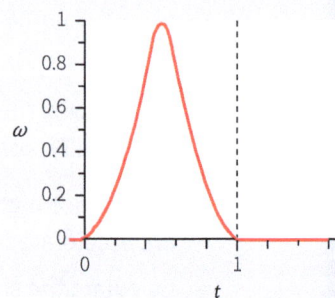

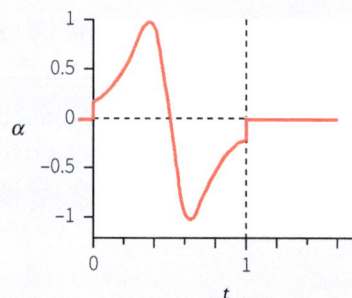

Rotational kinetic energy, work, and power

High torques / large amounts of work are required for some theme park rides, swing bridges, cranes, and so on.

Work done can be found from the area under a graph of torque against angular displacement.

Frictional torque must be considered in rotating machinery, as it reduces efficiency.

Some **power** is expended in opposing frictional torque.

Mechanics	linear	rotational
Kinetic energy	$E_k = \frac{1}{2}mv^2$	$E_k = \frac{1}{2}I\omega^2$
Work done	$W = Fd$	$W = T\theta$
Power	$P = Fv$	$P = T\omega$

Angular momentum and impulse

angular momentum $(N\,m^2\,s^{-1}) = I\omega$

Conservation of angular momentum
Angular momentum is only conserved if no external torque (not force) acts. For example:

- in sport, divers tuck themselves in, while skaters, dancers, and trampolinists move their arms in and out
- in clutches (in cars), shafts of differing MoIs (I_1, I_2) are brought together to rotate at the same speed ω = rotational dynamics collision:

$$I_1\omega_1 + I_2\omega_2 = (I_1 + I_2)\omega$$

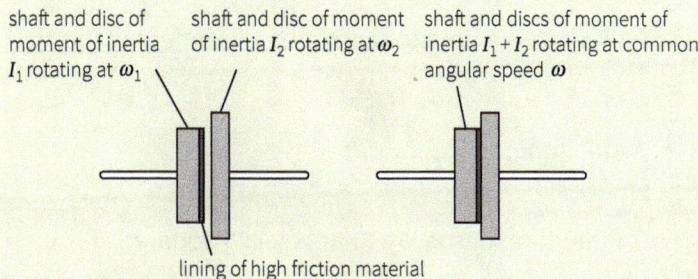

shaft and disc of moment of inertia I_1 rotating at ω_1 shaft and disc of moment of inertia I_2 rotating at ω_2 shaft and discs of moment of inertia $I_1 + I_2$ rotating at common angular speed ω

lining of high friction material

- angular impulse = $T\,\Delta t$, where T is constant
- angular impulse = change in angular momentum, $\Delta(I\omega)$, where T is constant
- angular impulse can be found from the area under a graph of torque against time.

Flywheels

Flywheels are used to 'store' energy.

$$E_k\,(J) = \frac{1}{2}I\omega^2$$

- The greater the MoI, the greater the speed, so the more energy stored.
- The maximum speed depends on the braking stress of the material of the flywheel.
- Examples include push-and-go cars, experimental vehicles with 'flywheel batteries' (charged by motor), regenerative braking / kinetic energy recovery systems (energy in braking is transferred to the flywheel to be used later).

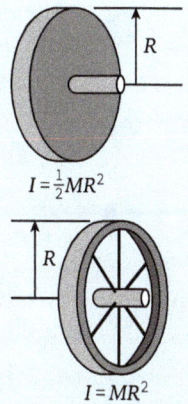

$I = \frac{1}{2}MR^2$

$I = MR^2$

Flywheels are used to smooth out fluctuations in rotational speed.

- They are fitted to the crankshaft of internal combustion engines / reciprocating steam engines.
- The torque provided by the engine varies with pressure / crank angle.
- The flywheel takes the engine over the 'dead centres' when there is no turning effect.
- The greater the MoI of the flywheel, the smaller the fluctuation in speed.
- A greater MoI can be produced without a substantial increase of weight using spoked wheels.

Difference in torque	Effect on flywheel
engine > load	accelerates
load > engine	decelerates

Flywheels are used in machine tools when a large work output is needed over a short time period – an electric motor brings the flywheel up to speed, then the flywheel stops suddenly to do work.

Learn the answers to the questions below, then cover the answers column with a piece of paper and write as many as you can. Check and repeat.

Questions	Answers
1 Which quantity in linear mechanics is analogous to moment of inertia in rotational mechanics?	mass
2 In what units do you measure angle, angular velocity, and angular acceleration?	radians, rad s^{-1}, rad s^{-2}
3 Which quantity in rotational mechanics is analogous to force in linear mechanics?	torque
4 How do you convert angular speed from rpm to radians per second?	$\text{rpm} \times \dfrac{2\pi}{60}$
5 How do you find the linear acceleration from angular acceleration?	$a = \alpha r$
6 How do you determine angular acceleration from a graph of angular velocity against time?	it is the gradient
7 Which has the biggest moment of inertia: a rod rotated about its centre, or the same rod rotated about one end?	same rod rotated about one end
8 In an experiment to measure the moment of inertia you use a torque to produce what?	angular acceleration
9 The area under which graph can be used to find the angular impulse?	torque against time
10 Under what condition is angular momentum conserved?	no external torque acts on the system
11 A spinning wheel comes to a halt because of friction. What needs to be determined to calculate the frictional torque?	initial speed; time to reach a halt; moment of inertia
12 What can be calculated from the area under a torque against angular displacement graph?	work done
13 What does rotational kinetic energy depend on?	moment of inertia, angular velocity squared
14 Why are flywheels usually not solid discs, but instead have spokes and a rim?	spoke and rim wheel has a larger moment of inertia for the mass, hence stores more energy
15 What does KERS stand for?	kinetic energy recovery system
16 Why are flywheels fitted in internal combustion engines?	smooth out fluctuations in the rotational speed of the crankshaft

(Put paper here)

17	Why are flywheels used in machine tools?	when the wheel stops over a short time it provides a large work output
18	What determines the maximum speed of a flywheel?	breaking stress of the material of the wheel
19	What are the units of angular momentum?	$N\,m^2\,s^{-1}$ (= $kg\,m^2\,rad\,s^{-1}$)
20	Why do skaters spin faster when they pull their arms in?	as angular momentum is conserved, pulling in their arms reduces their moment of inertia so their angular speed increases

Put paper here

🖩 Maths skills

Practise your maths skills using the worked example and practice questions below.

Rotational motion

The angular displacement θ of an object in radians is:

$$\theta = \frac{s}{r}$$

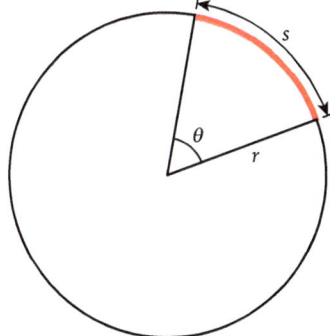

$$360° = 2\pi$$

The angular velocity ω is given by:

$$\omega = \frac{\text{angular displacement}}{\text{time}}$$
$$= \frac{\Delta\theta}{\Delta t}$$

Worked example

Question

A compact disc (CD) spins at 500 revolutions per minute.

Calculate the angular velocity of the CD.

Answer

First, we calculate the angular displacement of the CD. Each revolution of the CD is 2π radians; therefore, in 1 minute, the CD will rotate through $500 \times 2\pi$ rad:

$$\Delta\theta = 1000\pi\ \text{rad}$$

Next, convert the time into SI units:

$$\Delta t = 60\ \text{s}$$

Substituting these values into the equation for angular velocity:

$$\omega = \frac{\Delta\theta}{\Delta t} = \frac{1000\pi}{60}$$

$$\omega = 52\ \text{rad s}^{-1}$$

Practice

1. Calculate the angular velocity of the hour hand of an analogue clock.

2. The blades in a food blender rotate at 23 000 revolutions per minute.

   Calculate the angular velocity of the blades.

3. A car travels 200 m in 20 s. The tyres are 50 cm in diameter. Calculate the angular velocity of the tyres.

Practice

Exam-style questions

01 A grinding wheel is rotating at 100 revolutions per minute when it is switched off. When a piece of wood is held against the wheel, the angular deceleration of the wheel is 2.00 rad s^{-2}.

01.1 Calculate the time it takes for the wheel to stop. **[2 marks]**

time = _____ s

01.2 Describe how to find the force exerted by the wood on the wheel. **[2 marks]**

_____

_____

_____

> **! Exam tip**
> Remember that there are 2π radians in each revolution.

01.3 Calculate the number of revolutions that the wheel makes while it is slowing down to a stop. **[3 marks]**

revolutions = _____

01.4 The wheel is turned on again and reaches the same angular speed. When the wheel is turned off this time, the force applied is steadily increased.

Sketch a graph of angular speed against time for the wheel. **[2 marks]**

01.5 Explain **one** difference between the graph in **01.4** and a graph for the situation in **01.1**. **[2 marks]**

_____

_____

_____

02 An experiment to model a method to find the moment of inertia of satellites that are put into orbit is shown in **Figure 1**. The mass m is attached to a piece of string. The radius R of the central part of the cylinder around which the string is wound is 3.2 cm.

Figure 1

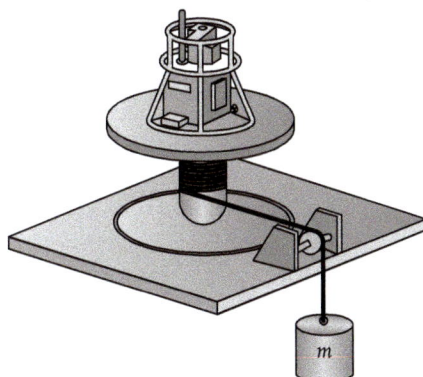

02.1 Suggest why it might be necessary to know the moment of inertia of a satellite. **[1 mark]**

_____

_____

02.2 The mass is held at different distances D above the floor. The time it takes the mass to hit the floor t is recorded in **Table 1**.

Table 1

D / m	t / s	
0.2	0.55	
0.4	0.82	
0.6	0.99	
0.8	1.18	
1.0	1.33	

> **! Exam tip**
>
> Remember that there are linear equations of motion that you can use as well.

Complete **Table 1**, and draw a linear graph using this data so that the acceleration of the mass can be deduced from the gradient.

Calculate the acceleration using the graph. **[6 marks]**

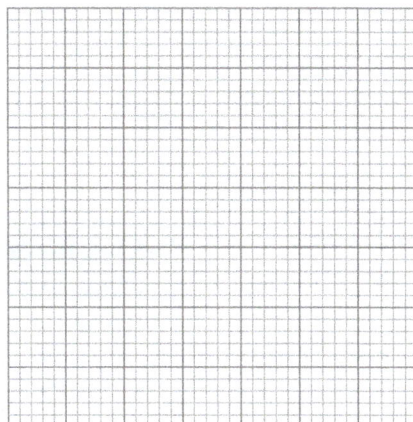

acceleration = _____ m s^{-2}

02.3 The relationship between the moment of inertia I of the object and the acceleration of the mass m is:

$$I = m R^2 \left(\frac{g}{a} - 1\right)$$

Calculate the moment of inertia of the satellite. The mass m is 100 g.

[2 marks]

moment of inertia = _____ kg m^2

02.4 There is an error in the calculation: the actual radius is larger than that of the cylinder because of the string that is wound around it.

Suggest **one** other source of error in this experiment, and explain its effect on the value of the moment of inertia. **[2 marks]**

_____

_____

_____

_____

_____

_____

03 A single piston engine contains a flywheel that is connected to a rotating crankshaft.

03.1 Suggest **one** reason why the flywheel is used. **[1 mark]**

03.2 The flywheel engages with a second disc that forms part of the clutch.

State the conditions under which angular momentum is conserved in a collision. **[1 mark]**

03.3 The flywheel is spinning with an angular velocity of 530 rad s^{-1} and has a moment of inertia of 0.21 kg m^2. The clutch plate is spinning with an angular velocity of 240 rad s^{-1} and has a moment of inertia of 0.15 kg m^2. The two wheels then come into contact and lock together.

Calculate the final angular velocity in rad s^{-1} of the combined system. **[2 marks]**

03.4 The final angular velocity is 10% less than that calculated in **03.3**. When the wheels are brought into contact, there is a frictional torque that slows the wheels.

Calculate the impulse exerted on the flywheel and state the unit. State **one** assumption that you have made. **[4 marks]**

> **! Exam tip**
>
> You should learn the main reasons for using a flywheel and the situations in which they are used.

04 The head of a grass strimmer has 80 g of plastic line wound around a spool inside a line head, as shown in **Figure 2**.

Figure 2

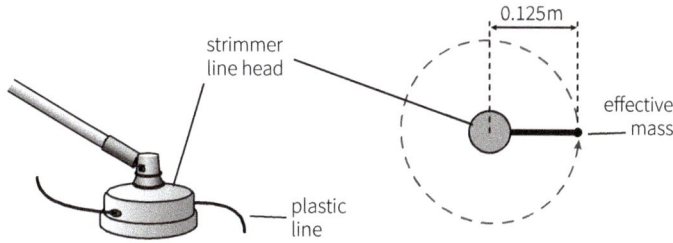

04.1 State the mathematical relationship between angular velocity and linear velocity. **[1 mark]**

04.2 The effective mass of the strimmer speeds up from rest to 2200 revolutions per minute in 0.25 s.

Calculate the maximum linear velocity in m s^{-1} of the mass. **[2 marks]**

04.3 Calculate the power in W delivered to the mass. **[2 marks]**

04.4 The effective mass is 1.3 g.

Calculate the moment of inertia in kg m^2 of the effective mass.

$$\text{moment of inertia} = \frac{1}{2} m R^2$$ **[2 marks]**

> **! Exam tip**
>
> Read the information in the question to see which equations you need to apply.

04.5 The grass exerts a force of 6.50 N on the effective mass. Calculate the change in angular speed in rad s^{-1} of the mass, assuming that the power of the motor is constant. **[2 marks]**

05 A figure skater enters a spin with arms outstretched. The skater can be is modelled as a cylinder with two rods attached, as shown in **Figure 3**.

Figure 3

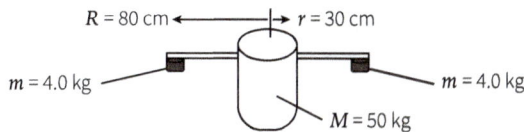

05.1 The moment of inertia of the cylinder is $\frac{1}{2} m R^2$, where m is the cylinder's mass and R is the cylinder's radius.

Calculate the moment of inertia in kg m^2 of the cylinder. **[1 mark]**

> **! Exam tip**
>
> Make sure that you use standard units when taking measurements from diagrams.

05.2 Calculate the total moment of inertia in kg m^2 of the skater. Assume the 'arms' do not contribute moment of inertia, but the 'hands' do. **[3 marks]**

05.3 The skater enters the spin with a rotational speed of 2.2 rad s^{-1}.

Calculate the angular momentum of the model skater with this rotational speed. **[2 marks]**

05.4 The skater now puts their arms by their side.

Without calculations, describe what happens to the moment of inertia, angular momentum, and speed of the skater. State **one** assumption you have made. **[4 marks]**

05.5 The skater comes to a halt after spinning for 24 seconds.

Calculate the torque exerted on the skater. **[2 marks]**

06 A kinetic energy recovery system (KERS) in a racing car system uses a flywheel to recover the energy that would have been dissipated when the car decelerates. During braking, a flywheel is accelerated.

06.1 The mass m of the flywheel is 15 kg, and it has a radius R of 20 cm. It is accelerated to a rotational velocity 1920 rad s^{-1}, during braking. The moment of inertia of the wheel is $\frac{1}{2}mR^2$.

Calculate the stored energy in J. **[2 marks]**

06.2 The car has a mass of 660 kg and goes into the corner at a speed of or 67 m s^{-1}.

Calculate, using your answer to **06.1**, the deceleration of the car in m s^{-2} if the 'braking' takes 3.3 s. **[4 marks]**

06.3 The flywheel adds considerable weight to the racing car. A racing team considers using a capacitor, or bank of capacitors, as storage. An alternative system uses a capacitor that charges.

Suggest how the capacitor could be charged during braking. **[2 marks]**

06.4 Calculate the potential difference in V to which a 1.0 F capacitor would need to be charged to store the same energy as the flywheel. Comment on your answer. **[3 marks]**

06.5 Most high-value capacitors (1 F, 5 F, 10 F) have an operating p.d. of less than 20 V.

Suggest and describe what would happen to one of these capacitors charged to the pd calculated in **06.4**. **[2 marks]**

07 The Joint European Torus (JET) is a facility that includes a tokamak, which is a prototype fusion reactor. The device uses a strong magnetic field to confine hot plasma, which is an ionised gas. The tokamak is a doughnut-shaped device, with D-shaped copper coils wrapped around it.

07.1 Explain why a current needs to flow in the plasma for it to be contained in the magnetic field. **[2 marks]**

07.2 To produce the plasma, the pulse needs a peak power of 1000 MW. Show, using the information in **Table 2**, that the facility needs two flywheels to produce the pulse. **[3 marks]**

Table 2

Mass of flywheel	775×10^3 kg
Diameter of flywheel	9 m
Top speed of wheel	24 rad s⁻¹
Duration of pulse for peak power	4.2 s

07.3 The JET flywheels are designed to work for up to 100 000 pulses.

Suggest and explain why there is a limit to the number of times the flywheels can be used. **[2 marks]**

07.4 The magnetic field that contains the plasma is produced by copper-wound magnets, each of which carries a current of 51 MA. The melting point of copper is 1085 °C, the resistivity of copper is 1.72×10^{-8} Ω m, the density of copper is 8960 kg m⁻³, and the specific heat capacity of copper is 385 J °C⁻¹ kg⁻¹.

Calculate, by considering a 1 m length of wire, the minimum diameter of wire in m that could carry this current for the duration of the pulse without melting.

Comment on your answer. **[6 marks]**

> **! Exam tip**
>
> Multi-step questions should be set out clearly. You might want to sketch out the method before writing things out.

08 A student models a lower leg kicking a ball by attaching one end of a ruler to a stand so that the ruler can pivot about the joint and 'kick' a table tennis ball directly below the pivot. When the ruler is released from a horizontal position, the free end hits the table tennis ball, and the ruler continues to move about the pivot until it is momentarily stationary.

> **⊶ Synoptic links**
>
> 3.4.1.2   3.4.1.8

08.1 Distinguish between the centre of mass and the moment of inertia. **[1 mark]**

08.2 Describe and explain the motion of the ruler from when it is released, through being momentarily vertical, up to when it next becomes stationary. **[3 marks]**

> **! Exam tip**
>
> Always look at the number of marks as an indication of the level of detail or number of points to include in your answer.

08.3 The student uses a computer model to plot a graph of torque against angle.

Explain, by analogy with linear motion, what can be deduced from the graph. **[2 marks]**

08.4 The moment of inertia of a uniform ruler is $\frac{1}{3}mL^2$, where m is 65 g and L is 30 cm.

Calculate the angular speed in rad s⁻¹ of the ruler just before it hits the ball. **[3 marks]**

08.5 The table tennis ball has a mass of 5 g, and it leaves the table with a linear velocity of 2.5 m s⁻¹.

Calculate the angular speed in rad s⁻¹ of the ruler after the collision. **[3 marks]**

08.6 Suggest whether the difference in height that the ruler reaches after the collision would be noticeable. **[2 marks]**

⚙ Knowledge

26 Thermodynamics and engines

Indicator diagrams, work done, power, and efficiencies

Work done **on** a gas is the area under the compression curve.

Work done **by** a gas is the area under the expansion curve.

Net work is the area enclosed by the loop:

$$\text{time for one cycle} = \frac{1}{\text{cycles per second}}$$

The area between induction/exhaust strokes is negative but it is negligible.

Calculating power

- indicated power = area of $p-V$ loop $\times$ 0.5 (number of cycles per second) $\times$ number of cylinders (because there is one power stroke for every two revolutions of crankshaft)

- output or brake power = P_{out} = torque at output shaft $\times$ angular velocity = $T\omega$

- friction power = indicated power − brake power

Calculating efficiencies

- $\text{mechanical} = \dfrac{\text{brake power}}{\text{indicated power}}$

- $\text{thermal} = \dfrac{\text{indicated power}}{\text{input power}}$

- $\text{overall} = \dfrac{\text{brake power}}{\text{input power}}$

First law of thermodynamics

Law of **conservation of energy:**

$$Q = \Delta U + W$$

where:

Q = energy transferred by heating (positive if transferred in to system; negative if transferred away from system)

ΔU = change in internal energy of the system (postive if internal energy increases; negative if internal energy decreases)

W = work done (positive if work done by system; negative if work done on system).

A **system** is a region in space containing a quantity of gas or vapour.

- In an open system, the gas or vapour can flow in and out of the region.

- In a closed system, gas or vapour cannot flow in and out, but the boundary can move.

Non-flow processes

A **process** is a change from one state to another; a **state** is defined by its pressure p, volume V, and temperature T (thermodynamic temperature in K). Non-flow processes involve closed systems.

Ideal gas equation:

$$pV = nRT$$

where: n = number of moles R = ideal gas constant.

Type	No change in	What happens	Gas law
isothermal (impossible unless process very slow)	temperature	p changes as gas expands or contracts	$pV = \text{constant}$ $p_1 V_1 = p_2 V_2$
adiabatic	heat in or out	T changes as work done by / on gas	$pV^\gamma = \text{constant}$ $p_1 V_1{}^\gamma = p_2 V_2{}^\gamma$
constant pressure	pressure	V changes as gas heated	$W = p\,\Delta V$
constant volume	volume	p changes as gas is heated or cooled	$p \propto T$

The p–V diagram

Isothermal: $\Delta U = 0$

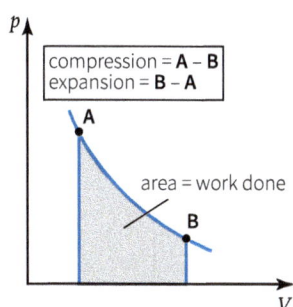

compression = **A – B**
expansion = **B – A**

area = work done

Compression:

$$-Q = -W$$

Heat transferred out of the gas, so work is done on the gas.

Expansion:

$$Q = W$$

Heat transferred to the gas, so work is done by the gas.

Adiabatic: constant Q

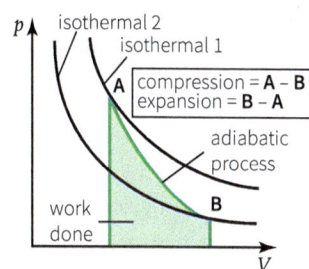

isothermal 2
isothermal 1
compression = **A – B**
expansion = **B – A**
adiabatic process
work done

Compression:

$$-W = \Delta U$$

Work done on the gas increases the temperature.

Expansion:

$$W = -\Delta U$$

Work done by the gas decreases the temperature.

Isothermal: constant volume

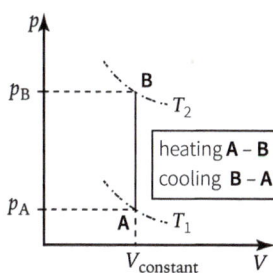

heating **A – B**
cooling **B – A**

Heating:

$$Q = \Delta U$$

Heat transferred into the gas increases the temperature.

Cooling:

$$-Q = -\Delta U$$

Heat transferred out of the gas decreases the temperature.

Adiabatic: constant pressure

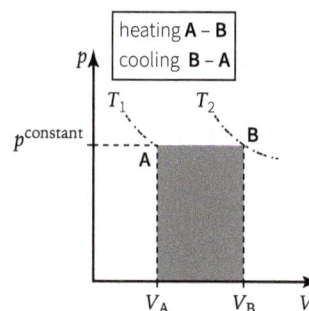

heating **A – B**
cooling **B – A**

Heating:

$$W = p\,(V_2 - V_1)$$

Heat transferred to the gas, so work is done by the gas.

Cooling:

$$W = p\,(V_1 - V_2)$$

Heat transferred out of the gas, so work is done on the gas.

The graph for adiabatic processes is steeper than for isothermal processes.
In a cyclic process, work done per cycle = area of loop.

Engine cycles

In a real engine, energy is transferred from the fuel supplied at a fuel flow rate with a certain calorific value, to produce work done by each cycle:

input power = calorific value × fuel flow rate

In a theoretical engine, the same air is continuously taken through the cycle (called an air standard).

Petrol engine cycle

- induction: p constant, V increases, air and petrol drawn into cylinder as piston moves out
- compression: V decreases, T increases, p increases as piston moves in, spark ignites fuel at end of stroke so T increases, p increases at constant V
- expansion: V increases, p decreases, work is done as piston moves out, exhaust valves open
- exhaust: p constant (atm), piston moves in expelling burnt gases

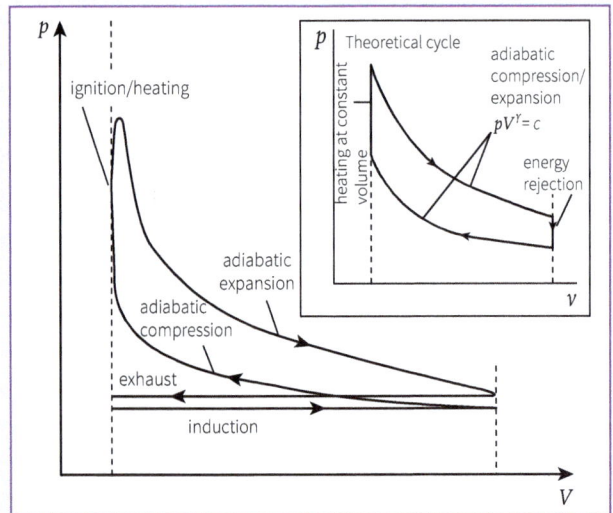

Diesel engine cycle compared with petrol engine

- induction: air only drawn into cylinder
- compression: fuel injected into cylinder, T increases to ignite mixture
- expansion: no difference
- exhaust: no difference

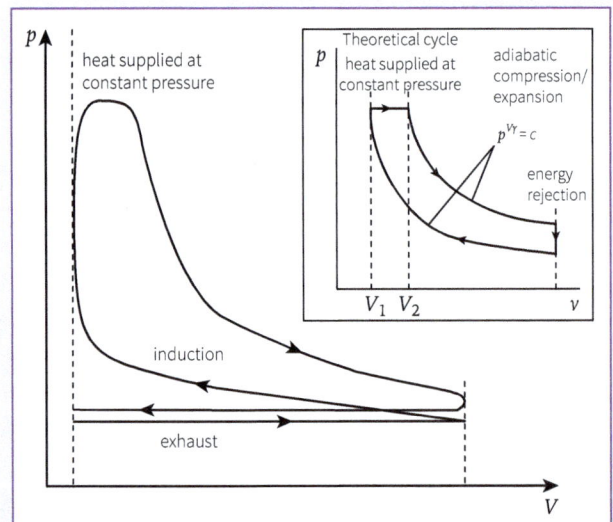

Theoretical efficiency (10:1 compression) = 60%. In reality, this is about 30% because of:

- rounded corners due to time for valves to open/close; these processes are not instantaneous
- induction/exhaust strokes
- overcoming friction; pumping oil/coolant around engine
- fuel not being completely burned
- heat transfer, so processes are not adiabatic.

Second law of thermodynamics and heat engines

Ideal heat engines

- use a difference in temperature to do work
- cannot work only by the first law of thermodynamics: $W_{out} = Q_H$; friction losses reduce W_{out}.

Second law of thermodynamics

A heat engine needs to operate between a source and a sink.

$$\text{efficiency } \eta = \frac{W_{out}}{Q_H} = \frac{Q_H - Q_c}{Q_H} = 1 - \frac{Q_c}{Q_H}$$

$$\text{maximum theoretical efficiency} = \frac{T_H - T_c}{T_H}$$

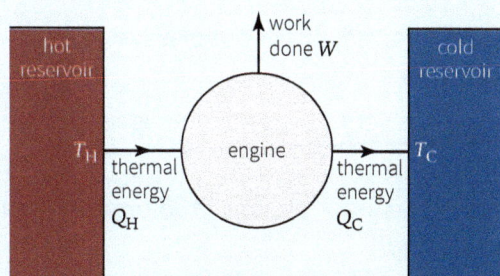

Practical heat engines

These are less efficient because:

- T_C is above absolute zero (the temperature should be as cold as possible)
- T_H is limited by the high-temperature properties of materials of the engine (it should be as hot as possible)
- real engines reject energy by heating the atmosphere.

Combined heat and power (CHP) schemes

- Conventional power stations use a steam/gas turbine to drive an alternator.
- Maximum theoretical efficiency is about 60%, but efficiency is usually nearer 35% in practice.
- Thermal energy is used for space heating of houses, businesses, and factories.

Reversed heat engines

A reversed heat engine does work to produce a difference in temperature – how good it is at this is measured by a **coefficient of performance** (not an efficiency, so this can be greater than one), which is the ratio of useful heating or cooling to input work.

Examples include heat pumps and refrigerators.

A heat pump operates between a hot temperature and the atmosphere; a refrigerator operates between a cold temperature and the atmosphere.

Refrigerator

$$COP_{REF} = \frac{Q_c}{W} = \frac{Q_c}{Q_H - Q_c} = \frac{T_c}{T_H - T_c}$$

Heat pump

$$COP_{HP} = \frac{Q_H}{W} = \frac{Q_H}{Q_H - Q_c} = \frac{T_H}{T_H - T_c}$$

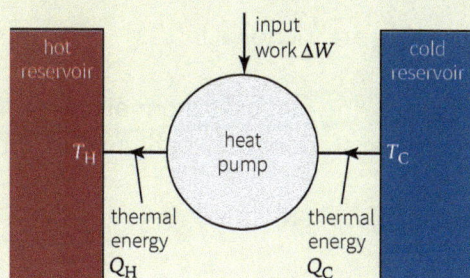

Retrieval

Learn the answers to the questions below, then cover the answers column with a piece of paper and write as many as you can. Check and repeat.

	Questions		Answers
1	What is the first law of thermodynamics?		Q (heat) = ΔU (internal energy) + W (work)
2	What is a system?		region of space containing a quantity of gas or vapour
3	What is the unit of thermodynamic temperature?		Kelvin
4	What is an adiabatic process?		no energy transferred into or out of a system by heating
5	When is a process approximately isothermal?		process happens very slowly so there is no change in temperature
6	How do you calculate the work done compressing a gas at constant pressure?		work done = pressure × change in volume, or area under a $p-V$ graph
7	What is the difference between a $p-V$ graph for an adiabatic process, and a $p-V$ graph for an isothermal process?		adiabatic graph is steeper
8	What two factors affect the energy supplied to an internal combustion engine?		calorific value of the fuel, fuel flow rate
9	Name the four strokes of the petrol engine cycle.		induction, compression, expansion, exhaust
10	What is the approximate efficiency of a 4-stroke petrol engine?		~30%
11	How do you calculate the work done per cycle from an engine cycle graph?		area enclosed by the line showing the cycle
12	What is the brake power of an engine?		torque produced at a crankshaft at a particular angular velocity
13	What is the difference between a petrol engine and a diesel engine in terms of how the fuel enters the engine?		in a petrol engine fuel and air are drawn into a cylinder; in a diesel engine fuel is injected into the cylinder at the end of the compression stroke
14	What is a heat engine?		engine that uses a temperature difference to do work
15	Why can a heat engine not work only by the first law of thermodynamics?		there are frictional losses
16	What is a combined heat and power scheme?		a scheme that uses thermal energy produced by power stations to heat homes and businesses
17	For maximum efficiency, what should the temperature of the sink be in a heat engine?		absolute zero
18	What is a reversed heat engine?		engine that does work to produce a temperature difference

Put paper here

19 What is the coefficient of performance of a reversed heat engine related to?

ratio of useful heating or cooling to input work

20 What is the difference between a heat pump and a refrigerator?

heat pump operates between a hot temperature and the atmosphere; refrigerator between a cold temperature and the atmosphere

Put paper here

📱 Maths skills

Practise your maths skills using the worked example and practice questions below.

Isothermal and adiabatic changes in gases

An isothermal change is one in which the temperature of the gas does not change and $\Delta U = 0$, so pV = constant.

In an adiabatic change, no heat flows into or out of the system and $Q = 0$, so pV^{γ} = constant, where the power $\gamma = 1.4$ for air.

We can also use $pV = nRT$ to calculate changes in the pressure, volume, and temperature of a gas.

Worked example

Questions

A gas cylinder contains $200\,cm^3$ of air at a pressure of $100\,kPa$. The air is compressed adiabatically until the volume halves.

1 Calculate the pressure of the air after compression.

2 The temperature of the gas before compression was $20\,°C$. Calculate the temperature after compression.

Answers

1 After compression, new volume = $100\,cm^3$. pV^{γ} is constant, so $p_1 V_1^{\gamma}$ = constant = $p_2 V_2^{\gamma}$

$p_1 = 100\,kPa$; $V_1 = 200\,cm^3$; $V_2 = 100\,cm^3$

$100 \times 200^{1.4} = p_2 \times 100^{1.4}$

$p_2 = \dfrac{100 \times 200^{1.4}}{100^{1.4}} = 264\,kPa$

2 $pV = nRT$

$\dfrac{p_1 V_1}{T_1} = nR = \dfrac{p_2 V_2}{T_2}$

$T_1 = (20 + 273)\,K = 293\,K$

Rearranging for T_2:

$T_2 = \dfrac{p_2 V_2 T_1}{p_1 V_1}$

$T_2 = \dfrac{264 \times 100 \times 293}{100 \times 200} = 387\,K$

Practice

1 A diatomic gas at $7\,°C$ has an initial pressure of $150\,kPa$ and a volume of $300\,cm^3$. It is compressed adiabatically until its pressure is $600\,kPa$.

Calculate the final pressure and temperature of the gas.

$\gamma = 1.4$ for a diatomic gas

2 A gas is compressed isothermally from an initial volume of $0.0\,m^3$ to a final volume of $0.04\,m^3$. The initial pressure is $150\,kPa$.

Calculate the final pressure of the gas.

3 A monatomic gas has an initial volume of $50\,cm^3$ at a pressure of $200\,kPa$. The gas is compressed adiabatically to a volume of $5.0\,cm^3$.

Calculate the new pressure of the gas.

$\gamma = 1.67$ for a monatomic gas

Exam-style questions

01 An ideal gas goes through a process that produces an indicator diagram as shown in **Figure 1**.

Figure 1

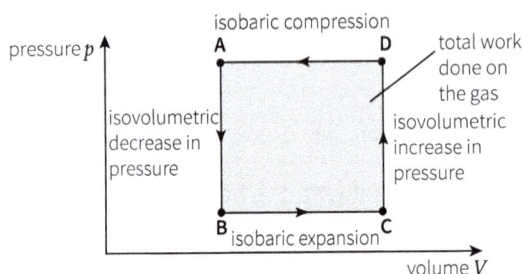

01.1 Point **A** is at a pressure of 5.0×10^5 Pa, a volume of 2×10^{-4} m^3, and a temperature of 100 °C.

Calculate the number of moles of gas in the container. **[2 marks]**

> **! Exam tip**
>
> Remember to use the thermodynamic temperature in Kelvin.

moles of gas = _____ mol

01.2 Point **C** is at a pressure of 1.0×10^5 Pa and a temperature of 0 °C.
Show that the new volume is approximately 10^{-3} m^3. **[2 marks]**

01.3 Compare the changes to pressure, temperature, and volume in the transition from point **A** to **B** with the changes to pressure, temperature, and volume in the transition from from point **C** to **D**. **[2 marks]**

_____

_____

_____

01.4 Compare the work done from point **B** to **C** with the work done from point **D** to **A**. Include calculations as part of your answer. **[3 marks]**

_____

_____

_____

_____

02 A cylinder of an ideal gas initially has a volume of 2×10^{-3} m^3. There are 3 mol of gas in the cylinder, and the initial pressure of the gas is 3×10^6 Pa. This is state **A**.

The gas is then brought at constant pressure to state **B**, where the volume is $4\times10^{-3}\,m^3$. The gas is then brought at constant volume to a new pressure in state **C**. The gas is then brought at constant temperature back to state **A**.

02.1 Calculate the pressure in state **C**. State any assumptions you make.

[2 marks]

pressure = _____ Pa

02.2 **Figure 2** shows the position of state **B**.

Complete **Figure 2** to show the complete cycle, labelling states **A** and **C**. Annotate **Figure 2** to show the direction of the cycle. [4 marks]

Figure 2

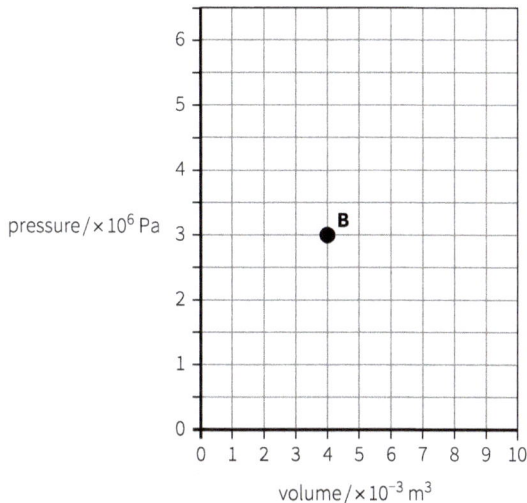

pressure / × 10⁶ Pa

volume / × 10⁻³ m³

02.3 Explain whether this cycle represents a heat pump or a refrigerator.

[2 marks]

_____

_____

_____

> **! Exam tip**
>
> If work is done on the system then W is negative; if work is done by the system then W is positive.

02.4 Explain why the transition from state **C** to **A** may involve heat transfer despite it happening at a constant temperature. [2 marks]

_____

_____

_____

02.5 Calculate the maximum efficiency of a heat engine operating between the two temperatures of the gas in this cycle. [2 marks]

maximum efficiency = _____ %

03 A student researches petrol and diesel engines. In both engines, the air in the cylinders is compressed adiabatically. In a cylinder with a volume of 240 cm^3 the initial pressure is 1.00×10^5 Pa. The gas is compressed adiabatically to a volume of 50.0 cm^3.

03.1 Explain in terms of the first law of thermodynamics what is meant by an adiabatic change.

Use these to explain how the change in internal energy of an ideal gas is related to the work done by the gas during an adiabatic change. **[3 marks]**

03.2 Calculate the new pressure of the gas in Pa.
$\gamma = 1.4$ **[2 marks]**

Exam tip

Remember that all calculations must be done in SI units: kg, m, s, and if pressure is in Pa then volume must be in m^3.

03.3 Sketch a simplified indicator diagram of a theoretical petrol engine and a theoretical diesel engine. The loop between the induction and exhaust cycles can be ignored. Include the following labels: compression, expansion, heat addition, and heat rejection.

· Describe how your diagrams indicate the difference between the two engines. **[4 marks]**

03.4 The information about two types of engine is shown in **Table 1**. It is claimed that:

- the frictional losses in engine **B** are greater than those in engine **A**
- engine **A** has an overall efficiency that is 1.4 times the overall efficiency of engine **B**.

Use **Table 1** to evaluate these claims. **[6 marks]**

Exam tip

In an extended answer remember to address each bullet point, and explicitly show how you use the information you are given.

Table 1

Property	Engine A	Engine B
number of cylinders	4	4
number of cycles per minute	4250	5360
work done per cycle / J	430	307
brake power / kW	35	27
energy per kg of fuel / MJ kg^{-1}	38	42
fuel flow rate / $\times 10^{-3}$ kg s^{-1}	2.8	2.6

04 Engines that propel container ships are some of the largest in the world. Some data relating to one such engine are given in **Table 2**.

Table 2

Quantity	Value
brake power per cylinder	74 MW
torque measured at output shaft	6.5 MN m
fuel flow rate	3.43 kg hr^{-1}
fuel calorific value	38 MJ kg^{-1}

04.1 Calculate the rotational speed at the output shaft of the engine in rpm. **[2 marks]**

04.2 Compare the rate at which energy is supplied to the engine of the container ship with the rate at which energy is supplied to the engine of a diesel car.

calorific value of diesel fuel = 46 MJ kg^{-1}; fuel flow rate in a diesel car = 0.62 kg hr^{-1} **[2 marks]**

04.3 A diesel engine has an efficiency of approximately 35% when used in a car.

Calculate the brake power in W of a diesel engine with this efficiency. **[2 marks]**

04.4 Compare the torque produced by the diesel engine with the torque produced by the container ship engine. Assume the rotational speed of the output shaft is the same.

Suggest **one** reason for any difference, including the effect of the assumption that you made. **[4 marks]**

05 The temperature inside a refrigerator −3 °C, and the temperature outside is 24 °C.

05.1 Calculate the efficiency of the refrigerator if the 1.2 kW electric motor is 50% efficient. **[2 marks]**

05.2 Calculate the maximum coefficient of performance, and the energy transferred to the room in kJ s^{-1}. **[3 marks]**

05.3 **Figure 3** shows the pressure–volume (p–V) changes for one cycle of the working substance of a refrigerator.

Figure 3

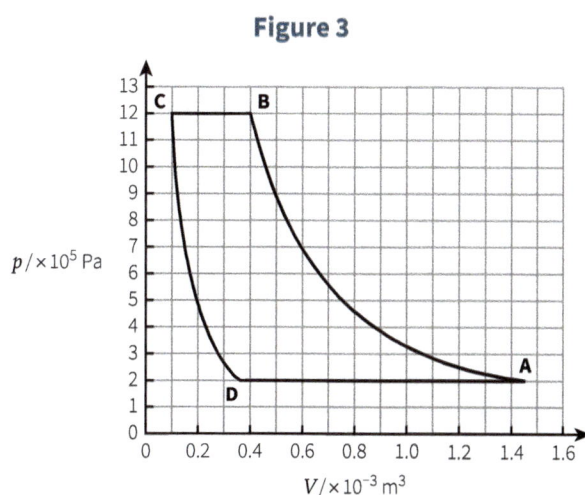

$p / \times 10^5$ Pa

$V / \times 10^{-3}$ m^3

Identify the sections of the graph where:
- the air inside the refrigerator is being cooled
- the motor is doing work on the refrigerant
- the refrigerant is condensing. **[3 marks]**

05.4 Describe **one** practical reason why the coefficient of performance calculated in **05.2** can not be achieved in practice. **[1 mark]**

06 Air source heat pumps are sometimes called reverse-cycle air conditioners. They transfer energy between the inside and outside of a building. The typical temperatures inside and outside a building in different seasons are shown in **Table 3**.

Synoptic links

3.5.1.3   3.6.2.1

Table 3

Season	Average temperature inside the building/°C	Average temperature outside the building/°C
winter	20	10
summer	24	31

06.1 Compare the coefficient of power when the device operates as a heat pump with the coefficient of power when the device operates as a refrigerator. **[2 marks]**

06.2 The coefficient of performance for a heat pump is greater than when the same heat pump is working as a refrigerator. State why this is not the case for **06.1**. **[1 mark]**

06.3 A backup heater for winter consists of a length of resistance wire coiled in a pipe. The water comes in from the outside of the building and is heated to a temperature of 60 °C.

Calculate the maximum flow rate in $m^3 s^{-1}$ achievable by a heater with a power of 230 kW.
density of water = 1000 kg m^{-3};
specific heat capacity of water = 4.2 kJ kg^{-1}°C^{-1} **[4 marks]**

06.4 A safety device in the heater uses a thermistor to monitor the water temperature.

Explain how a thermistor could be used in a potential divider circuit to shut down the heater if the water gets too hot. **[3 marks]**

! Exam tip

When you are asked to describe a circuit, it is useful to draw a diagram.

07 A student models a pogo stick by compressing a spring with a mass on top of it. When the mass is released, it is projected upwards to a maximum height h_{max}. The data are shown in **Table 4**.

Synoptic links

3.4.1.8   3.4.2.1

Table 4

Column A Compression/cm	Column B Maximum height/cm	Column C
0.0	0.0	
0.5	2.0	
1.0	8.1	
1.5	18.2	
2.0	32.4	

07.1 Determine a heading for Column **C** that would enable the student to plot a graph giving a straight line.

Write the labels on each axis of the graph.

Describe how the student could use the gradient of this graph to calculate the spring constant of the spring in the model. Show your reasoning. **[5 marks]**

! Exam tip

There will be two options for producing a straight line graph, and you will be given credit for whichever one you describe.

07.2 The student deduces that the spring constant of the spring used in their model is $80\,\text{N m}^{-1}$.

Calculate, using the data in **Table 4**, the maximum possible velocity of a $50\,\text{g}$ mass when it is released.

State at which point the mass has this maximum speed. You may use a sketch in your answer. **[4 marks]**

07.3 A student of mass of $45\,\text{kg}$ uses a real pogo stick with a spring that can be compressed by four times the compression of the spring used in the model.

Calculate the spring constant in N m^{-1} that would be needed for the student to travel at the same speed as the mass in the model. **[1 mark]**

07.4 A different type of pogo stick has a piston that compresses a cylinder of air, instead of a spring.

Suggest whether a piston filled with air at atmospheric pressure when compressed would return to its original length as the spring does in the first design. Explain your answer using the first law of thermodynamics. **[3 marks]**

08 In a diesel engine, the gas is compressed to a temperature such that the fuel-air mixture injected will ignite.

08.1 The ignition temperature of a diesel fuel is approximately $70\,°\text{C}$. Assume that air taken into the cylinder has a temperature of $20\,°\text{C}$ when the volume is V_1, and compressed to a volume V_2.

For an adiabatic change in an ideal diesel engine $T_1 V_1^{\gamma-1} = T_2 V_2^{\gamma-1}$. For air, $\gamma = 1.4$.

Show that the ratio $\dfrac{V_1}{V_2}$ is about 1.5. **[2 marks]**

08.2 In real diesel engines, the ratio $\dfrac{V_1}{V_2}$ is much larger than 1.5. Suggest **two** reasons why. **[2 marks]**

08.3 In a petrol engine, the petrol-air mixture is ignited by a spark at the spark plug, near the end of the compression stroke, resulting in a sudden increase in temperature and pressure at almost constant volume. The spark is produced by an induction coil, which produces a p.d. of thousands of volts. Air breaks down at an electric field strength of $10\,\text{kV cm}^{-1}$. The gap across which the spark is produced is $0.8\,\text{mm}$.

Calculate the p.d. in V needed for air to conduct.

Suggest a reason for the difference between this p.d. and that produced by the coil. **[3 marks]**

08.4 The induction coil is powered from the car battery, which has a voltage of $12\,\text{V}$.

Suggest how this constant p.d. can be used to produce a p.d. of thousands of volts. **[3 marks]**

Synoptic links

3.7.3.2   3.7.5.6

Exam tip

You do not need to know the construction of a petrol or diesel engine, but you should be aware of what happens to the gas in a cylinder of each engine during the four-stroke cycle.

⚙ Knowledge

27 The discovery of the electron

Cathode rays

In a **Crookes discharge tube:**

- gases at low pressure conduct
- a magnet deflects the positive column
- a paddle wheel placed in the tube rotates, which are negatively charged.

J. J. Thomson showed that cathode rays are the same – they are electrons.

Glowing gases

- The gas glows with a characteristic colour.
- The potential difference applied is sufficiently high to produce a field that ionises atoms.
- Electrons accelerated towards the anode cause excitation of gas molecules, which is seen as a positive column of glow discharge.
- Positive ions move toward the cathode where they recombine to give a negative glow.

Thermionic emission of electrons

An electron beam is produced by:

- heating the cathode (a filament with a current flowing through it due to low p.d. V_1)
- accelerating the beam using an anode at high p.d.
- evacuating the glass tube so there are no collisions between electrons and gas molecules,

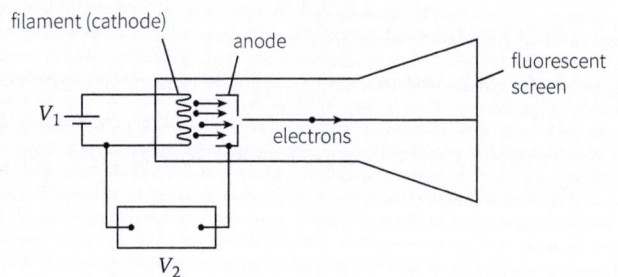

Work is done on an electron accelerated through a high p.d. V_2 so that $eV = \frac{1}{2}mv^2$

If v is much less than c, $v = \sqrt{\dfrac{2eV}{m}}$

Principle of Millikan's determination of the electronic charge

Millikan determined electronic charge e to be 1.6×10^{-19} C using his **oil drop experiment.**

With a uniform electric field between a positively charged top plate and a negatively charged bottom plate, a charged oil drop was held stationary.

$$F_e = QE = \frac{QV}{d} = mg$$

With no electric field between the plates, the oil drop moves at a terminal (steady) speed because of an upwards viscous drag F_d.

Stokes' law

- $F_d = 6\pi\eta rv$ (η = viscosity of air)
- for a spherical droplet $m = \rho \times \frac{4}{3}\pi r^3$
- weight $= mg = \frac{4}{3}\pi\rho g r^3$
- $\frac{4}{3}\pi\rho g r^3 = 6\pi\eta rv$, so $r^2 = \dfrac{9\eta v}{2\rho g}$

Hence radius and mass can be found from values of v, p, g, η.

So, $Q = \dfrac{mgd}{v}$, which was found to be a multiple of 1.6×10^{-19} C.

Electric charge is **quantised**, meaning it only has values which are integer multiples of e.

Thomson's measurements

Thomson found that $\frac{e}{m}$ for the electron was much larger than that of the hydrogen ion (proton).

He deduced that electrons were much less massive than hydrogen – they are subatomic particles.

Deflection

By a magnetic field

Diagram	Equations	What to measure
magnetic field out of diagram (at 90° to the plane of the diagram) electron gun Bev v electron • electron beam moves in a circle • no work is done because the force is perpendicular to the velocity	• $F_{magnetic} = B\,q\,v$, here $q = e$ • circular motion: $F = \dfrac{m\,v^2}{r}$ • $\dfrac{m\,v^2}{r} = B\,e\,v$ • $v^2 = \left(\dfrac{B\,e\,r}{m}\right)^2$ • $eV = \dfrac{1}{2}\,m\,v^2$ • $eV = \dfrac{1}{2}\,m\left(\dfrac{B\,e\,r}{m}\right)^2$	Find specific charge: $$\dfrac{e}{m} = \dfrac{2v}{B^2\,r^2}$$ Measure: • anode V • magnetic field B • radius of circle r

By magnetic and electric fields

Diagram	Equations	What to measure
cylindrical anode cathode magnetic field into page F_E F_M filament electron beam evacuated tube V • electron beam is undeflected in magnetic and electric fields • the 'crossed' fields mean forces cancel	• $F_{magnetic} = B\,e\,v$ • $F_{electric} = e\,v$ • $E = \dfrac{V_{plates}}{\text{plate separation } d}$ • $v = \dfrac{E}{B} = \dfrac{V_{plates}}{B\,d}$ To find v: • $F = \dfrac{m\,v^2}{r} = B\,e\,v$ • $\dfrac{m\,v}{r} = B\,e$	Find: $\dfrac{e}{m} = \dfrac{v}{B\,r}$ Measure: • V_{plates} • d • B • r

By an electric field

Diagram	Equations	What to measure
filament anode $+V$ evacuated tube d y L $0V$ high voltage • electron beam is deflected by an electric field • deflection is due to a vertical force • separately measure v using a magnetic field as well to straighten the beam	• $F_{electric} = e\,V$ • $E = \dfrac{V_{plates}}{\text{plate separation } d}$ • $a = \dfrac{F}{m} = \dfrac{eV_{plates}}{m\,d}$ To find a: • deflection $y = \dfrac{1}{2}\,a\,t^2$ • $a = \dfrac{2y}{t^2}$ • $t = \dfrac{L}{v}$ To find v: • $v = \dfrac{E}{B} = \dfrac{V_{plates}}{B\,d}$	Find: $\dfrac{e}{m} = \dfrac{da}{V_{plates}}$ Measure: • d • V_{plates} • y • L • B

Learn the answers to the questions below, then cover the answers column with a piece of paper and write as many as you can. Check and repeat.

Questions	Answers
1 Under what conditions do gases conduct electricity?	low pressure, high voltage
2 How can you show that there are particles moving from the cathode to the anode in a discharge tube?	paddle wheel inserted between the anode and the cathode rotates
3 Which type of ions cause the negative glow near the cathode of a discharge tube?	positive ions
4 Why is a cathode ray tube evacuated?	so there are no collisions between air molecules and electrons
5 How did scientists establish the sign of the charge on cathode rays?	using direction the ray is deflected by a magnetic field
6 What is thermionic emission?	emission of electrons from a cathode that is heated by passing a current through it
7 What is specific charge?	$\dfrac{\text{charge}}{\text{mass}}$
8 What is the angle between the beam direction and movement of electrons when using a magnetic field to determine the specific charge of an electron?	$90°$
9 How much work is done on an electron that is accelerated to an anode at a potential V?	$eV \,(= W)$
10 How much work is done when an electron moves in a circle due to a magnetic field?	none
11 How can a beam moving through a magnetic field be made to travel undeflected?	use magnetic field to exert an equal and opposite force to the beam
12 How can the speed of an electron in a beam be determined using a magnetic field?	centripetal force $= \dfrac{mv^2}{r} =$ magnetic force $(B\,eV)$
13 When an electric field is used to find the specific charge on the electron, how do you find the acceleration of the electron?	using the time in the field (from length and speed), and the vertical deflection y
14 What was the particle with the largest specific charge before Thompson found $\dfrac{e}{m}$ for the electron?	hydrogen ion
15 What made Thomson deduce that the electron was a subatomic particle?	it had a mass that was much smaller than that of hydrogen
16 What is the force on an electron in a field E?	$F = e\,E$

Put paper here

17	In Millikan's oil drop experiment is the upper plate positively or negatively charged?	positively charged
18	When an electric field is used to hold an oil drop stationary, what does the upwards electric force equal?	weight, mg
19	What two features of an oil drop can be deduced from the drop moving between charged plates when there is no electric field?	radius and mass
20	What did Millikan deduce from his oil drop experiment?	electric charge is quantised / only occurs in integer multiples of 1.6×10^{-19} C

Put paper here

Maths skills

Practise your maths skills using the worked example and practice questions below.

Thermionic emission

When a metal filament is heated, thermionic emission occurs.

This effect is used in cathode ray tubes. The emitted electrons are accelerated through a potential difference. The work done accelerating the electrons is the charge multiplied by the potential difference, eV. The electrons gain $\frac{1}{2}mv^2$ kinetic energy, so:

$$eV = \frac{1}{2}mv^2$$

This equation is used to calculate the specific charge of the electron:

$$\frac{e}{m} = \frac{v^2}{2V}$$

Worked example

Question

A beam of electrons is produced by thermionic emission. The electrons are accelerated towards an anode through a potential difference of 4.2 kV. The speed of the electrons after acceleration is $3.9 \times 10^7 \, \text{m s}^{-1}$.

Calculate the specific charge of an electron.

Answer

Convert values to SI units:

$V = 4.2 \, \text{kV} = 4200 \, \text{V}$

Then substitute into the equation:

$$\frac{e}{m} = \frac{v^2}{2V} \frac{e}{m}$$

$$= \frac{(3.9 \times 10^7 \, \text{m s}^{-1})^2}{2 \times 4200 \, \text{V}}$$

specific charge $= 1.8 \times 10^{11} \, \text{C kg}^{-1}$

Practice

1 Electrons are emitted by thermionic emission from a cathode. The potential difference between the cathode and the anode is 30 V.

Calculate the speed of the electrons as they pass through the anode.

2 Calculate the work done in accelerating thermionic electrons to a speed of $3.0 \times 10^7 \, \text{m s}^{-1}$.

3 The work done accelerating a thermionic electron between a metal filament and an anode is 1.2×10^{-17} J.

Calculate the potential difference between the filament and the anode.

01 William Crookes showed that a gas at low pressure conducts electricity when a potential difference is applied across the ends of the tube containing the gas. He observed a 'positive column' near the anode.

01.1 State the charges on the anode and the cathode. **[1 mark]**

_____

_____

01.2 Describe the observation that showed that the particles moving through the gas when it conducts are charged. **[1 mark]**

_____

_____

01.3 Crookes observed a 'negative glow' near the cathode.

Explain why the gas near the cathode glowed. **[2 marks]**

_____

_____

_____

_____

_____

_____

01.4 Suggest why the term 'cathode ray' and not 'anode ray' was used. **[1 mark]**

_____

_____

01.5 J.J. Thomson used different gases in the tube and observed that the negative particles were the same.

Suggest **two** methods that Thompson could use to confirm that the particles were the same. **[4 marks]**

_____

_____

_____

_____

_____

_____

_____

> **! Exam tip**
>
> Particles are characterised by their mass and their charge.

02 **Figure 1** shows a cathode ray oscilloscope.

Figure 1

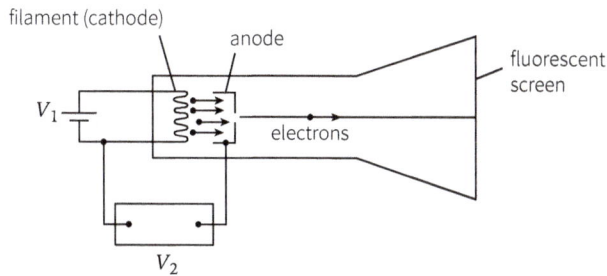

02.1 Explain why the tube needs to be evacuated. **[1 mark]**

_____

_____

02.2 There are two potential differences, V_1 and V_2, shown in **Figure 1**.

Explain why each p.d. is needed. **[3 marks]**

_____

_____

_____

_____

02.3 The output of the high-voltage supply is 4500 V.

Calculate the speed of the electrons as they leave the slit. **[2 marks]**

speed = _____ $m\,s^{-1}$

02.4 When the beam hits the screen, a dot of light is seen.

Suggest and explain what happens to the dot on the screen when the accelerating p.d. is decreased. **[4 marks]**

_____

_____

_____

_____

_____

_____

_____

> **! Exam tip**
>
> The number of marks indicates the level of detail needed. Check the verbs used in the question.

03 J.J. Thomson measured the specific charge of particles emitted in a cathode ray tube. He concluded that the particles were subatomic particles, which we call electrons.

03.1 Explain why Thomson's measurement of the specific charge of the electron was significant. **[3 marks]**

Figure 2

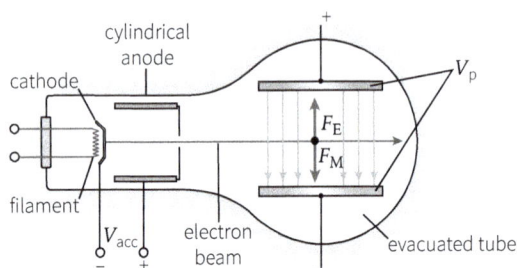

03.2 **Figure 2** shows a method of measuring the specific charge. A magnetic field is produced and applied to the tube to cancel out the deflection by the charged plates.

Describe and explain the orientation of the magnetic field that would produce an undeflected beam. **[2 marks]**

03.3 The accelerating potential difference V_{acc} can be increased.

Explain how and why the p.d. between the plates V_p should be changed to continue to produce an undeflected beam. **[2 marks]**

03.4 The distance between the plates is d and the magnetic field strength is B.

Show that a graph of V_{acc} against V_p^2 has a gradient of $\dfrac{m}{ed^2 B^2}$. **[3 marks]**

03.5 The gradient of the line is calculated as $(8.30 \pm 0.35) \times 10^{-15}$ kg $C^{-1} m^{-2} T^{-2}$. The distance between the plates is 15.0 cm, and the magnetic field strength is 0.250 T.

Calculate the charge to mass ratio of the electron, and the uncertainty in the value. **[3 marks]**

> **! Exam tip**
>
> Uncertainty can be expressed as a percentage. Alternatively, you can incorporate the uncertainty and recalculate the answer.

04 Millikan's oil drop experiment uses two parallel plates, between which there are charged droplets of oil. There is a potential difference between the plates that can be adjusted.

Figure 3

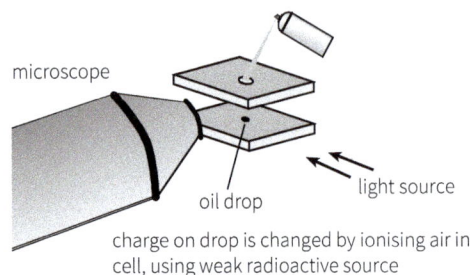

charge on drop is changed by ionising air in cell, using weak radioactive source

When there is no p.d. between the plates, an oil drop between them falls with a constant velocity of 4 mm s^{-1}.

04.1 Explain why the oil drop falls with a constant velocity. **[2 marks]**

04.2 Derive an expression for the mass of the droplet in **04.1** in terms of the velocity v of the drop, the radius r, the viscosity of the air η, and the density of the oil ρ. **[3 marks]**

> **! Exam tip**
>
> When deriving equations, make each of the steps in the derivation clear.

04.3 A drop is held stationary in the field between the plates by applying a p.d. of 4700 V. The distance between the plates is 12.0 mm. The mass of the drop is 3.2×10^{-11} g.

Deduce the number of excess electrons on the drop. **[3 marks]**

04.4 Millikan's reported value of the charge on the electron was $1.5924(17) \times 10^{-19}$ C. It has been suggested that one reason for the difference was that he used an incorrect value for the viscosity of air.

Suggest and explain the difference between the value of viscosity that he used and the correct value. **[2 marks]**

05 William Crookes developed a tube that he could evacuate to a very low pressure. He applied a high potential difference across the tube and made observations.

05.1 Explain why the discussion of these experiments at the time did not involve the idea of electrons. **[1 mark]**

05.2 Cathode rays were produced using two different methods. The Crookes tube method is known as the 'cold cathode' method, whereas thermionic emission is known as the 'hot cathode' method.

> **! Exam tip**
>
> Extended answer questions should be clearly laid out, showing answers to each bullet point.

Explain why. You should:

- describe the production of cathode rays in a Crookes tube, and why it uses a 'cold cathode'
- describe the production of cathode rays by thermionic emission, and why it uses a 'hot cathode'. **[6 marks]**

05.3 To show that particles were emitted from the cathode, Crookes adapted his tube.

Describe the adaptation that he made and explain how it showed that particles were emitted. **[2 marks]**

05.4 In both methods, the cathode rays produced light when they hit a fluorescent screen.

Suggest and explain whether the brightness of the light emitted will vary in the same way with the accelerating potential difference. **[1 mark]**

06 Scientists are able to detect fractions of the electronic charge e on a charged sphere in the electric field between two plates.

06.1 Suggest where fractions of the electronic charge might come from. **[1 mark]**

> **Synoptic links**
>
> 3.2.1.6   3.7.5.2   3.2.1.3

06.2 The scientists used a niobium sphere of mass 7×10^{-5} g, which compares to the mass of an oil drop that has a mass of about 3×10^{-11} g.

The potential difference used to keep the oil drop stationary between the two plates is 3000 V.

Deduce the p.d. in V required to keep the niobium sphere stationary. State any assumptions you make. **[3 marks]**

! Exam tip

Remember that there are particles smaller than protons and neutrons.

06.3 The niobium spheres were made to oscillate between the plates. They were held between the plates using a magnetic field, not an electric field.

Suggest why the spheres had to be made to oscillate. **[1 mark]**

06.4 The initial charge on the niobium sphere was very large, equivalent to 105 electrons.

To reduce this charge, the scientists used a radioactive source that produced positrons.

The source was held very close to the sphere.

Suggest why this reduced the charge on the sphere. **[2 marks]**

06.5 The scientists deduced that the niobium sphere had a fractional charge.

Suggest why Millikan was not able to detect this fractional charge.

[1 mark]

07 The specific charge of the electron can be determined by several different methods.

07.1 Compare the method of determining $\frac{e}{m}$ by deflecting electrons in a magnetic field with the method of determining $\frac{e}{m}$ by deflecting electrons in an electric field. **[4 marks]**

⚙ Synoptic links

3.4.1.3   3.7.5.2   3.6.1.1

07.2 A teacher uses the deflection of a tennis ball in a gravitational field as an analogy for the deflection of the electron in an electric field.

The tennis ball leaves a desk horizontally at a velocity v.

The ball hits a wall 10 m away at a distance of 50 cm below the horizontal.

Calculate the initial horizontal velocity in m s^{-1} of the ball. **[3 marks]**

! Exam tip

When you make comparisons you need to state both similarities and differences.

07.3 In the experiment described in **07.2**, the ball falls because of its weight.

Explain why the effect of the weight of the electron is negligible in the experiments that you described in **07.1**. **[1 mark]**

07.4 A cyclotron is a particle accelerator that accelerates a beam of charged particles using a high frequency alternating potential difference, as shown in **Figure 4**.

Figure 4

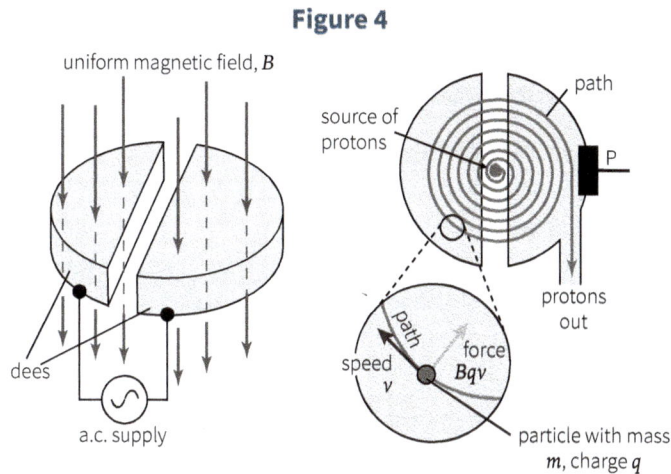

Suggest why the magnetic field does not speed up the charged particles. **[1 mark]**

07.5 A cyclotron has a resonance frequency of

$$f = \frac{qB}{2\pi m}$$

This is the frequency of the alternating p.d. required to make the charged particles move in a circle.

Derive using ideas about circular motion, the equation above.

[3 marks]

08 X-rays can be produced when accelerated electrons hit a target and transfer energy to it.

08.1 Suggest how X-rays can be emitted from the screen in a Crookes tube. **[3 marks]**

08.2 The wavelength of an X-ray is of the order of the size of an atom.

Determine an estimate for the energy in J of an X-ray photon.

[3 marks]

08.3 Calculate the potential difference in V through which an electron must be accelerated to produce an X-ray with this wavelength.

[2 marks]

08.4 An induction coil is a type of transformer and is often characterised by the length of spark it can produce.

Suggest how an induction coil produces a spark. **[3 marks]**

08.5 A typical Crookes tube requires a p.d. of 10 kV.

Calculate the electric field in $V\,m^{-1}$ that is needed to make the air conduct if the induction coil produces a spark across a 1 cm gap.

[1 mark]

> **Synoptic links**
>
> 3.2.2.2   3.2.1.3
> 3.7.5.4   3.7.3.2

> **Exam tip**
>
> Remember that X-rays are part of the electromagnetic spectrum.

Knowledge

28 Wave-particle duality and special relativity

Theories of light

Prior to Young's double slit experiment, there were two competing theories about light: Newton's corpuscular theory and Huygens' wave theory. Newton's theory was preferred because he was the more famous scientist.

Theory	Details	Reflection	Refraction	Cannot explain
Newton's corpuscular theory (1670s)	light is made of 'corpuscles' = particles which are perfectly elastic, rigid, weightless, and travel in straight lines	corpuscles bounce off surfaces	corpuscles attracted to boundary, travel faster in denser media (incorrect)	diffraction, interference, polarisation
Huygens' wave theory (1690s)	light is made of waves, with wave fronts constructed from secondary wavelets	wavelets reflected	waves travel slower in denser media (correct)	n/a

Young's double slit experiment
Young investigated interference in this experiment in 1801.

- Light that goes through two slits forms a pattern of bright and dark fringes on a distant screen.
- Waves superpose; either constructively to form a bright fringe, or destructively to form a dark fringe.
- Interference cannot be explained using the corpuscular theory of light.

Wave theory only began to be accepted in the 1850s because corpuscular theory could not explain polarisation.

The discovery of photoelectricity

A **black body** emits all wavelengths of radiation possible at that temperature.

An **ultraviolet catastrophe** is predicted (most energy should be emitted at short wavelengths), but this doesn't match the data – the peak occurs at shorter wavelengths, explained by **Wien's law**.

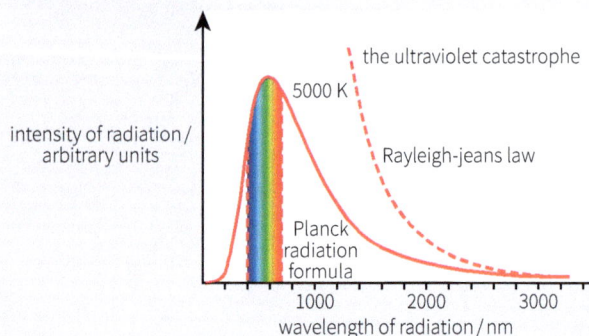

Planck and quanta
Planck suggested that:

- all radiation is emitted or absorbed in **quanta** (now called **photons**)
- photons are not corpuscles
- the energy of the quantum depends on the (single) frequency of the radiation $E = hf$
- there is no catastrophe because higher frequency photons carry more energy.

The observations of the **photoelectric effect** were explained by Einstein using photons. Both wave and particle explanations are needed, depending on the phenomenon being observed.

Explaining the photoelectric effect

Observation	Explanation in wave theory
more intense radiation means more photoelectrons emitted per second	more energy 'accumulated' by metal, so more electrons more photons absorbed by metal, so more electrons
The following three observations are not explained by the wave theory, only using quanta	
minimum frequency below which no electrons emitted, however intense the radiation	minimum frequency = minimum energy of photon = **work function**
instantaneous emission of electrons	electron emitted as soon as photon absorbed
emitted electrons have a range of kinetic energy for a given metal f up to a maximum	$E_k = hf - \Phi$ (Φ = work function)

Work function can be found from a graph of frequency and stopping potential.

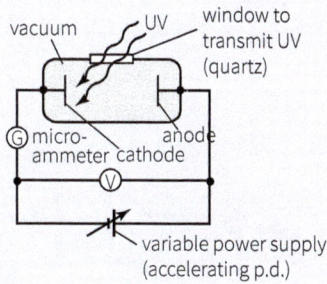

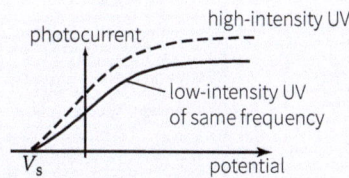

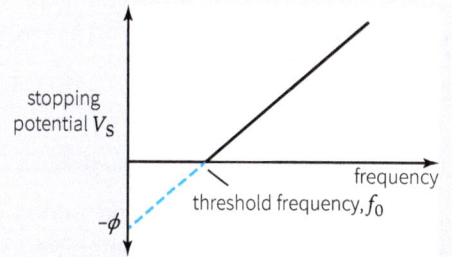

Theories of light – electromagnetic waves (part 1)

Fizeau's determination of the speed of light
Previously, the speed of light was thought to be infinite. Fizeau spun a toothed wheel with n teeth a distance d between mirrors. When spinning at frequency f, light just gets through the gap, so $c = 4\,d\,n\,f$.

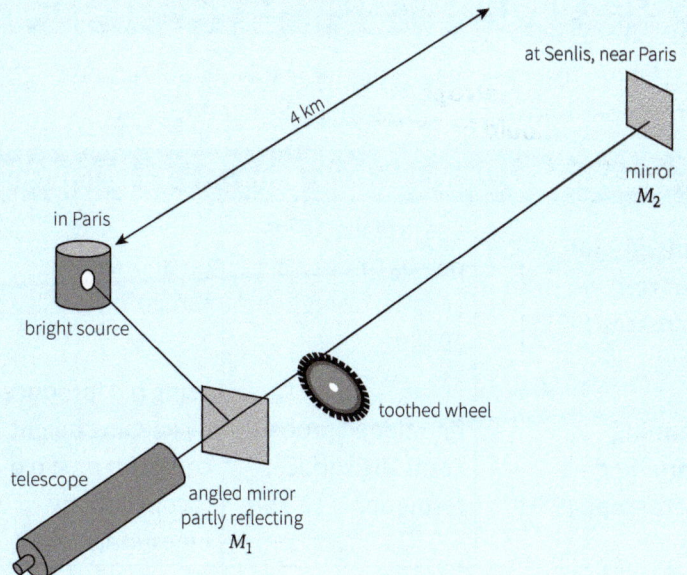

Einstein's theory of special relativity

An **inertial frame of reference** is one moving at a constant velocity (relative to other frames).

The two postulates of Einstein's theory of special relativity (consequences of Maxwell's equations) are:

- physical laws have the same form in all inertial frames

- the speed of light in free space is invariant (as shown in the Michelson–Morley experiment).

Wave–particle duality

Matter waves were proposed first by de Broglie's hypothesis so that $m v \times \lambda = h = $ Planck constant.

Evidence from electron diffraction was proposed later – electrons accelerated to a potential difference V:

$$\tfrac{1}{2} m v^2 = eV$$

$$\lambda = \frac{h}{\sqrt{2\, m\, eV}}$$

In low-energy electron diffraction experiments, increasing V ($V = 150\,V$ gives $\lambda = 10^{-10}\,m$):

- increases energy, hence increases the speed of the electrons

- decreases the de Broglie wavelength

- produces rings in electron diffraction pattern, which are brighter with smaller radius ($n \lambda = d \sin \theta$, $d = $ atom spacing).

The Michelson–Morley experiment

A Michelson–Morley interferometer tests for **aether**; light should travel more slowly when the Earth is moving 'against' the aether than with it. A difference in speed would be seen as a shift in the interference pattern. In reality, no difference is seen, so there is no ether – the speed of light is not affected by the Earth's motion, it is constant.

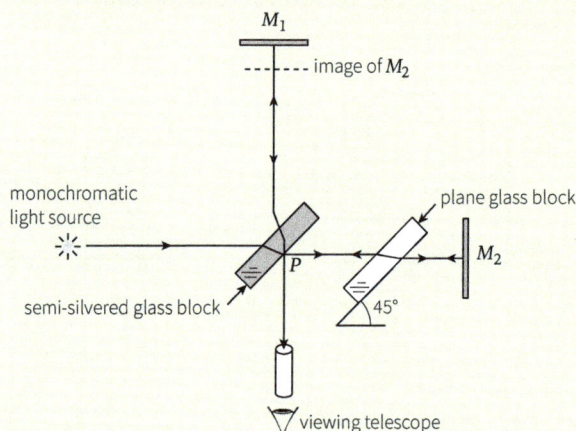

Electron microscopes

Electron microscopes have a higher resolving power than light microscopes because $\lambda_{electron}$ is much less than $\lambda_{visible}$.

Microscope	Mode of operation	Image formed	Focused with	Detail limited by
transmission electron microscope (TEM)	beam of electrons passes through a sample; resolution = $10^{-10}\,m$	on fluorescent screen	magnetic lenses – deflection depends on speed	lens aberration; speed of electrons varies slightly
scanning tunnelling microscope (STM)	fine-tipped probe scans the surface; resolution = $10^{-12}\,m$	using p.d. produced by keeping height constant; or, if p.d. is constant, by measuring height	n/a	resolution of measuring instruments

Theories of light – electromagnetic waves (part 2)

Discovery of electromagnetic waves and speed

Hertz discovered radio waves using:

- an induction coil and capacitor to produce them
- a spark gap in a wire loop/dipole to detect them.

He found:

- the speed by finding the wavelength (by producing standing waves)
- the frequency (using a rotating mirror) and $c = f\lambda$
- they were polarised by a change in signal strength when he rotated the dipole.

Maxwell used his equations for electricity and magnetism to predict the existence of self-propagating waves of oscillating electric and magnetic fields (**electromagnetic waves**) in 1864; before, there was no experimental evidence for them.

His theory suggested that:

$$\text{speed } c = \frac{1}{\varepsilon_0 \, \mu_0}$$

where:

ε_0 = permittivity of free space – relates to electric field strength due to a charged object in free space

μ_0 = permeability of free space – relates to magnetic flux density due to a current-carrying wire in free space.

Consequences of the invariance of the speed of light

Effect	Equation	Definitions	Evidence
time dilation	$t = \dfrac{t_0}{\sqrt{1 - \dfrac{v^2}{c^2}}}$	t_0 = proper time (stationary with respect to the clock) t = time as measured by observer with speed v relative to the clock – time runs more slowly	muons decay – more muons detected at the bottom of a mountain than would be expected from their half-life; if $v \approx c$, time dilated and length contracted so they go further
length contraction	$l = l_0\sqrt{1 - \dfrac{v^2}{c^2}}$	l_0 = proper length (stationary with respect to the clock) l = length as measured by observer with speed v relative to clock – time runs more slowly	
mass and energy	$E = \dfrac{m_0 c^2}{\sqrt{1 - \dfrac{v^2}{c^2}}}$	at zero speed, $E = m_0 c^2$ or rest energy at speed v, $E_k = m c^2 - m_0 c^2$, so $m = \dfrac{m_0}{\sqrt{1 - \dfrac{v^2}{c^2}}}$	Bertozzi measured kinetic energy of electrons at different speeds (electrons heated aluminium plate) – within 10% of predicted values

Graphs of variation of mass and kinetic energy with speed show how mass and energy increase with speed.

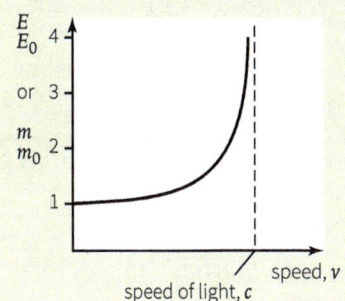

Retrieval

Learn the answers to the questions below, then cover the answers column with a piece of paper and write as many as you can. Check and repeat.

Questions	Answers
1 Name **two** processes that Newton's corpuscular theory explained.	reflection; refraction
2 Name **three** processes that Newton's corpuscular theory could not explain.	diffraction; interference; polarisation
3 What was produced in Young's double slit experiment?	bright and dark fringes on a screen
4 Why was Newton's theory accepted despite not explaining some observations?	Newton was a respected scientist
5 What is a black body?	object that emits all wavelengths possible at a particular temperature
6 What is the ultraviolet catastrophe?	most energy should be emitted at short (ultraviolet) wavelengths
7 What is a photon?	a quantum of light
8 What does the number of photoelectrons emitted per second from a metal surface depend on?	intensity of light (number of photons per second)
9 What **two** things determine the maximum kinetic energy of a photoelectron ejected from a metal surface?	frequency of the incident photon and work function of the metal
10 How did Hertz first detect radio waves?	seeing them make a spark across a gap in a loop
11 In low energy electron diffraction experiments what happens to the de Broglie wavelength as the accelerating p.d. increases?	decreases
12 For the rings produced in electron diffraction, what two observations would be made as the accelerating p.d. increases?	rings are brighter; diameter is smaller
13 What focuses electrons in a transmission electron microscope?	magnetic fields
14 How does a scanning tunnelling microscope work?	fine-tipped conducting probe scans the surface by applying a p.d.
15 What did the Michelson–Morley experiment disprove the existence of?	aether
16 What is an inertial frame of reference?	frame of reference moving with a constant velocity / not accelerating

Put paper here

17	In addition to the laws of physics having the same form in all frames of reference, what is the second postulate of special relativity?	speed of light is invariant / the same in all frames of reference
18	The observation of which particles provided evidence for time dilation and length contraction?	muons
19	What happens to the mass of a particle as it moves close to the speed of light?	significantly increases
20	How did Bertozzi show that the energy of electrons moving close to the speed of light is relativistic?	measured speed of the electrons and the increase in temperature of a metal plate that the electrons hit

Put paper here

🖩 Maths skills

Practise your maths skills using the worked example and practice questions below.

Travelling near the speed of light	Worked example	Practice

Travelling near the speed of light

When objects travel at speeds close to the speed of light c they experience time dilation and length contraction.

For an object travelling at speed v:

$$t = \frac{t_0}{\sqrt{1 - \frac{v^2}{c^2}}}$$

where t is the time measured by an observer and t_0 is the proper time.

Also:

$$l = l_0 \sqrt{1 - \frac{v^2}{c^2}}$$

where l is the length measured by an observer and l_0 is the proper length

Worked example

Question

A muon is travelling at a speed of $0.997c$.

Calculate, in the laboratory frame of reference, the distance travelled by the muon in proper time 2.2×10^{-6} s.

Answer

Calculate the time t in the laboratory frame of reference.

$$t = \frac{2.2 \times 10^{-6}}{\sqrt{1 - \frac{(0.997c)^2}{c^2}}}$$

$$t = \frac{2.2 \times 10^{-6}}{\sqrt{1 - (0.997)^2}} = \frac{2.2 \times 10^{-6}}{0.0774015}$$

$t = 2.84 \times 10^{-5}$ s.

Convert $0.997c$ to SI units, and use distance = speed × time to calculate how far the muon has travelled.

$s = (0.997 \times 3.0 \times 10^8 \, \text{m s}^{-1}) \times 2.84 \times 10^{-5}$ s

$\quad = 8500$ m

Practice

1. A muon is travelling through the atmosphere at a speed of $0.999c$.

   Calculate, in the laboratory frame of reference, the distance travelled by the muon in proper time 3.0×10^{-6} s.

2. A pion, with a mean lifetime of 26 ns at rest, is accelerated to a speed of $0.995c$.

   Calculate the mean lifetime of the pion in the laboratory.

3. In a movie, a spaceship of length 1.0 km flies over a planet at a speed of $0.7c$. Calculate the length of the ship as measured from the planet as it flies past.

01 One of the earliest attempts to measure the speed of light involved two people standing on hills several miles apart with lanterns. It was extremely difficult to produce a reliable value for the speed of light with this method.

01.1 Explain why this method is inconclusive in producing a value for the speed of light, using estimates and calculations. **[2 marks]**

_____

_____

_____

_____

01.2 A more accurate method was found in 1849 when Fizeau set up a toothed wheel that rotated in the path of a beam of light.

Figure 1

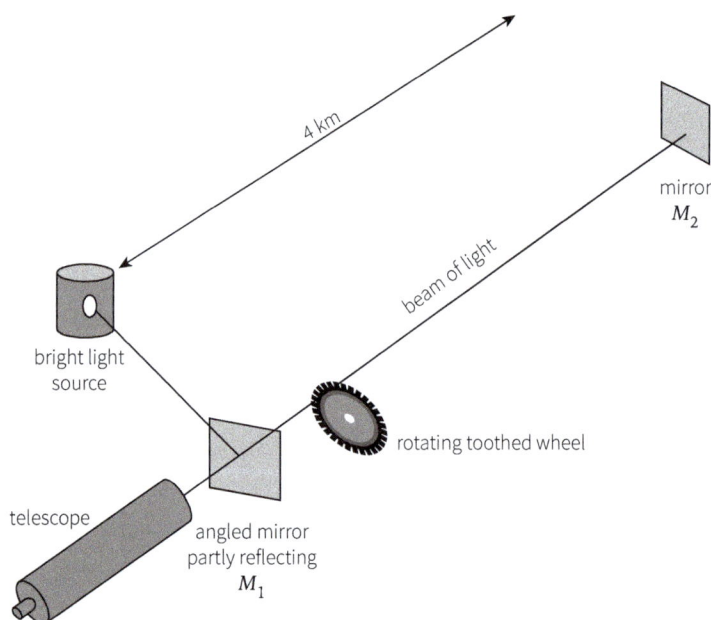

Explain how the equipment in **Figure 1** enabled Fizeau to find a precise value of the time it took light to travel. **[5 marks]**

_____

_____

_____

_____

_____

_____

01.3 In his experiment, Fizeau used a distance of 8630 m between the two mirrors, and the wheel had 720 cogs.

Calculate the frequency of rotation of the wheel in rotations per second using the time it takes the light to travel and the current value of the speed of light. **[3 marks]**

frequency of rotation = _____ rotations

01.4 The Michelson–Morley interferometer also used a partially reflecting mirror.

Compare the use of this mirror in attempting to detect aether with the use of the mirror in Fizeau's method. **[2 marks]**

_____

_____

_____

> **(!) Exam tip**
>
> Remember that comparison questions expect you to say what is the same and what is different in the two scenarios.

02 Electromagnetic waves with a frequency of 1 GHz can be produced using a dipole. A student set up transmitting and receiving dipoles as shown in **Figure 2**. The receiving dipole is connected to a microammeter. Both dipoles are made of metal.

Figure 2

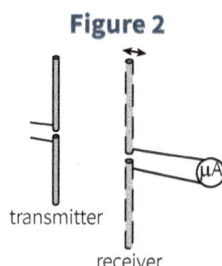

transmitter

receiver

02.1 Suggest and explain what the transmitter needs to be connected to in order to produce electromagnetic waves. **[3 marks]**

_____

_____

_____

_____

02.2 Explain how the waves produce a signal in the receiver. **[2 marks]**

_____

_____

_____

02.3 When Hertz first carried out experiments he found that the reading on the microammeter connected to the receiver decreased to zero as he rotated the dipole so that it was at 90° to the transmitter.

Explain why. **[2 marks]**

_____

_____

_____

02.4 A student put a reflector behind the receiver. They moved the reflector until a standing wave was produced between the transmitter and the reflector. They measured the distance between two positions where the reading on the meter was a minimum as 15 cm.

Calculate the speed of the electromagnetic waves. **[3 marks]**

> **! Exam tip**
>
> Remember that you can use the positions of nodes, or antinodes, in a standing wave to deduce wavelength.

speed = _____ $m\,s^{-1}$

02.5 Maxwell's theory suggested that electromagnetic waves travel with a speed of:

$$c = \sqrt{\frac{1}{\varepsilon_0 \mu_0}}$$

The radio waves produced by the dipole have a frequency that is approximately a million times greater than that of visible light.

Suggest how, according to Maxwell, light and radio waves can travel at the same speed. **[2 marks]**

_____

_____

_____

03 A significant turning point in the development of scientists' understanding of the nature of light came from observations of black-body radiation.

03.1 Describe a black body. **[1 mark]**

03.2 Sketch a graph to show the prediction, according to classical theory, of the energy intensity of light emitted from a very hot object against wavelength.

Explain the problem caused in classical physics by the graph that you have drawn. **[3 marks]**

03.3 The solution to the problem in **03.2** was to postulate that radiation can only be emitted or absorbed in quanta.

Explain how this suggestion overcame the problem in **03.2**.
[2 marks]

03.4 A student read an article that claims 'a force equivalent to 1 kg of light hits your face every year'.

Suggest a connection between the solution suggested in **03.3** and the statement in the article. **[3 marks]**

! Exam tip

Remember that this chapter is about wave–particle duality, so you need to discuss how particles can show wave behaviour.

04 In an experiment with a vacuum photocell, electromagnetic radiation with different wavelengths is shone on a photocell (**Figure 3**).

Figure 3

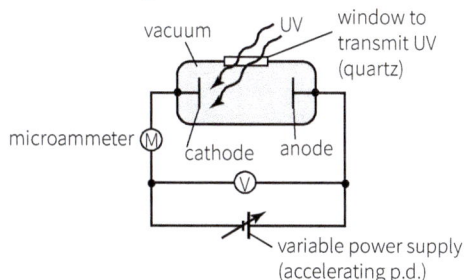

04.1 A student sets the p.d. to zero and measures the current on the microammeter as a function of the intensity of the light.

Sketch a graph of current against intensity. **[2 marks]**

! Exam tip

Sketched graphs do not have to show numbers on axes, but should have labelled axes with units.

04.2 Suggest and explain whether the graph sketched in **04.1** can be explained by the wave theory of light. **[2 marks]**

04.3 Suggest how the equipment in **Figure 3** can be adapted to apply a p.d. to just stop the electrons ejected from the photocathode from reaching the anode. **[2 marks]**

04.4 The new experimental arrangement described in **04.3** is used to collect data to deduce the work function of the metal of which the cathode is made.

Calculate the work function in J of the metal using the data in **Table 1**. **[3 marks]**

Table 1

Frequency / $\times 10^{14}$ Hz	Stopping potential / V
6.00	0.50
9.00	1.75

04.5 Suggest how the data could be used to produce a result that cannot be explained using the wave theory of light. **[2 marks]**

05 A scanning tunnelling microscope (STM) can be used to produce images of atoms.

05.1 State the maximum distance in m of the STM probe above the surface of the sample. **[1 mark]**

05.2 Explain why electrons can be observed to 'tunnel' through small gaps. **[2 marks]**

05.3 Compare the two modes of use of an STM. **[4 marks]**

05.4 In electron microscopy, scientists need to apply a 'relativistic correction' when the electron velocity becomes very high. The accelerating potential difference is adjusted to produce the same electron wavelength as would be achieved without a relativistic effect.

Explain how and why the potential difference needs to be adjusted. **[2 marks]**

> **! Exam tip**
>
> Remember that tunnelling is linked to the wave nature of matter.

06 A teacher demonstrates a visible analogy of electron diffraction. They ask students to look at a point source of green light through a thin gauze. The students rotate the gauze through 360°. This represents the diffraction of electrons through the polycrystalline structure of a very thin layer of metal foil through which electrons are transmitted.

When electrons are diffracted through a thin metal foil, concentric rings are seen on a screen.

> **Synoptic links**
>
> 3.3.2.2   3.8.1.5   3.8.1.6

06.1 In the electron diffraction experiment, the electrons are accelerated to a potential of 3.0 kV.

Calculate the wavelength in m of the electron.

State **one** assumption that you need to make. **[4 marks]**

06.2 The angles between the rings observed in the visible light spectrum are approximately the same as that in the electron diffraction experiment. The wavelength of the green light is 540 nm.

Determine an estimate for the grating spacing in m for the electrons. Use this estimate to calculate an approximate value for the spacing between the threads in the gauze.

State any assumptions that you make. **[4 marks]**

06.3 The teacher changes the colour of the visible light to red.

State and explain the effect on the pattern of maxima observed through the gauze. **[2 marks]**

> **! Exam tip**
>
> You can use equations in your explanation of phenomena.

06.4 Describe and explain what should be done in the electron diffraction experiment that would be analogous to changing the colour of the light as described in **06.3**. **[2 marks]**

06.5 Electron diffraction can be used to investigate the size of nuclei.

Sketch the graph of intensity against angle for the diffraction of electrons by a nucleus. **[2 marks]**

06.6 An electron diffraction pattern suggests a nucleus has a radius of 6.6×10^{-15} m.

Calculate the nucleon number of the nucleus.
$R_0 = 1.1 \times 10^{-15}$ m **[3 marks]**

07 Prior to Young's double slit experiment, there were two competing theories about light: Newton's corpuscular theory and Huygens' wave theory.

07.1 Suggest why Newton's theory of light was not rejected for a very long time. **[2 marks]**

07.2 In 1850, light was discovered to travel slower in water than in air.

Explain the significance of this observation to the debate about which theory of light was correct. **[2 marks]**

07.3 In 1933, Blackett and Occhialini observed pair production in a cloud chamber.

Describe what happens in pair production, and explain why it was not necessary to consider a wave theory explanation for pair production. **[2 marks]**

07.4 The work of Blackett and Occhialini had to be validated by the scientific community.

Describe this process. **[1 mark]**

Synoptic links
3.2.1.3   3.2.2.4

Exam tip
Ensure that your answers to each bullet point are clearly organised.

08 An atom of gold has no colour, but many atoms of gold together have a golden colour. Rutherford found an approximate value for the nucleus of an atom using the distance of closest approach of an alpha particle fired at a thin piece of gold foil.

08.1 Calculate the distance of closest approach in m of a 5.0 MeV alpha particle to a gold nucleus.
atomic number of gold = 79 **[3 marks]**

08.2 Suggest why alpha particles cannot be used to work out the size of the atom. **[1 mark]**

08.3 The inner electrons in a gold atom can have a velocity of 0.58c.

Calculate the relativistic energy in J of the electrons in this orbit. **[2 marks]**

08.4 The transition of electrons between the inner and outer electrons of a gold atom would (without relativistic effects) involve photons with a wavelength in the ultraviolet region of the electromagnetic spectrum.

Suggest the effect of relativistic energy on the energy spacing of levels that means gold is seen as a golden colour. **[3 marks]**

08.5 Muons are detected at an observatory at the top of a mountain and again at a lower observatory 1500 m below the upper observatory.

Show that, if the muons are travelling approximately 1.7 times faster than the electrons in **08.3**, then 75% of muons passing the upper observatory will reach the lower observatory.
half-life of muons at rest = 2.2 µs **[3 marks]**

Synoptic links
3.2.2.2   3.8.1.5

Exam tip
You will need to find the half-life of the *moving* muons to answer **08.5**.

⚙ Knowledge

29 Discrete semiconductor devices

MOSFET

A **metal-oxide semiconducting field-effect transistor** (MOSFET) is used to switch on large p.d. devices using a small p.d., and for switching and amplifying electronic signals.

- A MOSFET is a transistor that consists of two p-n junctions, making a **source**, **drain**, and **gate**
- it has an SiO_2 layer between the gate and other electrodes
- the SiO_2 layer means very high (infinite) resistance/impedance
- it draws zero current from circuit.

A p.d. at the gate:

- produces an electric field in the p-type material
- creates an n-type channel – a current flows between the drain and the source.

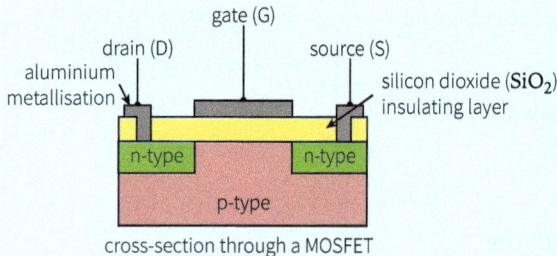

cross-section through a MOSFET

Photodiodes

Photodiodes are used to detect electromagnetic radiation and generate a p.d.

A photodiode:

- has a p-n junction that is exposed (via a transparent window)
- absorbs photons
- produces a current and p.d. because charge carriers are released (**dark current** of about 500 pA flows when there is no light)
- has a typical response of $0.6\,A\,W^{-1}(= 600\,\mu A\,mW^{-1})$.

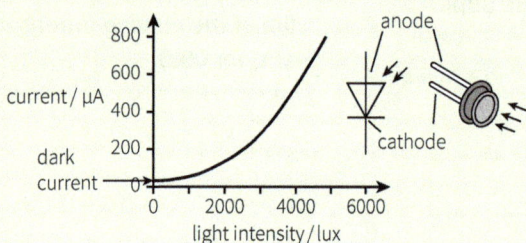

MOSFET as a switch

Operating a MOSFET as a switch

- If the V_{GS} is greater than V_{th} (1 V–2.5 V), the transistor turns on a small p.d. (for example, from a logic system) that can turn on a high-power circuit such as a heater/solenoid.
- If the output current is large, no current is drawn from the logic circuit.

Operating a MOSFET in a circuit

- A current still flows when the gate p.d. is zero; it varies with temperature.
- V_{th} will fall (by $8\,mV\,K^{-1}$) as the temperature increases.
- If the MOSFET is used to switch on an inductor, then a protective diode is needed.
- MOSFET is susceptible to electrostatic charge build up at the gate, so a (high) resistor is connected between G and the ground.

V_{DS}	p.d. between drain and source
V_{GS}	p.d. between gate and source
I_{DSS}	drain current when 'off'
V_{th}	threshold p.d. to switch on

MOSFET input characteristics

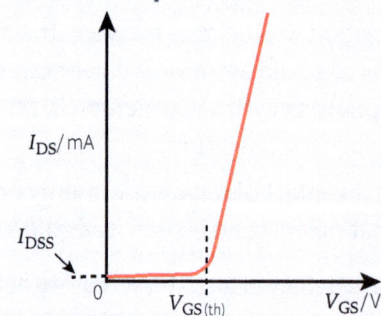

MOSFET output characteristics

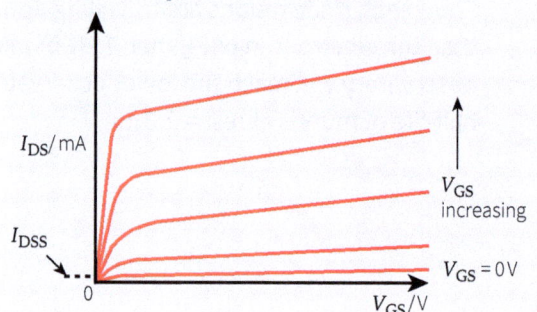

Zener diode

- Forward direction: anode positive with respect to the cathode, so V_Z is positive.
- Reverse direction: anode negative with respect to the cathode, so V_Z is negative.
- Breakdown p.d. for Zener diode is designed to be smaller than normal diode.
- used as a **voltage reference**, where its reverse breakdown characteristic provides a stable voltage across the diode over for a range of currents flowing though it.

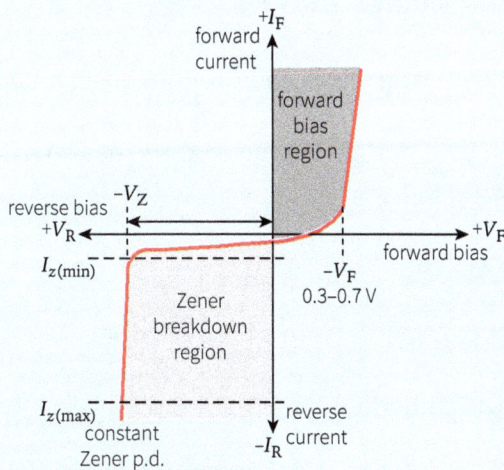

Constant p.d. source

A Zener diode can be used with a resistor to provide a constant p.d. so V_{out} stays at V_Z whatever the input voltage, as long as:

- $I_Z \geq -5\,mA$, and constant
- no current is drawn from the output of the circuit.

Hall effect sensor

A Hall effect sensor is used to measure the magnitude of a magnetic field.

A p.d. is produced when a current-carrying 'slab' of semiconducting material is in a magnetic field – the p.d. is proportional to the magnetic field strength B.

- A magnet embedded in a spinning component (e.g., car engine) produces a changing p.d. in the Hall effect sensor; this is used in **tachometers**.
- As the sensor is tilted, the component of B changes; this is used to monitor attitude.
- As the sensor is moved from the magnet, B changes; this is used in contactless switches.

One advantage of a Hall effect sensor is that they have no contact bounce – mechanical switches can bounce up and down before contact.

Photovoltaics and photoconductives

A photodiode can be operated in a **photoconductive** or a **photovoltaic** mode.

Photoconductive mode (producing current) is reverse-biased:

- $I_D \propto$ light intensity.
- Response time smaller than photovoltaic mode.
- Used in electronics, and as detector in optical systems, and with scintillator to detect atomic particles.

Photovoltaic mode (producing p.d.) is forward biased:

- In solar cells = photodiodes with lots of exposed junctions in photovoltaic mode (hence solar PV).

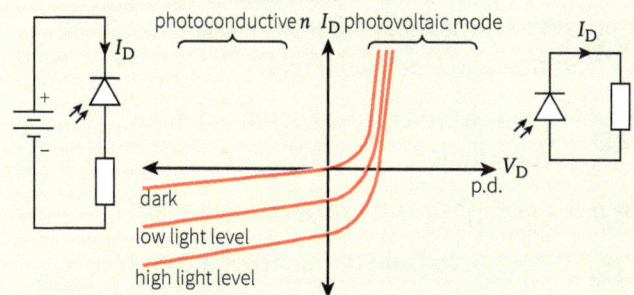

Spectral response

The type of semiconductors used in p-n junctions determines the range of wavelengths over which a photodiode responds. Diodes can be designed to absorb photons of different wavelengths depending on their use.

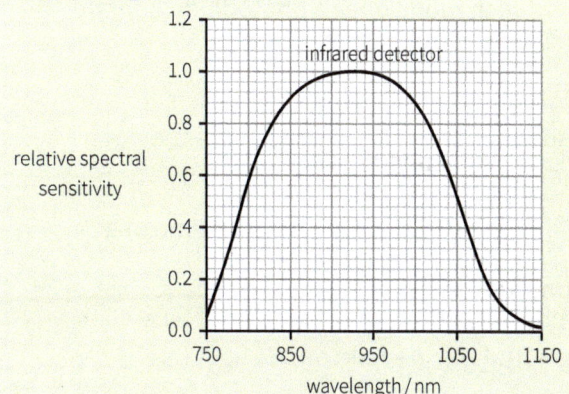

Retrieval

Learn the answers to the questions below, then cover the answers column with a piece of paper and write as many as you can. Check and repeat.

	Questions	Answers
1	Name the three connections on a MOSFET device.	source (S), drain (D), gate (G)
2	How much current does a MOSFET draw from a circuit connected to the gate/drain?	zero
3	Apart from amplification, what is the main use of a MOSFET?	switch on a circuit needing a large current using a small p.d. (at the gate)
4	When is a protective diode needed in the use of a MOSFET?	transistor is using the MOSFET to turn on an inductor
5	What is the main difference between a Zener diode and a normal diode?	Zener diode is designed to break down at a specific, low, reverse p.d.
6	What is a typical breakdown p.d. for a Zener diode?	about 5 V
7	What is the main use of a Zener diode?	constant p.d. source (with a fixed resistor)
8	What are the two terminals of a Zener diode called?	anode and cathode
9	What is the minimum Zener current?	reverse current above which the breakdown pd becomes almost independent of the current
10	What does a photodiode do?	absorbs photons and produces a current and p.d.
11	What is the dark current in a photodiode?	current that flows when there is no light incident on it
12	In which mode does a photodiode operate as photovoltaic?	forward bias mode
13	When a photodiode is used to detect subatomic particles, which other device is required?	scintillator
14	In which mode is the response time of a photodiode shorter?	reverse
15	How does the current produced by a photodiode depend on the light intensity?	current is proportional to the light intensity
16	What determines the spectral response of a photodiode?	type of semi-conductors from which the p–n junction is made
17	A slab of semiconducting material will generate a p.d. across two parallel sides under which two conditions?	slab carries a current and is in a magnetic field
18	How does the p.d. produced in the Hall effect depend on the magnetic field strength?	p.d. is proportional to the magnetic field strength

Put paper here

19 What can happen in a mechanical switch that does not happen with a Hall effect switch?

20 How is a Hall effect probe used in a tachometer?

Put paper here

switch can 'bounce' and not make complete contact when pressed

magnet is attached to spinning part of the engine, and the Hall effect sensor tracks speed from a changing p.d.

Maths skills

Practise your maths skills using the worked example and practice questions below.

Zener diodes as a constant-p.d. source	Worked example	Practice
Zener diodes are marked with their Zener p.d. V_Z and their maximum power rating P_Z. These values can be used to calculate the maximum current I_{max} flowing through the diode and the series resistor R_s that should be used in the circuit to ensure that there is a current flowing through the diode. $$I_{max} = \frac{P_z}{V_z}$$ For a circuit with a supply p.d. V_s, the minimum value of resistor in series with the diode is given by: $$R_{min} = \frac{(V_s - V_z)}{I_{max}}$$	**Question** A Zener diode is marked 4.3 V, 2.0 W. It is connected in a circuit with a 12 V power supply. Calculate the maximum current and minimum value of the series resistor for this diode. **Answer** Substituting the values into the first equation: $$I_{max} = \frac{2.0\,W}{4.3\,V}\ 0.47\,A$$ $V_s = 12\,V;\ V_Z = 4.3\,V$ Substituting the values into the second equation: $$R_{min} = \frac{12\,V - 4.3\,V}{0.47\,A}$$ $R_{min} = 16\,\Omega$	**1** A Zener diode is marked 5.1 V, 500 mW. Calculate the maximum current for this diode. **2** A 2.7 V Zener diode is used in a circuit through which a maximum current of 10 mA flows. Calculate the power rating of the Zener diode. **3** A Zener diode is used in a circuit to produce a constant pd of 10 V from a 12 V power supply. The maximum current from the power supply is 100 mA. Calculate the minimum value of resistor that should be used in the circuit.

01 **Figure 1** shows a MOSFET used in a circuit to switch on a motor.

Figure 1

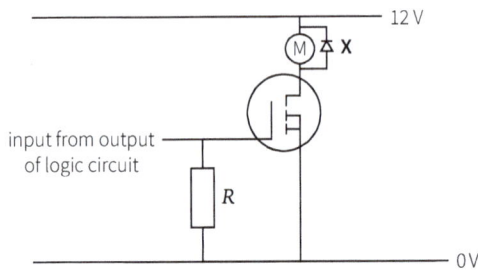

input from output of logic circuit

01.1 Annotate **Figure 1** to show the gate (G), drain (D), and source (S).

[1 mark]

01.2 Explain why a MOSFET is useful as a switch in a logic circuit.

[2 marks]

_____

_____

_____

01.3 Explain why resistor R is needed in the circuit. **[1 mark]**

_____

_____

01.4 State the name of component **X** in the circuit and explain why it is needed. **[3 marks]**

_____

_____

_____

_____

02 Streetlights are designed to come on in the dark. **Figure 2** shows a simple circuit that controls a streetlight.

Figure 2

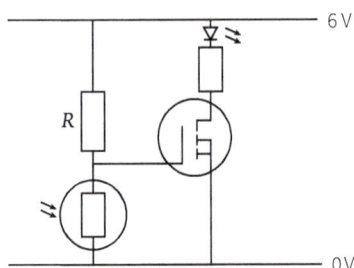

> ! **Exam tip**
>
> Make sure you know the basic structure of each device.

> ⊙ **Synoptic link**
>
> 3.5.1.5

02.1 The MOSFET has a threshold p.d. of 2.2 V.
Describe what this means. **[1 mark]**

_____

_____

02.2 Explain why the LED is switched on when it becomes dark. **[3 marks]**

_____

_____

_____

_____

02.3 The LED is to be switched on when the resistance of the LDR is 100 kΩ.

Determine a suitable value for resistor R. **[3 marks]**

> **! Exam tip**
>
> The LDR and resistor R are a simple potential divider arrangement.

resistance = _____ kΩ

03 This question is about Zener diodes.

03.1 Compare the characteristics of a diode with a Zener diode. **[4 marks]**

03.2 A Zener diode is used to produce a stabilised 4.3 V from a 9.0 V source.

Draw the Zener diode correctly placed between **X** and **Y** in **Figure 3**.
[2 marks]

> **⊗ Synoptic links**
>
> 3.5.1.1   3.5.1.4

Figure 3

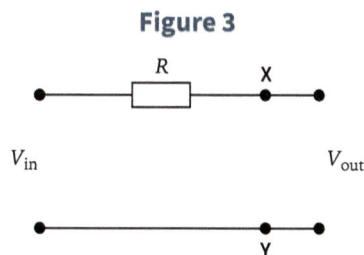

V_{in} V_{out}

> **! Exam tip**
>
> Compare questions mean you need to consider similarities as well as differences.

03.3 The maximum power rating of the Zener diode is 1.3 W.

Calculate the maximum current in A through the Zener diode.
[1 mark]

03.4 Calculate the minimum resistance in Ω of the resistor R for safe working with the diode. **[2 marks]**

04 The circuit diagram in **Figure 4** has been designed so that a Zener diode is used to provide a constant p.d. source across a load resistor of 440 Ω.

Figure 4

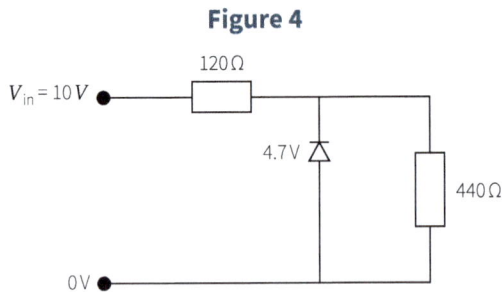

$V_{in} = 10\,V$, 120 Ω, 4.7 V, 440 Ω, 0 V

04.1 Calculate the current in A through the load resistor. **[2 marks]**

04.2 Determine the current in A through the 120 Ω resistor. **[2 marks]**

04.3 Calculate the current in A through the Zener diode. **[1 mark]**

04.4 **Figure 4** does not give the power rating for the Zener diode. A choice of two is available: 250 mW and 500 mW.

Explain which Zener diode should be used. **[2 marks]**

05 A photodiode used in a smoke detector has a peak sensitivity of 850 nm. In the smoke detector, a beam of infrared light is directed towards one side. When smoke enters the enclosure, it scatters the beam causing some light to fall on the photodiode.

05.1 State the bias direction of a photodiode in a circuit when it is used in photoconductive mode. **[1 mark]**

05.2 Sketch a graph of relative spectral intensity against wavelength for this photodiode. **[3 marks]**

05.3 The photodiode has a sensitivity of 0.62 A W^{-1}. The light sensitive area of the photodiode is 1 mm^2.

Calculate the photocurrent in A if the intensity of light falling on the photodiode is 10 W m^{-2}. **[3 marks]**

05.4 Explain how the photodiode can be used to trigger an alarm when there is smoke. You may use a diagram in your answer. **[4 marks]**

06 A Hall effect proximity sensor can be used to adjust airbag settings in cars based on the position of the car seats.

06.1 Describe how the sensor can give an output dependent on the seat position. **[3 marks]**

06.2 The magnetic flux density varies from 3.5×10^{-4} T to 0.014 T. The sensitivity of the device is quoted as 9 mV per mT.

Calculate the output p.d. range in mV. **[2 marks]**

Synoptic links

3.5.1.1   3.5.1.4

Exam tip

Remember your basic circuit rules – the diode and the resistor are in parallel.

Exam tip

Look at the units – how can you obtain a value for power from W m^{-2} and m^{-2}?

Exam tip

A diagram may help you organise your thoughts – when do you want the p.d. to be high? When it is dark or light?

06.3 The Earth's magnetic flux density varies from 25 to 65 μT.

Calculate the maximum p.d. output in mV due to the Earth's magnetic field. Explain whether this will **affect the function of the** sensor circuit. **[2 marks]**

Exam tip

Compare this pd output to the minimum needed for the chair position.

07 **Figure 5** shows how a Hall effect sensor could be used, in principle, to measure the speed of a car and the distance it has travelled.

Figure 5

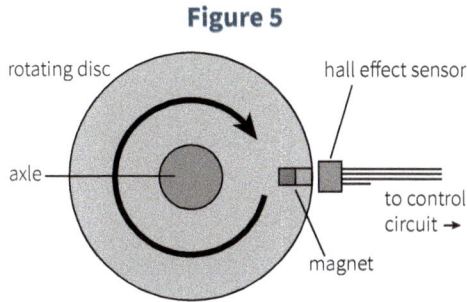

07.1 Explain how data from this arrangement could be used to determine the distance travelled. **[3 marks]**

07.2 The diameter of the disk is 5.0 cm and the pulses are 6.5 ms apart.

Calculate the speed of the car in $m\,s^{-1}$ at that point in the journey. **[2 marks]**

Exam tip

$Speed = \dfrac{distance}{time}$, but how can you calculate the distance?

07.3 Suggest how the accuracy of the distance and speed measurements could be improved. **[1 mark]**

08 **Figure 6** shows a photodiode circuit. The photodiode has a sensitivity of $0.6\,A\,W^{-1}$ and a dark current of 5 nA.

Figure 6

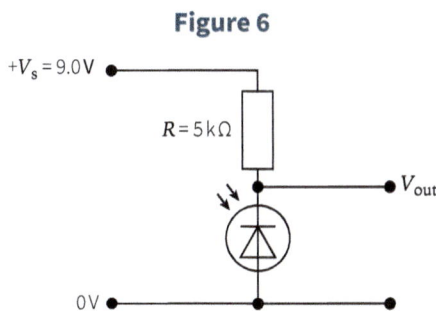

Synoptic links

3.5.1.1   3.5.1.4

08.1 State the direction of the photodiode bias. **[1 mark]**

08.2 State what is meant by dark current. **[1 mark]**

08.3 Calculate the current in A in the diode when light of power 0.2 mW falls on the active area of the photodiode. **[1 mark]**

08.4 V_S is 9.0 V and R is 5 kΩ.

Calculate the output p.d. V_{out} in V when 0.2 mW falls on the active surface. **[2 marks]**

Exam tip

If in doubt, look at the units $A\,W^{-1}$ and W – you need to find A.

Knowledge

30 Analogue and digital signals

Analogue and digital signals

Bits and bytes

An analogue signal can have any value, but a digital signal has two states, such as 1/0, on/off, high/low, T/F.

Bit means a **b**inary dig**it**, usually indicated by a 0 or a 1; there are 8 bits in a byte.

analogue signal $\rightarrow$ ADC $\rightarrow$ 000, 000, 001, 100,

Comparing analogue and digital signals

Both acquire **noise** when transmitted due to stray electrical signals interfering with the signals, but noise can overwhelm analogue signals.

Analogue sensors generate analogue p.d.s (microphones, Hall effect probes, photodiodes, pH/moisture/pressure/lambda (oxygen) sensors) or by changing p.d.s as a result of changing resistance (thermistors, accelerometers, strain gauges).

Analogue-to-digital conversion (ADC)

To convert an analogue signal to a digital signal:

- sample it, taking measurements at a frequency called the sampling rate of the amplitude of the signal (in V) to the nearest voltage level (minimum sampling rate for audio about $2 \times 20\,\text{kHz} = 40\,\text{kHz}$)

- number of voltage levels = resolution
- each value is converted to a binary number, equivalent to two voltage levels = **quantisation**
- analogue signal is reconstructed from bits using digial-to-analogue (DAC) conversion.
- best quality signals have high sampling rate and low resolution

Advantages and disadvantages of digital sampling

Advantages: easier to process, store, reproduce; signal less susceptible to noise; many signals can be sent down the same transmission line.

Disadvantages: reconstruction not exactly the same; low sampling rate means higher frequencies can be lost (= **aliasing**); requires more bandwidth than analogue.

Pulse code modulation: many digital signals can be sent at the same time down the same transmission line using a parallel-to-serial converter to send bits from many signals.

Analogue signal processing

LC resonance filters

A parallel circuit containing an inductor (coil of wire on a core) L, and a capacitor (parallel plates separated by a dielectric) C, will resonate.

- It is analogous to a mass–spring system.
- The frequencies in a signal can be changed; a signal can be 'filtered' using an LC circuit.
- An LC circuit can be tuned to receive a radio signal via an aerial.

An energy (voltage) response curve shows the variation of the output p.d. with frequency.

Quality factor or $Q = \dfrac{f_0}{f_B}$; if the peak is narrow, Q is large and only a narrow range of frequencies 'pass' (low-pass or high-pass).

f_0 = resonant frequency of filter

f_B = bandwidth (range of frequencies) of filter at the half power points (power $\propto V^2$, so $\frac{1}{2}$ power at $\frac{1}{\sqrt{2}}V_{\text{peak}}$).

Equation	Mass–spring system	LC circuit	Component
Inertial component	mass m	inductance L	$L \equiv m$
'Elastic' component	spring constant k	capacitance C (variable)	$C \equiv \dfrac{1}{k}$
Resonant frequency	$f_0 = \dfrac{1}{2\pi}\sqrt{\dfrac{k}{m}}$	$f_0 = \dfrac{1}{2\pi\sqrt{LC}}$	

Ideal operational amplifiers

An ideal operational amplifier (op-amp) can perform mathematical operations:

- it needs a power supply, $\pm V_S$
- it has inverting and non-inverting inputs
- in an open loop (no feedback), $V_{out} = A_{OL}(V_+ - V_-)$, where A_{OL} = open-loop gain
- an ideal op-amp has infinite input resistance (it draws no current) and zero output current.

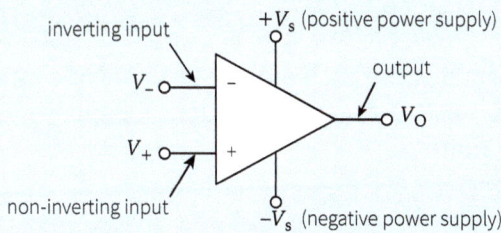

It can be used as a comparator – if $(V_+ > V_-)$, $V_{out} = +V_S$, but if $(V_+ < V_-)$, $V_{out} = -V_S$.

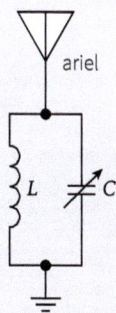

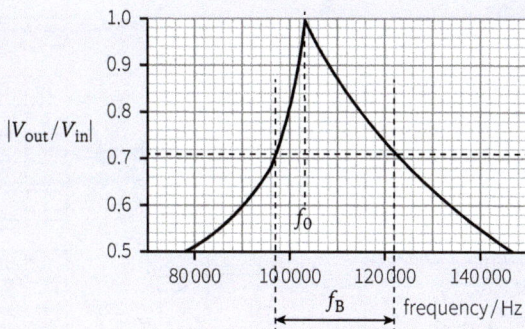

Op-amp uses

Inverting amplifier

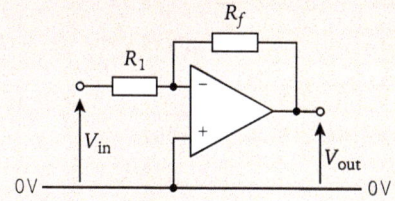

$$\frac{V_{out}}{V_{in}} = -\frac{R_f}{R_{in}}$$

- Non-inverting input = virtual earth, at 0 V;
- Used in virtual earth analysis, $V_{out} = -V_{R_f}$

Non-inverting amplifier

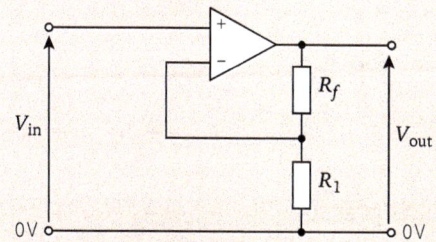

$$\frac{V_{out}}{V_{in}} = 1 + \frac{R_f}{R_1}$$

- No virtual earth
- Provides infinite resistance – buffer, acts as voltage follower.

Summing amplifier

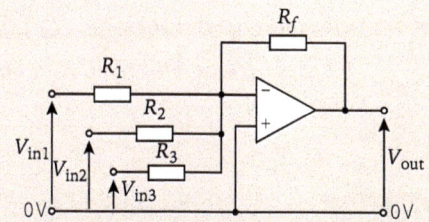

$$V_{out} = -R_f\left(\frac{V_1}{R_1} + \frac{V_2}{R_2} + \frac{V_3}{R_3}\right)$$

- Use, for example, in an audio mixer.

Difference amplifier

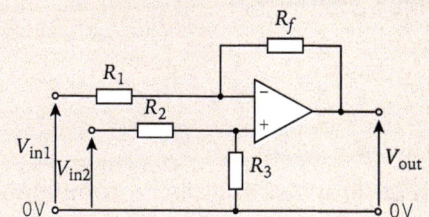

$$V_{out} = (V_+ - V_-)\frac{R_f}{R_1}$$

- Use, for example, in noise cancellation, EEG machines.

Real operational amplifiers

Real operational amplifiers may:

- not have an input resistance that is infinite
- not have infinite open-loop gain
- have a limited frequency response.

For any given device:

gain × bandwidth = constant

Different types of counter

- binary counter – outputs taken together form binary numbers
- modulo-n counter – basic counter with a logic driving the reset pin
- BCD counter – binary coded decimal; counts to 10, used in digital displays
- Johnson counter – has 10 outputs which turn on in sequence; decade counter

Astables

The astable is an oscillator that can provide a clock pulse. Clock (pulse) rate (frequency) depends on pulse width t_W

period $t_p = t_{on} + t_{off}$

duty cycle $= \dfrac{t_{on}}{t_p} \times 100\%$

A charging or discharging RC circuit + NOT gate produces a square pulse.

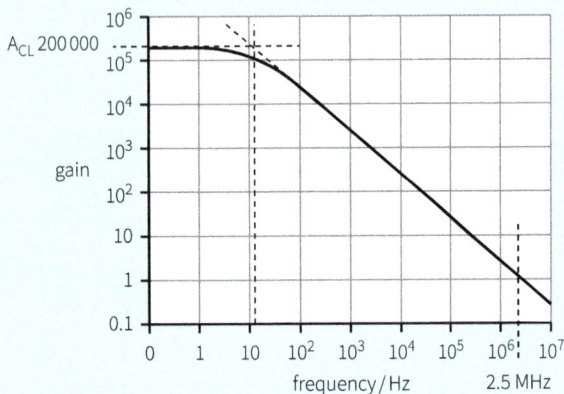

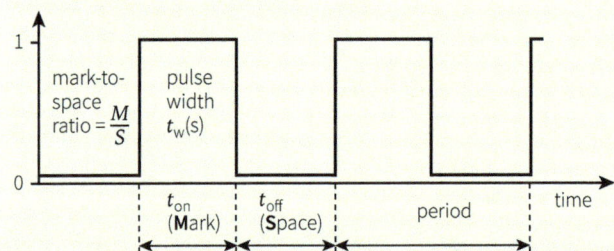

Sequential logic

Counting circuits are used for non-static inputs, and have many outputs.

Input	Detail
clock	input for pulses that are counted
enable	has to be high for counter to work
reset	when high, counter is set to zero
up/down	counts up when high, counts down when low

Boolean algebra and truth tables

Boolean algebra can be related to truth tables and logic gates.

Basic logic symbols

NOT $\triangleright$ $\bar{A}$ = not A

AND $\mathcal{D}$ A . B = A and B

OR $\mathcal{D}$ A + B = A or B

A **truth table** shows outputs for combinations of inputs.

Gates can be combined to perform different functions; for example, NAND, NOR, and EOR.

NAND gate:
output is NOT ON when inputs **A** AND **B** are ON

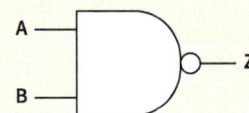

A	B	Z
0	0	1
0	1	1
1	0	1
1	1	0

NOR gate:
output is NOT ON if inputs **A** OR **B** are ON

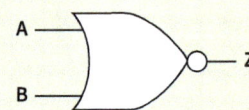

A	B	Z
0	0	1
0	1	0
1	0	0
1	1	0

EOR gate:
output is ON if inputs **A** OR **B** are ON, but not both

A	B	Z
0	0	0
0	1	1
1	0	1
1	1	0

Principles of data communication systems

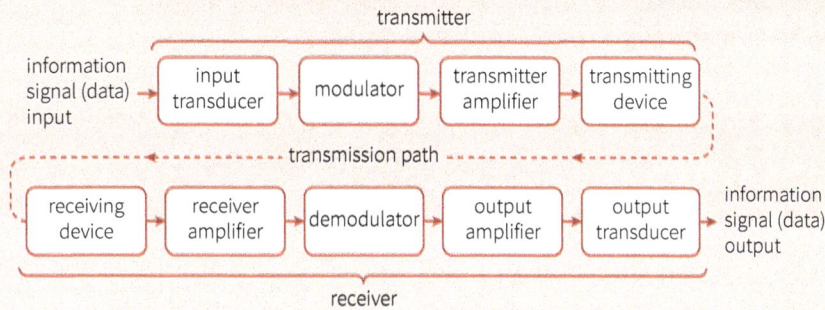

Transmission

If a very high frequency carrier wave (radio/microwave) is **amplitude modulated** (AM) or **frequency modulated** (FM), then it is:

- sent directly as a ground wave; wave can be diffracted by hills / around the Earth to beyond the line of sight
- reflected/refracted from the ionosphere (can get reflections from ground = interference)
- sent to/from a satellite in low earth orbit, or geostationary orbit; up-link (to satellite) at different frequency from down-link (from satellite) so that amplified down-link does not overwhelm up-link received by satellite (receivers would be de-sensitised).

The carrier wave has:

- bandwidth AM $= 2 f_m$ where f_m = max frequency in signal, typically $f_m = 4$ kHz, channels > 9 kHz apart
- bandwidth FM $= 2 (\Delta f + f_m)$, where Δf = frequency deviation, typically $f_m = 15$ kHz, $\Delta f = 75$ kHz, channels > 0.5 MHz apart
- traffic rates can be increased by **time-division multiplexing**; slotting packets of data between each other, then reconstructing (used in telephone networks).

Transmission path	Principle	Use	Power $\frac{1}{2}$ in	Cost	Range	Bandwidth	Security	Affected by noise?
metal wire (co-axial cable, twisted pair cable, plain copper)	p.d. applied	televisions, computer networking	co-axial – 5 m; twisted pair –12.5 m	low	short	low – medium	low	yes – a lot
optic fibre	pulses of light sent	networking, multimedia	1 km	high	medium	very high	high	no
em wave / satellites	modulated carrier wave sent	wifi, wireless, multimedia	n/a	varies	long	very high	very low	not if digital

⟷ Retrieval

Learn the answers to the questions below, then cover the answers column with a piece of paper and write as many as you can. Check and repeat.

	Questions		Answers
1	What is the difference between an analogue and a digital signal?	Put paper here	analogue signal can have any value; digital signal can only be, for example, on/off, 1/0, high/low, T/F
2	What is sampling in the context of communication?	Put paper here	measuring the value/amplitude of an analogue signal at certain time intervals
3	What is the resolution of an analogue to digital conversion?		smallest difference between the value of a sample, which depends on the number of voltage levels
4	What is noise in the context of communication?	Put paper here	additional signal acquired by a signal when it travels, due to stray electrical signals
5	What is pulse code modulation?	Put paper here	process of converting a parallel signal to a series of bits so many signals can be sent at the same time down the same transmission line
6	In an *LC* resonance filter, what are the inductor and capacitor analogous to?	Put paper here	inductor is analogous to the mass; capacitor is analogous to the spring
7	Why does a tuning circuit contain a variable capacitor?		enables you to change the resonance frequency to that of the carrier wave frequency
8	In an *LC* circuit energy response curve, what is the quality factor?	Put paper here	ratio of the peak frequency to the bandwidth
9	What is the open-loop gain of an ideal operational amplifier?	Put paper here	infinite
10	What is the input resistance of an ideal operational amplifier?		infinite
11	Name **four** types of use of an operational amplifier.	Put paper here	inverting, non-inverting, summing, difference
12	In Boolean algebra, what is the difference between 'A . B' and 'A + B'?		'A . B' means 'A AND B'; 'A + B' means 'A OR B'
13	What is the difference between a BCD counter and a Johnson counter?	Put paper here	BCD counter produces an output that counts to 10; Johnson counter has 10 outputs that turn on in sequence
14	Which type of external circuit determines the clock rate of an astable?		*RC* (resistance/capacitor) circuit
15	What does modulation mean in communication?	Put paper here	to add an information signal to a (high frequency) electromagnetic carrier wave
16	What is the difference between AM and FM?		AM means amplitude modulated; FM means frequency modulated

17	How can ground waves be used in communication beyond the line of sight?	they are diffracted around the Earth / hills
18	Why is the down-link frequency of a signal from a satellite different from the frequency of the up-link?	so that the satellite receiver is not de-sensed
19	Which has a larger bandwidth, AM or FM?	FM
20	What is time-division multiplexing?	dividing a signal into packets, slotting packets of data between each other to send more down a transmission line

Put paper here

Maths skills

Practise your maths skills using the worked example and practice questions below.

LC resonance filters

LC resonance filters are often used to remove unwanted frequencies when processing analogue signals.

In a resonance circuit, the maximum output pd occurs at a frequency f_0 given by:

$$f_0 = \frac{1}{2\pi\sqrt{LC}}$$

where L is the inductance of the inductor and C is the capacitance of the capacitor.

Worked example

Question

A filter is designed to reduce interference from a mains supply with a frequency of 50 Hz. The filter uses an inductor with an inductance of 2.0 H.

Calculate the value of capacitor that should be used to filter the mains supply.

Answer

The resonant frequency of the circuit should be the same as the frequency that is to be filtered.

$f_0 = 50\,\text{Hz}; L = 2.0\,\text{H}$

Rearranging the equation to make C the subject and then substituting the values in gives:

$$C = \frac{1}{L} \times \left(\frac{1}{2\pi f_0}\right)^2 = \frac{1}{2.0} \times \left(\frac{1}{100\pi}\right)^2$$

$$= 5.1 \times 10^{-6}\,\text{F}$$

Practice

1. A resonance circuit has an inductor of 8.0 H and a capacitor of 1.2 µF.

   Calculate the resonant frequency of the circuit.

2. A filter is designed to filter out frequencies with a peak of 160 Hz.

   Calculate the value of capacitor that should be used with an inductor of inductance 100 mH.

3. A 1.0 µF capacitor is used in a circuit that has a resonant frequency of 4.1 kHz.

   Calculate the inductance of the inductor in the circuit.

Practice

Exam-style questions

01 Part of an audio recording is shown in **Figure 1a**.

Figure 1a

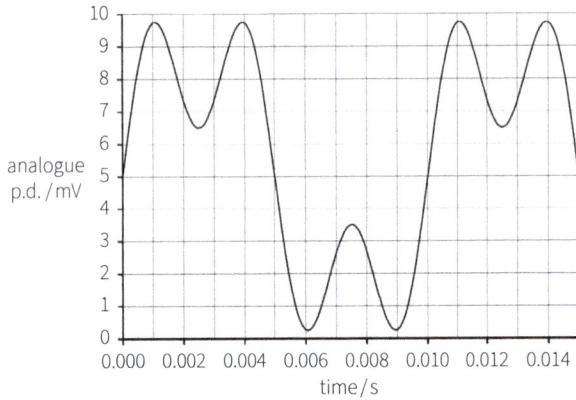

Figure 1b

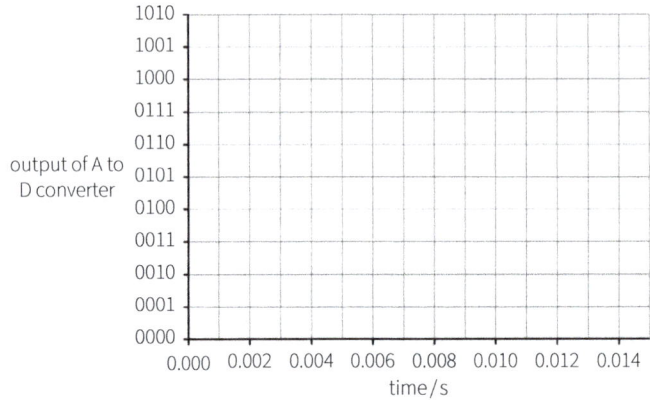

01.1 The signal is sampled at a rate of 500 Hz.

Complete the output of the analogue to digital conversion on **Figure 1b**. **[2 marks]**

01.2 Comment on the sampling rate used in **01.1**. **[2 marks]**

_____

_____

_____

> (!) **Exam tip**
>
> You should clearly distinguish between the effects of sampling rate and resolution on the quality of a signal.

01.3 The audio recording lasts for 3 minutes.

Calculate the storage required in bytes. **[3 marks]**

storage = _____ B

01.4 Suggest **one** change, apart from to the sampling rate, that could improve the accuracy of the reproduced signal.

State **one** disadvantage of the change you suggest. **[2 marks]**

_____

_____

_____

01.5 Explain why the transmission system for transferring the audio recording incorporates a parallel-to-serial converter. **[2 marks]**

_____

_____

_____

02 An engineer is selecting components to make a first-stage filter for a simple radio receiver.

02.1 Sketch a circuit that could enable the engineer to do this. **[2 marks]**

02.2 The circuit contains a 2.3 mH inductor.

Determine the required variable capacitor from **Table 1** if the circuit needed to receive a frequency of 1.1 MHz. Show your working.

[3 marks]

Table 1

Capacitor	Capacitor range / pF
A	1–20
B	20–40
C	40–60

capacitor = _____

02.3 The Q value for the circuit is 50.

Sketch the response curve for the circuit on **Figure 2**.

Explain the shape of the graph using a calculation. **[4 marks]**

Figure 2

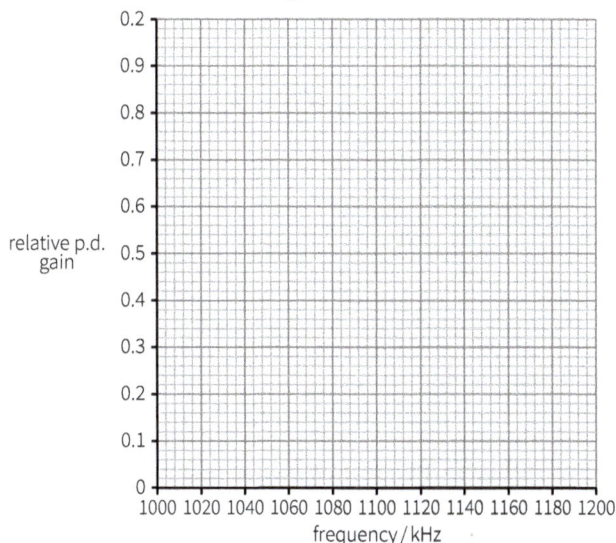

02.4 The equation for the resonant frequency of an LC circuit can be explained using a mass–spring analogy.

Explain:

- the analogy between the values of the components in an LC circuit and those in a mass–spring system
- how the analogy explains the resonant frequency, and how the energy is stored.

You may wish to use equations in your answer. **[6 marks]**

_____

_____

_____

_____

_____

_____

_____

_____

_____

Exam tip

If a question says 'You may wish to' then you should do what it says.

03 The inputs and outputs to a difference amplifier are show in **Table 2**. The supply potential difference was ± 24 V.

Table 2

Inverting input p.d. / V	Non-inverting input p.d. / V	Output p.d. / V
0.0	0.0	0.0
0.0	1.0	4.0
0.0	2.5	10.0
1.0	0.0	−4.0
2.5	0.0	−10.0
3.0	3.0	0.0
−3.0	3.0	24.0

03.1 Describe **one** similarity and **one** difference between a circuit where an operational amplifier is being used as a differential amplifier and a circuit where it is being used as a comparator. **[2 marks]**

03.2 State how you know from **Table 2** that the operational amplifier is being used as a differential amplifier and not as a comparator. **[1 mark]**

03.3 The input resistance R_{in} to both inputs is 10 kΩ.

Deduce the value of the feedback resistance R_f in kΩ. **[2 marks]**

$$V_{out} = (V_+ - V_-)\frac{R_f}{R_{in}}$$

03.4 The circuit is adapted to detect irregular heart rhythms. The inputs come from electrodes placed on the human body. They are of the order of mV, but the electrodes also pick up a 50 Hz signal from the mains. To reduce the effect of this noise, a third 'indifferent' electrode is placed somewhere else on the body.

Explain why the difference amplifier is a good choice to produce a system of monitoring irregularities in heart rhythms. **[3 marks]**

04 A summing amplifier can be used to change a digital signal to an analogue signal, as shown in **Figure 3**.

Figure 3

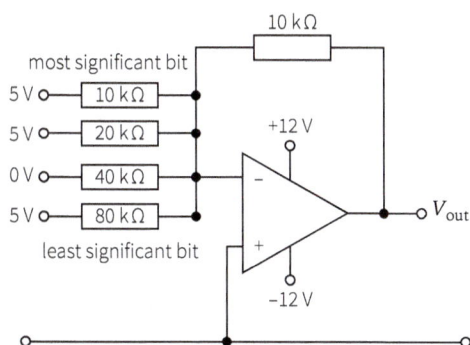

04.1 Write the binary number that forms the input of the digital-to-analogue converter. **[1 mark]**

04.2 Calculate the output p.d. in V. **[2 marks]**

$$V_{out} = -R_f \left(\frac{V_1}{R_1} + \frac{V_2}{R_2} + \frac{V_3}{R_3} + ... \right)$$

04.3 A student wishes to use another amplifier circuit to convert the output of this circuit to the binary number. They want to use a feedback resistor of the same value.

Suggest a type of circuit they can use to turn the output of this circuit into the binary number, and suggest a value for the input resistance. **[3 marks]**

$$\frac{V_{out}}{V_{in}} = -\frac{R_f}{R_{in}}$$

04.4 State the **two** principles that you need to use to show that $\frac{V_{out}}{V_{in}} = -\frac{R_f}{R_{in}}$ for the amplifier used in **04.3**. **[2 marks]**

05 A person with a key can gain access to a high security building at night. A person with a swipe card can only gain access during the day. Assuming the conditions described, give the following inputs to a logic gate system as shown in **Table 3**.

Table 3

Input	Condition	State
A	valid card swiped	1
B	night time	0
C	correct key used	1

05.1 Write the Boolean equation expressing the conditions described.

[1 mark]

05.2 Draw the logic circuit for the Boolean equation in **05.1**.   [2 marks]

05.3 Annotate any intermediate inputs and complete the truth table (**Table 4**) for the circuit in **05.2**.   [3 marks]

Table 4

A	B	C	D	E	F
0	0	0			
0	1	0			
1	0	0			
1	1	0			
0	0	1			
0	1	1			
1	0	1			
1	1	1			

05.4 Once the person has met the access requirements, the door closes for 10 seconds while a camera takes a photograph, and then opens. The person can see the time counting down on a display.

Suggest which type of counter should be used for this function. Describe how the circuit can be adapted to enable the door to open after 10 seconds.   [3 marks]

> **! Exam tip**
>
> You should look for patterns in truth tables that confirm the arrangement of the gates.

06 A string of festive lights has variety of different modes. In one mode, the brightness of each lamp gradually increases to a maximum value, then turns off for a short period before the sequence repeats.

06.1 A circuit with a capacitor and a resistor can be used to produce a p.d. that increases from zero to a maximum value for this mode.

Sketch a circuit with a power supply, resistor, capacitor, and voltmeter. The p.d. reading on the meter should increase from zero when the circuit is switched on.   [2 marks]

> **⊛ Synoptic link**
>
> 3.7.4.4

06.2 The resistor in the circuit has a resistance of $330\,k\Omega$.

Determine an estimate for the value of the capacitance in farads that would produce a noticeable increase in brightness as described.

[2 marks]

06.3 In another mode, the lights flash on and off. The duration that the lights are on is the same as the duration that they are off. The bulbs go on and off four times in 10 seconds.

Calculate the clock rate in Hz and the duty cycle of the pulses being used to power the lights.   [4 marks]

> **! Exam tip**
>
> You should be clear about the difference between t_{on} and t_{p}.

06.4 A pulse-generating circuit uses an RC circuit and a NOT gate to produce pulses of period $1.3\,RC$.

Deduce the value of the capacitance in farads in this circuit that could be used to produce the stream of pulses described in **06.3**. Assume that the resistance of the resistor is $330\,k\Omega$.   [2 marks]

07 A student designs an experiment to find the Young modulus of the material of a wire sample.

07.1 Describe the measurements the student would need to take, and how to use those measurements to determine the Young modulus. **[6 marks]**

Synoptic links

3.4.2.2   3.5.1.3   3.5.1.5

07.2 The student decides to use a different method to measure the change in the length of the wire. They connect the wire in the circuit shown in **Figure 4**. The wire is then attached and stretched as before.

Figure 4

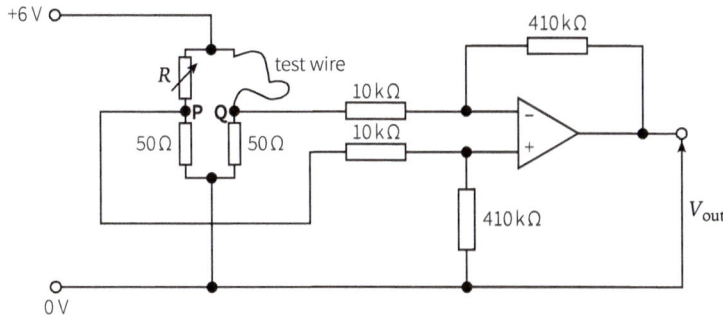

Explain how R can be adjusted so that the output p.d. is zero. **[3 marks]**

07.3 The wire is put under tension so that the reading on the voltmeter measuring the output p.d. is 12.5 V.

Calculate the change in potential in V at the inverting input. **[2 marks]**

$$V_{out} = (V_+ - V_-)\frac{R_f}{R_{in}}$$

Exam tip

In synoptic questions, it is important to clearly identify information that you are given so you can work out which equations to use.

07.4 When the output p.d. was zero, the potential at point **P** was 5.688 V.

Calculate the change in resistance in Ω of the wire. **[4 marks]**

08 A microphone contains a coil and a magnet. It is used to record a conversation.

Synoptic links

3.3.1.1   3.3.2.2   3.7.5.4

08.1 Explain how the microphone converts a sound wave into a changing p.d. **[2 mark]**

08.2 The conversation needs to be sent securely between two buildings.

State the names of the **transmitting device**, **transmission path**, and **receiving device** that would be required to send the message.

State a reason your choice over one other transmission path. **[4 marks]**

08.3 A less sensitive signal needs to be sent to a remote area using a satellite. Explain why the up-link frequency must be different from the down-link frequency. **[3 marks]**

08.4 People living in the remote area can receive a radio station when they tune their radios to AM frequencies. One of the stations can be found at a frequency of 1548 kHz. Calculate the wavelengths in m of the waves.

Discuss whether AM waves are appreciably diffracted by the hills over which they travel. **[2 marks]**

08.5 A second radio station can be found at 1557 kHz. Explain why this is the closest that the carrier frequencies can be. **[3 marks]**

⚙️ Knowledge

31 Required practicals summary

Assessment of practicals

During your physics course you will carry out 12 Required Practicals. Your physics teacher will assess your ability to plan and implement practical work. In your exams you will answer questions that indirectly assess your use and application of the scientific method, data analysis skills, and knowledge of the use of practical apparatus. These questions may be based on practicals that you have already done, or you may be expected to apply your understanding to questions about a practical you haven't seen before. Paper 3 contains questions on data analysis, but you may also be asked about the measurement aspects of practical work in Papers 1 and 2. At least 15% of the marks for the course will be on practical skills.

Assessed practical skills

You will be assessed on four areas of practical skills in exams:

Independent thinking
- Solve problems set in a practical context.
- Apply scientific knowledge to a practical context.

Application of scientific methods and practices
- Comment on experimental design and evaluate scientific methods.
- Present data appropriately.
- Evaluate results and draw conclusions, with reference to measurement uncertainties and errors.
- Identify variables, including control variables.

Application of mathematical concepts
- Plot and interpret graphs.
- Process and analyse data.
- Consider margins of error, accuracy, and precision of data.

Instruments and equipment
- Know how to use a wide range of experimental and practical instruments, equipment, and techniques.

Apparatus and techniques

In your exam, you might be asked about the apparatus and techniques listed in the table.

a	use of analogue apparatus to record a range of measurements including length, temperature, pressure, force, angle, and volume; interpolation between scale markings on apparatus
b	use of digital instruments, including multimeters, to obtain measurements including time, current, potential difference, resistance, and mass
c	methods to increase the accuracy of measurements
d	use of a stopwatch or light gates for timing
e	use of callipers and micrometers for measuring small distances; use of digital or Vernier scales
f	construction of circuits from circuit diagrams
g	design and checking of circuits
h	use of a signal generator and oscilloscope
i	generating and measuring waves of different wavelengths using a microphone and loudspeaker, ripple tank, vibration transducer, or microwave source
j	use of lasers or other light sources to investigate characteristics of light
k	use of ICT, including data loggers, to collect data, and software to process data or model physical situations
l	use of ionising radiation, including detectors

RP 1 – Variation of the frequency of stationary waves on a string

In this practical you create stationary waves on a string, and investigate the effect of changing the string length, tension, and mass per unit length on the frequency of the wave produced.

Key knowledge and skills

- ☐ using analogue apparatus to obtain measurements, including length
- ☐ using digital instruments to obtain measurements, including mass
- ☐ using methods to increase the accuracy of measurements
- ☐ generating and measuring waves using a vibration transducer

Example practical

When an elastic string is vibrated at different frequencies, at some frequencies a stationary wave forms. The frequency of the wave depends on the length, tension, and mass per unit length of the string. The length and tension of the string can be varied using one string, and the mass per unit length can be varied by using different thicknesses of string.

Tips

- Use the fundamental (first harmonic) frequency when measuring the frequency of the stationary wave – this is the longest wavelength wave that 'fits' on the string.

- Repeat measurements by repeating the whole practical, then use your values to calculate a mean value of frequency.

RP 2 – Investigation of interference effects for light

In this practical you investigate the interference of light through a double slit (Young's double slit experiment) and through a diffraction grating.

Key knowledge and skills

- ☐ using analogue apparatus to record lengths and angles
- ☐ using a laser or light source to investigate characteristics of light

Example practical

When a beam of light passes through the double slit or diffraction grating the wave undergoes interference and a pattern of light and dark spots can be seen on a screen. The position of the spots depends on both the slit separation and the wavelength of the light.

Tips

- Make measurements that are as large as possible for separation of bright spots (e.g., $n = 2$) to minimise experimental uncertainty.

- Use a darkened room to reduce the lowest level of light that can be detected in experiments relying on the human eye.

RP 3 – Determination of g by a free-fall method

In this practical you investigate the motion of a moving object, and determine a value for the acceleration due to gravity.

Key knowledge and skills

- [ ] using analogue apparatus to record lengths and angles
- [ ] using methods to increase the accuracy of measurements
- [ ] using light gates for timing
- [ ] using a data logger and ICT software to measure time, velocity, or acceleration directly

Example practical

An object moving under free-fall has only force due to gravity acting on it. A metal ball bearing can be considered to be in free-fall. The time t that the ball takes to fall a known distance s can be measured using a pair of light gates connected to suitable software. Using the equation $s = \frac{1}{2}gt^2$, a graph of distance against time^2 can be plotted to obtain a value for g.

Tips

- When plotting graphs, do not include the origin in your points if you have not measured that value.
- All objects that fall through air experience air resistance, but using a metal ball bearing makes this force negligible.

RP 4 – Determination of the Young modulus

In this practical you investigate the elastic properties of a metal wire using a tensile force.

Key knowledge and skills

- [ ] using analogue apparatus to record lengths and distances
- [ ] using methods to increase the accuracy of measurements
- [ ] using micrometers to measure small distances, with digital or Vernier scales

Example practical

When a tensile force is applied to a length l of metal wire, such as copper or steel, the wire will extend by a small amount. The extension e depends on the length and cross-sectional area of the wire, and the force applied. A Vernier scale is often used to measure this extension because it tends to be small.

Tips

- Use safety goggles when loading and unloading thin wires to protect against eye damage if the wire snaps.
- In between scale readings, for example, to 0.5 mm on the millimetre scale of a ruler.
- When measuring the diameter of a metal wire take multiple measurements along the wire, including at right-angles to each other, then calculate the mean.

RP 5 – Determination of the resistivity of a wire

In this practical you will use a circuit diagram to build an electrical circuit, then use this circuit to determine the resistivity of a wire.

Key knowledge and skills

- [ ] using analogue apparatus to measure lengths and distances
- [ ] using digital instruments, including multimeters, to measure current, potential difference, and resistance
- [ ] constructing circuits from circuit diagrams using d.c. power supplies, cells, and a range of circuit components
- [ ] checking the functioning of electrical circuits

Example practical

A length of constantan wire is connected into an electrical circuit. The current I through the wire will depend on the length of the wire l and the potential difference V across the wire. The resistance of the wire can be calculated using $R = \frac{V}{I}$. Using the equation $R = \frac{\rho}{A}l$, where A is the cross-sectional area of the wire, a graph of R against I can be plotted and the gradient used to determine the resistivity ρ of the wire.

31

Tips

- Remember that conventional current and the movement of electrons are in opposite directions in a circuit.
- Connect the circuit for as little time as possible to avoid a temperature rise in the wire – this would alter its resistance.

RP 6 – e.m.f. and internal resistance of electric cells and batteries

Ideal cells are assumed to have no internal resistance. In this practical you investigate the behaviour of a real cell by modelling it as an e.m.f. source and internal resistance in series.

Key knowledge and skills

- [ ] using digital instruments to measure current and potential difference including ammeters, voltmeters, or multimeters
- [ ] constructing circuits from circuit diagrams using d.c. power supplies, cells, and a range of circuit components
- [ ] checking the functioning of electrical circuits

Example practical

A cell is connected in series with a variable resistor and ammeter (or multimeter). The resistor is used to vary the current I in the circuit. A voltmeter connected across the cell is used to measure the p.d. V. Using the equation $V = \varepsilon - I r$, where ε is the e.m.f. and r the internal resistance of the cell, a graph of V against I can be plotted. The gradient is the internal resistance of the cell, and the intercept on the y-axis is the emf.

Tips

- Batter are formed from multiple cells connected in series.
- Connect the circuit for as little time as possible during each reading to avoid a temperature rise in the wire – this would alter its resistance.

RP 7 – Investigation into simple harmonic motion

Oscillating systems that undergo simple harmonic motion are mathematically similar to each other and show the same displacement, velocity, and acceleration behaviour. You will investigate a simple pendulum and a mass on a spring, and compare their motions.

Key knowledge and skills

- [ ] using analogue apparatus to record measurements, including length and distance
- [ ] using digital instruments to record measurements for time and mass
- [ ] using methods to increase the accuracy of measurements, such as timing over multiple oscillations and the use of fiducial markers
- [ ] using a stopwatch for timing

Example practical

Simple harmonic motion can be investigated using a simple system that oscillates about an equilibrium point, such as a mass on a spring. The period of the oscillation is related to the mass m and the spring constant k of the spring. Using the equation $T^2 = \frac{4\pi^2}{k} m$ and plotting a graph of T^2 against m gives a value for the spring constant.

Tips

- The term 'light' string or spring means that you can ignore the effect of its mass on the oscillations.
- Place a fiducial marker on the apparatus to give a reference point for measuring the motion of the oscillating object.
- Timing many oscillations then dividing the time by the number of oscillations improves the accuracy of your measurement.

RP 8 – Investigation of Boyle's law and Charles's law for a gas

The relationships between the pressure, temperature, and volume of a fixed mass of gas can be investigated. In the Boyle's law practical, the pressure in the gas is altered and the volume measured. In the Charles's law practical, the temperature of the gas is varied and the volume measured. You may make an estimate of absolute zero using your data from the Charles's law practical.

Key knowledge and skills
☐ using analogue apparatus to record measurements of length, temperature, and pressure

Example practical
One method of measuring the pressure in a fixed mass of gas uses a round bottomed flask containing gas (air) placed in a water bath. Heating the water bath increases the temperature of the gas. Pressure can be measured using a Bourdon gauge. A graph of pressure against temperature crosses the x-axis, and can be used to estimate a value for absolute zero by extrapolating to find T when pressure is zero. This extrapolation is usually done by calculation using the equation of the straight line, rather than a scaled graph where the x-axis extends to approximately $-280\,^\circ$C.

Tips
- Always give the conditions when investigating a gas law, such as a fixed mass of gas and constant temperature for Boyle's law.
- Two main sources of error are the limited range of temperatures obtainable in a school laboratory, and the large effect of small changes of gradient on the value of the x-intercept.

RP 9 – Investigation of the charge and discharge of capacitors

Investigating the charging and discharging of a capacitor allows you to observe exponential increase and decay. You may use data-logging probes to measure rapid changes in values of current and potential difference.

Key knowledge and skills
☐ using digital instruments, including electrical multimeters, to measure p.d. and time

☐ using a stopwatch to measure time

☐ construct circuits from circuit diagrams using d.c. power supplies, cells, and a range of circuit components

☐ check the functioning of electrical circuits

Example practical
When a capacitor is connected in series with a switch and a variable resistor, the capacitor can be charged by closing the switch. When the switch is opened, the p.d. V across the capacitor will decrease and this can be measured over time t

The p.d. is given by the equation $V = V_0 e^{-t/RC}$, where V_0 is the initial p.d., R is the resistance of the variable resistor, and C is the capacitance of the capacitor. The value of RC is called the time constant of the circuit and has units of seconds.

Tips
- Connect electrolytic capacitors with the correct polarity to prevent them overheating; polarity may be indicated by the different 'leg' lengths.
- You should be able to convert an exponential equation into the equation of a straight-line by taking natural logarithms.

RP 10 – How the force on a wire varies with flux density, current, and length of wire

When a current flows through a wire in a magnetic field, the wire experiences a force. You can use a top pan balance to measure the magnitude of this force.

Key knowledge and skills
☐ using analogue apparatus to measure length

☐ using digital instruments to measure current and mass

☐ constructing circuits from circuit diagrams using d.c. power supplies, cells, and a range of components

Example practical

A length of copper wire is held between the poles of two magnets that are placed on a top pan balance. The copper wire is in a series circuit with a resistor and d.c. power supply. When current flows through the wire, a magnetic field is generated around it. This magnetic field interacts with the magnetic field of the two magnets and the wire experiences a force. However, because the wire is fixed in place, it remains stationary and the top pan balance moves instead. This can be seen as a change in the reading of the mass m on the balance. Using the equation $m = \dfrac{BL}{g}I$, a graph of m against I can be plotted.

Tips

- The copper wire must be parallel to the length of the magnets; if not, the force it experiences will decrease.
- Connect the circuit for as little time as possible to avoid a temperature rise in the wire – this would alter its resistance.

RP 11 – Effect on magnetic flux linkage of angle between search coil and magnetic field direction

A search coil is a small flat coil that has between 500 and 2000 turns of insulated wire. When a search coil is placed in a changing magnetic field, a small e.m.f. is generated in the coil. The e.m.f. can be viewed using an oscilloscope.

Key knowledge and skills

- ☐ using analogue apparatus to measure angles
- ☐ using digital instruments
- ☐ constructing circuits from circuit diagrams using d.c. power supplied, cells, and a range of circuit components
- ☐ using an oscilloscope, including volts / division

Example practical

As the angle θ of a search coil in a magnetic field B is changed, the flux linkage also alters. This is given by the equation flux linkage = $BAN\cos\theta$, where A is the area of the coil and N is the number of turns on the coil. The changing flux linkage causes the e.m.f. generated in the coil to change, which is displayed using an oscilloscope.

Tips

- It is sometimes easier to turn the oscilloscope timebase off, to allow measurement of the height of the trace.
- The number of turns will be marked on the search coil.

RP 12 – Investigation of the inverse square law for gamma radiation

In this practical you will learn how to safely handle a source of ionising radiation, and investigate the inverse square law for gamma radiation. You should always take measures to reduce your exposure to ionising radiation using 'as low as reasonably achievable' (ALARA) principles.

Key knowledge and skills

- ☐ using analogue apparatus to record distance
- ☐ using digital instruments such as a scalar and a stopwatch
- ☐ using ICT to model, collect, or process data
- ☐ handling and measuring ionising radiation, including using detectors

Example practical

Gamma radiation is electromagnetic radiation and obeys the inverse square law. Increasing the distance between a radioactive source and a detector, such as using a Geiger–Müller tube, will lead to a decrease in the measured activity of the source. However, when analysing the data from the experiment it is important to take into account the unknown distance e between the source and the point at which the gamma radiation is emitted from the holder. Intensity is therefore given by the equation $I = \dfrac{k}{(x + e)^2}$.

Tips

- Use safety precautions when handling radioactive sources; minimise the time the source is handled and maximise distance from the source.
- To plot log graphs when analysing data from radioactive practical work, practise using the 'ln' and 'e^x' functions on your calculator.

Knowledge

32 Unifying concepts

What is synoptic assessment?

Throughout physics there are 'key ideas' (unifying concepts) that are applied to topics across the subject. For example, the concept of fields can be applied to both objects with mass (gravitational fields) and objects with charge (electric fields). These big ideas allow physicists to simplify and explain topics that may appear unrelated.

Synoptic assessment tests your understanding of the key ideas in physics and the connections between different topics. You will be expected to apply the concepts that you have learned in one topic to solve problems in a different topic. These could be theoretical or practical topics. Synoptic assessment questions can be asked in any of the assessment components.

Strategies for synoptic questions

Here are some strategies to help you approach synoptic questions with confidence.

Look for the key ideas
Examiners will use unfamiliar contexts so that you can apply the knowledge you have. Don't be put off if you haven't learnt about the context in the question. When you read a question, first ask yourself 'what physics topic is this about?'

For example, the context might be about the motor in a child's toy. The physics topic could be current in an electrical circuit, Faraday's law, or Newton's laws of motion.

Identify relevant information
Underline key pieces of information. This could be the values of different variables, the aim of a practical set-up, or the physical mechanisms involved.

Choose the appropriate equations
Remember that you can use the information given in the Data, Formulae, and Relationships booklet if you need it.

Example of a synoptic theme

You might find it helpful to create a mind-map of different ideas in different topics, or which share similar mathematics.

For example, gravitational field strength can be linked to many other topics and practical assessments:

Learning outcome
gravitational field strength $$g = \frac{F}{m}$$
Links with other topics
scalar and vector quantities
equations of motion (suvat)
acceleration g of free fall
Practical Assessment Group 1
projectile motion
weight
free body diagrams
terminal velocity
gravitational potential energy
circular motion
simple harmonic motion for a mass on a spring
Practical Assessment Group 10
electric field strength $$E = \frac{F}{Q}$$

32

Example synoptic question

01 A metal string is used on a guitar as shown in the diagram. A 0.65 m length of the string between points **X** and **Y** is free to vibrate.

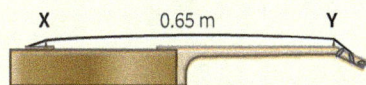

X   0.65 m   Y

01.1 When plucked by a guitarist, the wire vibrates at its fundamental frequency and produces sound waves of 254 Hz.

Calculate the wavelength in m of the fundamental mode of vibration of the wire. **[1 mark]**

01.2 The diameter of the wire is 0.26 mm. The density of steel is 7800 kg m^{-3}.

Show that μ, the mass per unit length of the wire, is approximately 4×10^{-4} kg m^{-1}. **[3 marks]**

01.3 Calculate the tension in N in the wire. **[2 marks]**

01.4 The steel wire will break if the stress is greater than 1.2×10^9 Pa. The guitarist wants to tighten the string so that the fundamental frequency is 500 Hz.

Discuss whether the string will break if the tension is increased to produce this fundamental frequency.

[3 marks]

Example synoptic question answer

The question is set in the context of a guitar, which you may not have not met before. However, you should be able to apply all of the physics content.

01.1 *The first part of the question is about stationary waves (topic 3.3.1). You should recall that, for a string fixed at both ends, the fundamental wavelength λ is twice the length of the string.*

$$\lambda = 2 \times 0.65 \text{ m} = 1.3 \text{ m}$$ [1]

01.2 *Use ideas about density (topic 3.4.2.1) and the data for the density of steel. You need to convert the diameter of the wire to SI units and use it to calculate the volume V of the wire.*

$$\text{volume of wire} = \text{cross-sectional area} \times \text{length}$$

$$V = \pi\left(\frac{d}{2}\right)^2 \times l$$

$$= \pi \left(\frac{2.6\times10^{-4}}{2}\right)^2 \times 0.65$$

$$= 3.45\times10^{-8} \text{m}^3$$ [1]

Now use the given value for density to calculate the mass of the wire.

$$\rho = \frac{m}{v}$$

$$\text{therefore, } m = \rho \times V$$

$$= 7800 \text{ kg m}^{-3} \times 3.34\times10^{-8} \text{ m}^3$$

$$= 2.69\times10^{-4} \text{ kg}$$ [1]

Use this mass to calculate μ.

$$\mu = \frac{m}{l} = \frac{2.69\times10^{-4}}{0.65} = 4.1\times10^{-4} \text{ kg m}^{-1} \approx 4\times10^{-4} \text{ kg m}^{-1}.$$ [1]

01.3 *Use the equation for the fundamental frequency for a vibrating string.*

$$f_0 = \frac{1}{2l}\sqrt{\frac{T}{\mu}}$$

$$\text{therefore, } T = \mu \, (2\,l f_0)^2$$ [1]

$$= 4.1\times10^{-4} \times (2 \times 0.65 \times 254)^2 = 45 \text{ N (to 2 s.f.)}$$ [1]

01.4 *The question now tests your knowledge of the Young Modulus (3.4.2.2). Recall that stress $= \frac{force}{area}$.*

$$\text{breaking force for wire} = \text{stress} \times \text{area}$$

$$= 1.2\times10^9 \text{ Pa} \times \pi \, (1.3\times10^{-4} \text{ m})^2 = 64 \text{ N}$$ [1]

$$T \propto f^2$$ [1]

If the frequency doubles then T would increase by a factor of 4. At 300 Hz the tension in the string would be 180 N and the string would break. [1]

Practice

33 Multiple-choice questions

Only **one** answer per question is allowed.

For each answer completely fill in the circle alongside the appropriate answer.

CORRECT METHOD ▣ WRONG METHODS ⊠ ◉ ⊜ ⊘

If you want to change your answer you must cross out your original answer as shown. ⊠

If you wish to return to an answer previously crossed out, ring the answer you now wish to select as shown. ⊗

01 A student makes measurements to determine the Young modulus of a wire. They measure the diameter, length, and extension of the wire, and the mass suspended from the wire. Each measurement has a 1% uncertainty.

Which measurement gives the greatest uncertainty to the final value of the Young modulus? **[1 mark]**

- A diameter ○
- B length ○
- C extension ○
- D mass ○

02 Which of the following is **not** an SI base unit? **[1 mark]**

- A ampere ○
- B second ○
- C coulomb ○
- D mole ○

03 An unstable nucleus **X** emits an alpha particle, followed by two beta particles by beta minus decay. A final nucleus **Y** is produced.

Which statement about nuclei **X** and **Y** is correct? **[1 mark]**

- A atomic masses of **X** and **Y** are the same ○
- B **X** and **Y** are isotopes of the same element ○
- C atomic mass of **Y** is greater than that of **X** ○
- D atomic number of **X** is greater than that of **Y** ○

04 A copper atom becomes a doubly-charged copper ion after the removal of two electrons from the atom $^{63}_{29}$Cu.

What is the magnitude of specific charge of the copper ion? **[1 mark]**

A 3.0×10^6 C kg^{-1} ○

B 1.5×10^6 C kg^{-1} ○

C 6.6×10^6 C kg^{-1} ○

D 1.9×10^{25} C kg^{-1} ○

05 A K^0, which has a quark content of d$\bar{s}$, reacts with a proton.

$$K^0 + p \rightarrow K^+ + n$$

What is the quark content of K$^+$? **[1 mark]**

A d$\bar{s}$ ○

B u$\bar{s}$ ○

C $\bar{u}$s ○

D $\bar{u}$d ○

06 When a proton and positive pion interact, a kaon and another strange particle can be produced.

$$\pi^+ + p \rightarrow K^+ + X$$

Which row in the table correctly shows some of the properties of X? **[1 mark]**

	Baryon number	Charge	Strangeness	
A	+1	+1	+1	○
B	+1	+1	−1	○
C	+1	−1	+1	○
D	0	+1	+1	○

07 Which of the following proves a wave is transverse? **[1 mark]**

A stationary waves ○

B polarisation ○

C superposition ○

D interference ○

08 Stationary sound waves are set up in a tube that is closed at one end. A node always forms at the closed end and an antinode at the open end. The first harmonic has a wavelength of 2.00 m.

Which of the following could also be a harmonic wavelength? **[1 mark]**

A 1.00 m ○

B 0.25 m ○

C 4.00 m ○

D 0.40 m ○

09 A diffraction grating has 500 lines per mm.

What is the angle of the third order spectral line when the incident light has a wavelength of 400 nm? **[1 mark]**

A   37°   ◯

B   35°   ◯

C   12°   ◯

D   0.6°   ◯

10 When a ray of light enters a glass block, it is refracted.

Which row in the table correctly identifies the changes to the light as it enters the glass block? **[1 mark]**

	Frequency	Velocity	Wavelength	
A	no change	increases	decreases	◯
B	decreases	decreases	decreases	◯
C	no change	decreases	decreases	◯
D	decreases	no change	increases	◯

11 **Figure 1** shows a uniform metal girder of weight W suspended by two cables. T gives the tension in the cable. The system is in equilibrium.

Figure 1

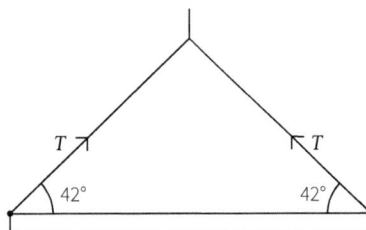

Which of the following shows the correct relationship between W and T? **[1 mark]**

A $W = 2T\sin 42$ ◯

B $W = T\cos 42$ ◯

C $W = T\sin 42$ ◯

D $W = 2T\cos 42$ ◯

12 A ball is dropped from rest onto the floor. **Figure 2** shows the velocity–time graph as the ball bounces.

Figure 2

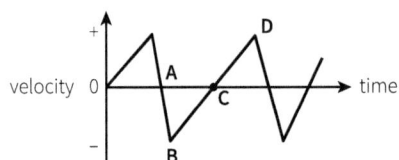

At which point does the ball reach its maximum height after the first bounce? **[1 mark]**

A   point **A**   ○

B   point **B**   ○

C   point **C**   ○

D   point **D**   ○

13   A person stands on scales in a lift. When the lift is stationary, the scales read 600 N. As the lift moves, the scales read 590 N.

Which statement describes how the lift is moving? **[1 mark]**

A   moving downwards at a constant velocity   ○

B   moving upwards and accelerating   ○

C   moving downwards and accelerating   ○

D   moving upwards at a constant velocity   ○

14   **Figure 3** shows the force on a steel ball when it is dropped and rebounds.

Figure 3

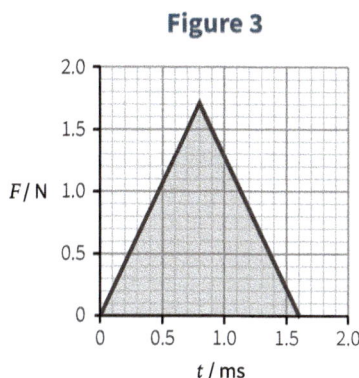

Which of the following graphs would have the same shape as **Figure 3**? **[1 mark]**

A   kinetic energy of the ball against time   ○

B   momentum of the ball against time   ○

C   displacement of the ball against time   ○

D   acceleration of the ball against time   ○

15   A student stretches a wire that has a cross-sectional area of $2.5 \times 10^{-7} \, m^2$ with a force of 10 N. The wire stretches from a length of 2 m to a length of 2.02 m.

What is the value of the Young modulus? **[1 mark]**

A $4 \times 10^7 \, N \, m^{-2}$ ○

B $4 \times 10^8 \, N \, m^{-2}$ ○

C $4 \times 10^9 \, N \, m^{-2}$ ○

D $4 \times 10^{10} \, N \, m^{-2}$ ○

16 Two materials, **X** and **Y**, are stretched up to a breaking point, shown by F_x and F_y in **Figure 4**.

Figure 4

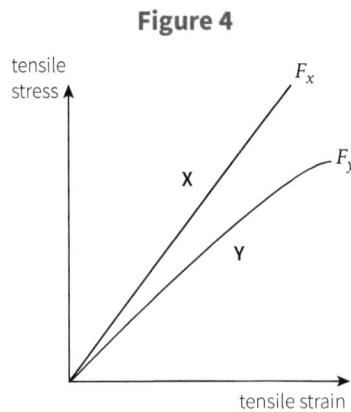

Which statement is correct? [1 mark]

A Young modulus of **X** has a smaller magnitude than that of **Y** ◯

B energy stored in **Y** just before it breaks is approximately double that stored in **X** ◯

C a sample of **X** with double the cross-sectional area of a sample of **Y**, but with the same length, would have a larger extension than a sample of **Y** ◯

D a sample of **Y** with the same cross-sectional area as a sample of **X**, but with double the length, would break at a smaller applied force than that applied to **X** ◯

17 **Figure 5** shows a simple circuit diagram.

Figure 5

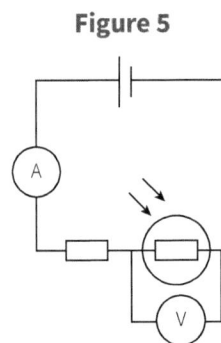

Which row in the table correctly describes how the ammeter and voltmeter readings change when it gets darker? [1 mark]

	Ammeter reading	Voltmeter reading	
A	decreases	stays the same	◯
B	stays the same	increases	◯
C	decreases	increases	◯
D	increases	increases	◯

18 Which graph correctly shows the p.d.-current characteristics for an Ohmic conductor? **[1 mark]**

A
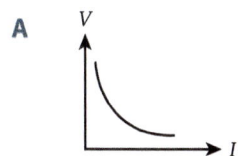
○

B
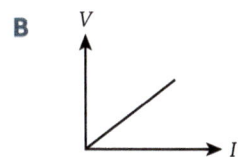
○

C
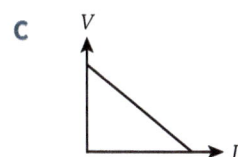
○

D
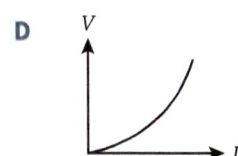
○

19 **Figure 6** shows the currents in the connectors meeting at a junction in a circuit.

Figure 6

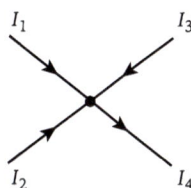

Which equation correctly shows the relationship between each of the currents? **[1 mark]**

A $I_1 + I_3 = I_2 + I_4$ ○

B $I_1 + I_2 + I_3 = I_4$ ○

C $I_1 - I_4 = I_2 + I_3$ ○

D $I_1 + I_2 = I_4 + I_3$ ○

20   **Figure 7** shows a circuit.

Figure 7

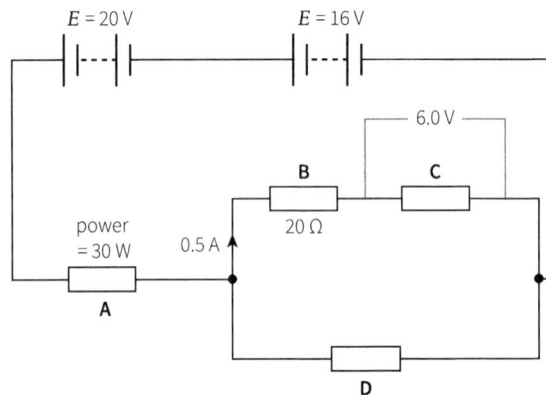

Which statement about the circuit is correct?   [1 mark]

A   current through resistor **A** is 1.5 A   ○

B   potential difference across resistor **D** is 6.0 V   ○

C   resistance of **C** is 6 Ω   ○

D   resistance of **D** is 20 Ω   ○

21   A proton in a particle accelerator moves in a circle of circumference 27 km.
The speed of the protons is approximately equal to the speed of light.

What is the magnitude of the force required to keep the protons moving in the circle?   [1 mark]

A 1.2×10^{-22} N   ○

B 2.0×10^{-19} N   ○

C 3.5×10^{-14} N   ○

D 2.2×10^{-13} N   ○

22   Two children are playing on a roundabout. Child **X** is standing on the outer edge of
the roundabout. Child **Y** is standing halfway between the outer edge and the centre.
The roundabout is spinning at a constant speed.

Which statement is correct?   [1 mark]

A   child **X** has a greater angular velocity than child **Y**   ○

B   child **X** has a greater linear velocity than child **Y**   ○

C   child **X** has a smaller angular velocity than child **Y**   ○

D   child **X** has a smaller linear velocity than child **Y**   ○

23   A student measures the time period of a mass on a spring to be 0.58 s. They pull the
spring down a distance of 3.5 cm and release it.

What is the maximum speed of the mass?   [1 mark]

A $0.38 \, \mathrm{m\,s^{-1}}$ ○

B $2.0 \, \mathrm{m\,s^{-1}}$ ○

C $6.0 \, \mathrm{m\,s^{-1}}$ ○

D $13 \, \mathrm{m\,s^{-1}}$ ○

24 A simple harmonic oscillator oscillates with a frequency of 3.4 Hz. The maximum kinetic energy of the oscillator is 5.2 J.

If the amplitude of the oscillation is doubled, what will the new maximum kinetic energy be? **[1 mark]**

A 2.6 J ⭕

B 5.2 J ⭕

C 10.4 J ⭕

D 20.8 J ⭕

25 Two liquids are heated by the same heater for the same length of time. Some data from the experiment are shown in **Table 1**.

Table 1

Material	Mass of material / kg	Temperature rise / °C
X	0.5	10
Y	2.0	5

What is the ratio of specific heat capacity of **X** to specific heat capacity of **Y**? **[1 mark]**

A 4 ⭕

B 2 ⭕

C 0.5 ⭕

D 0.25 ⭕

26 A student wants to estimate the value of absolute zero using measurements on a near ideal gas.

What graph should they plot? **[1 mark]**

A x-axis: T; y-axis: V ⭕

B x-axis: T; y-axis: V^{-1} ⭕

C x-axis: P; y-axis: V ⭕

D x-axis: P; y-axis: V^{-1} ⭕

27 The mass of the Earth is about 81 times the mass of the Moon. The radius of the Earth is about four times the radius of the Moon.

What is the ratio of gravitational field strength at the Earth's surface to gravitational field strength at the Moon's surface? **[1 mark]**

A 5.1 ⭕

B 20.3 ⭕

C 0.2 ⭕

D 2.3 ⭕

28 Which of the following statements becomes **incorrect** if you replace the word 'electric' with 'gravitational' and the words 'positive charge' with 'mass'? [1 mark]

    A   magnitude of electric potential is proportional to the size of the positive charge   ◯

    B   lines of equipotential join points of constant electric potential   ◯

    C   electric field strength due to a point positive charge obeys an inverse square law   ◯

    D   electric potential energy increases as you bring the positive charges closer together   ◯

29 Which of the following is the distance apart that two protons would have to be for the electrostatic force to be equal to the weight of one proton? [1 mark]

    A   14 cm   ◯

    B 4.2×10^{10} cm   ◯

    C 1.3×10^{-4} cm   ◯

    D   12 cm   ◯

30 **Figure 8** shows a point **P** near to two positive point charges of +10 nC and +30 nC.

Figure 8

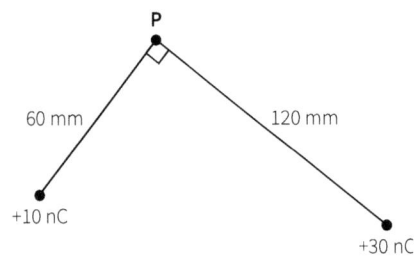

Which of the following is the magnitude of the resultant electric field at point **P**? [1 mark]

    A $24\,980\,N\,C^{-1}$ ◯

    B $31\,220\,N\,C^{-1}$ ◯

    C $18\,730\,N\,C^{-1}$ ◯

    D $43\,710\,N\,C^{-1}$ ◯

31 A student records values of current versus time as a capacitor charges through a resistor, as shown in **Figure 9**.

Figure 9

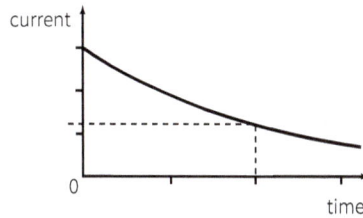

The value of the resistor is doubled.

Which statement is correct? [1 mark]

A initial current will double ○

B time it takes to charge will halve ○

C final charge on the capacitor will not change ○

D final potential difference across the resistor will double ○

32 A parallel plate capacitor is connected to a battery until it is fully charged. The distance between the plates is increased while the plates are still connected to the battery.

Which of the following are true? [1 mark]

A potential difference across the plates remains constant and the charge increases ○

B potential difference across the plates remains constant and the charge decreases ○

C charge remains constant and the potential difference across the plates increases ○

D charge remains constant and the potential difference across the plates decreases ○

33 A current-carrying wire of length l is placed perpendicular to a magnetic field of flux density B. When the current in the wire is I, the force on the wire is F.

Which of the following is the force felt by the wire when the current is $0.25I$, the magnetic flux density is $0.5B$, and the length of the wire is $2l$? [1 mark]

A $\frac{F}{4}$ ○

B F ○

C $\frac{F}{2}$ ○

D $4F$ ○

34 Charged particles create an ionisation track as they move through a bubble chamber.
Figure 10 shows the path of an electron in a bubble chamber.

Figure 10

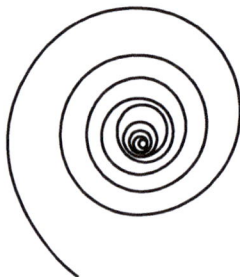

Which of the following can be deduced from **Figure 10**?   [1 mark]

A   magnetic field is acting out of the page   ○

B   the electron is annihilated   ○

C   magnetic field is acting into the page   ○

D   electron is moving anticlockwise   ○

35 The magnetic flux through a circular coil increases uniformly from 0 to B in time t.
The radius of the coil is r and the coil has N turns.

Which of the following is an expression for the maximum emf induced in the coil?   [1 mark]

A $\dfrac{B\,\pi r^2}{t}$ ○

B $\dfrac{NB}{t}$ ○

C $\dfrac{NB\,\pi r^2}{t}$ ○

D $\dfrac{NB\,\pi r^2}{2t}$ ○

36 **Figure 11** shows the variation of magnetic flux with time for a coil.

Figure 11

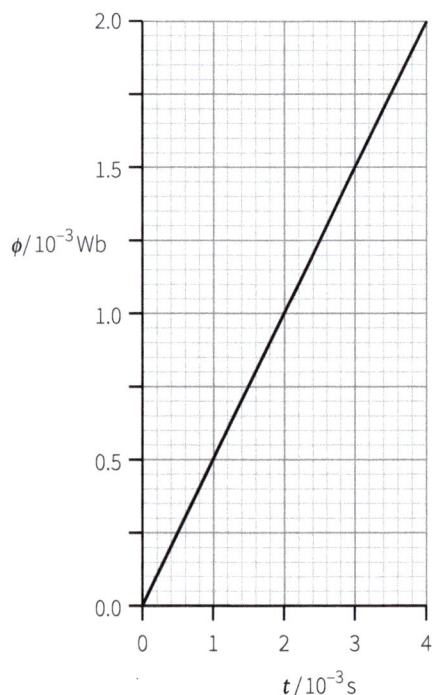

The e.m.f. induced in the coil is determined by multiplying the number of coils by which of the following? **[1 mark]**

A   area under the graph ◯

B   gradient of the graph multiplied by the area of the coil ◯

C   inverse of the gradient of the graph ◯

D   gradient of the graph ◯

37   A Geiger counter measures the count rate from a gamma source.
At a distance of 35 cm the count rate is 12 counts min^{-1}.

At what distance would the count rate be 450 counts min^{-1}? **[1 mark]**

A   0.9 cm ◯

B   3.1 cm ◯

C   5.7 cm ◯

D   32.0 cm ◯

38   An isotope of an element is unstable. The corrected count rate for the decay of the unstable element changes as shown in **Figure 12**.

Figure 12

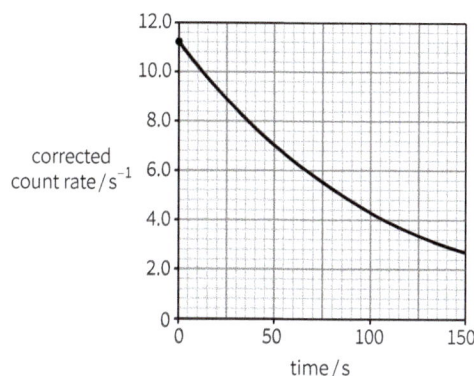

What is the decay constant of the isotope? **[1 mark]**

A $9.9 \times 10^{-3}\,s^{-1}$ ◯

B $101\,s^{-1}$ ◯

C $70\,s^{-1}$ ◯

D $2.8\,s^{-1}$ ◯

39 An electron moves from level $n = 3$ to $n = 1$ as shown in **Figure 13**.

Figure 13

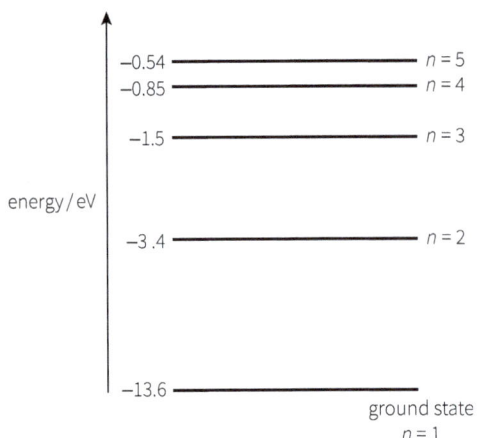

Which of the following statements is correct? **[1 mark]**

A one photon of frequency 1.8×10^{34} Hz is emitted ◯

B one photon of frequency 1.8×10^{34} Hz is absorbed ◯

C one photon of frequency 2.9×10^{15} Hz is emitted ◯

D one photon of frequency 2.9×10^{15} Hz is absorbed ◯

40 What is the relationship between the wavelength λ of an electron and the potential difference through which it is accelerated? **[1 mark]**

A $\lambda = \dfrac{\sqrt{2m\,eV}}{h}$ ◯

B $\lambda = \dfrac{h}{\sqrt{2m\,eV}}$ ◯

C $\lambda = h\sqrt{2m\,eV}$ ◯

D $\lambda = \dfrac{h}{2m\,eV}$ ◯

41 A diffraction grating is used to produce a line spectrum using light from a star. There are 1000 lines per mm in the grating.

What is the angular separation of lines with wavelengths of 570 nm and 610 nm? **[1 mark]**

A 2.0° ◯

B 2.8° ◯

C 3.0° ◯

D 3.5° ◯

42 A galaxy is a distance of 50 Mpc from the Earth.

If the Hubble constant is $73\,\text{km s}^{-1}\,\text{Mpc}^{-1}$, what is the recessional velocity? **[1 mark]**

A $0.68\,\text{km s}^{-1}$ ◯

B $1.46\,\text{km s}^{-1}$ ◯

C $123\,\text{km s}^{-1}$ ◯

D $3650\,\text{km s}^{-1}$ ◯

43 A lens in a pair of glasses has a focal length of −25 cm, and the lens is 18 mm in front of the eye.

Which of the following would be the correct prescription for a contact lens placed directly on the eye? **[1 mark]**

A   −4.0 D ⬜

B   0.04 D ⬜

C   −4.3 D ⬜

D   +4.3 D ⬜

44 The sound intensity from the dial tone of a phone is $1 \times 10^{-4}\,\text{W m}^{-2}$ at a distance of 5 cm.

Which of the following is the intensity at 80 cm? **[1 mark]**

A $2 \times 10^{-8}\,\text{W m}^{-2}$ ⬜

B $4 \times 10^{-7}\,\text{W m}^{-2}$ ⬜

C $3 \times 10^{-4}\,\text{W m}^{-2}$ ⬜

D $6 \times 10^{-6}\,\text{W m}^{-2}$ ⬜

45 Ultrasound in one material is incident on a boundary and enters a material of twice the acoustic impedance.

What is the percentage of ultrasound transmitted? **[1 mark]**

A   89% ⬜

B   33% ⬜

C   50% ⬜

D   11% ⬜

46 For 100 keV photons, the attenuation coefficient of lead is $6200\,\text{m}^{-1}$ and for aluminium it is $46\,\text{m}^{-1}$.

What is the thickness of aluminium that would be equivalent to 2 mm of lead? **[1 mark]**

A   270 mm ⬜

B   5.6 mm ⬜

C   0.015 mm ⬜

D   1.0 mm ⬜

Data sheet

DATA – FUNDAMENTAL CONSTANTS AND VALUES

Quantity	Symbol	Value	Units
speed of light in vacuo	C	3.00×10^8	ms^{-1}
permeability of free space	μ_0	$4\pi \times 10^{-7}$	Hm^{-1}
permittivity of free space	ε_0	8.85×10^{-12}	Fm^{-1}
magnitude of the charge of electron	e	1.6×10^{-19}	C
the Planck constant	h	6.63×10^{-34}	Js
gravitational constant	G	6.67×10^{-11}	Nm^2kg^{-2}
the Avogadro constant	N_A	6.02×10^{23}	mol^{-1}
molar gas constant	R	8.31	$JK^{-1}mol^{-1}$
the Boltzmann constant	k	1.38×10^{-23}	JK^{-1}
the Stefan constant	σ	5.67×10^{-8}	$Wm^{-2}K^{-4}$
the Wien constant	α	2.90×10^{-3}	mK
electron rest mass (equivalent to $5.5 \times 10^{-4}u$)	m_e	9.11×10^{-31}	kg
electron charge/mass ratio	$\dfrac{e}{m_e}$	1.76×10^{11}	Ckg^{-1}
proton rest mass (equivalent to $1.00728\,u$)	m_p	$1.67(3) \times 10^{-27}$	kg
proton charge/mass ratio	$\dfrac{e}{m_p}$	9.58×10^7	Ckg^{-1}
neutron rest mass (equivalent to $1.00867\,u$)	m_n	$1.67(5) \times 10^{-27}$	kg
gravitational field strength	g	9.81	Nkg^{-1}
acceleration due to gravity	g	9.81	ms^{-2}
atomic mass unit (1u is equivalent to 931.5 MeV)	u	1.661×10^{-27}	kg

ALGEBRAIC EQUATION

quadratic equation $\qquad x = \dfrac{-b \pm \sqrt{b^2 - 4ac}}{2a}$

ASTRONOMICAL DATA

Body	Mass/kg	Mean radius/m
Sun	1.99×10^{30}	6.96×10^8
Earth	5.97×10^{24}	6.37×10^6

GEOMETRICAL EQUATIONS

arc length	$= r\theta$
circumference of circle	$= 2\pi r$
area of circle	$= \pi r^2$
curved surface area of cylinder	$= 2\pi rh$
area of sphere	$= 4\pi r^2$
volume of sphere	$= \dfrac{3}{4}\pi r^3$

Particle Physics

Class	Name	Symbol	Rest energy/MeV
photon	photon	γ	0
lepton	neutrino	v_e	0
		v_μ	0
	electron	$e^\pm$	0.510999
	muon	$\mu^\pm$	105.659
mesons	π meson	$\pi^\pm$	139.576
		π^0	134.972
	K meson	$K^\pm$	493.821
		K^0	497.762
baryons	proton	p	938.257
	neutron	n	939.551

Properties of quarks
antiquarks have opposite signs

Type	Charge	Baryon number	Strangeness
u	$+\dfrac{3}{4}e$	$+\dfrac{1}{3}$	0
d	$-\dfrac{1}{3}e$	$+\dfrac{1}{3}$	0
s	$-\dfrac{1}{3}e$	$+\dfrac{1}{3}$	−1

Properties of Leptons

		Lepton number
Particles	$e^-, v_e; \mu^-, v_\mu$	+1
Antiparticles	$e^+, \overline{v}_e, \mu^+, \overline{v}_\mu$	−1

Photons and energy levels

photon energy
$$E = hf = \frac{hc}{\lambda}$$

photoelectricity
$$hf = \phi + E_{k\,(max)}$$

energy levels
$$hf = E_1 - E_2$$

de Broglie wavelength
$$\lambda = \frac{h}{p} = \frac{h}{mv}$$

Waves

wave speed $\qquad c = f\lambda \qquad$ period $\qquad f = \dfrac{1}{T}$

first harmonic
$$f = \frac{1}{2l}\sqrt{\frac{T}{\mu}}$$

fringe spacing $\quad w = \dfrac{\lambda D}{s} \quad$ diffraction grating $\quad d \sin\theta = n\lambda$

refractive index of a substance s, $n = \dfrac{c}{c_s}$

for two different substances of refractive indices n_1 and n_2,

law of refraction $\qquad n_1 \sin\theta_1 = n_2 \sin\theta_2$

critical angle $\qquad \sin\theta_c = \dfrac{n_2}{n_1}$ for $n_1 > n_2$

Mechanics

moments $\qquad Fd$

velocity and acceleration $\qquad v = \dfrac{\Delta s}{\Delta t} \qquad\qquad a = \dfrac{\Delta v}{\Delta t}$

equations of motion
$$v = u + at \qquad\qquad s = \left(\frac{u+v}{2}\right)t$$

$$v^2 = u^2 + 2as \qquad\qquad s = ut + \frac{at^2}{2}$$

force $\qquad F = ma$

force $\qquad F = \dfrac{\Delta(mv)}{\Delta t}$

impulse $\qquad F\Delta t = \Delta(mv)$

work, energy and power $\qquad W = Fs\cos\theta$

$$E_k = \frac{1}{2}mv^2 \qquad\qquad \Delta E_p = mg\Delta h$$

$$P = \frac{\Delta w}{\Delta t},\ P = Fv$$

$$\text{efficiency} = \frac{\text{useful output power}}{\text{input power}}$$

Materials

density $\rho = \dfrac{m}{v}$

Hooke's law $F = k\,\Delta L$

Young modulus $= \dfrac{\text{tensile stress}}{\text{tensile strain}}$

tensile stress $= \dfrac{F}{A}$

tensile strain $= \dfrac{\Delta L}{L}$

energy stored $E = \dfrac{1}{2}F\Delta L$

Electricity

current and pd	$I = \dfrac{\Delta Q}{\Delta t}$ $V = \dfrac{W}{Q}$ $R = \dfrac{V}{I}$
resistivity	$\rho = \dfrac{RA}{L}$
resistors in series	$R_T = R_1 + R_2 + R_3 + \cdots$
resistors in parallel	$\dfrac{1}{R_T} = \dfrac{1}{R_1} + \dfrac{1}{R_2} + \dfrac{1}{R_3} + \cdots$
power	$P = VI = I^2R = \dfrac{V^2}{R}$
emf	$\varepsilon = \dfrac{E}{Q}$ $\varepsilon = I(R + r)$

Circular motion

magnitude of angular speed	$\omega = \dfrac{v}{r}$ $\omega = 2\pi f$
centripetal acceleration	$a = \dfrac{v^2}{r}$ $\omega^2 r$
centripetal force	$F = \dfrac{mv^2}{r} = m\omega^2 r$

Simple harmonic motion

acceleration	$a = -\omega^2 x$
displacement	$x = A \cos(\omega t)$
speed	$v = \pm \omega \sqrt{(A^2 - x^2)}$
maximum speed	$v_{max} = \omega A$
maximum acceleration	$a_{max} = \omega^2 A$
for a mass-spring system	$T = 2\pi \sqrt{\dfrac{m}{k}}$
for a simple pendulum	$T = 2\pi \sqrt{\dfrac{l}{g}}$

Thermal physics

energy to change temperature	$Q = mc\Delta\theta$
energy to change state	$Q = ml$
gas law	$pV = nRT$ $pV = NkT$
kinetic theory model	$pV = \dfrac{1}{3} Nm \, (C_{rms})^2$
kinetic energy of gas molecule	$\dfrac{1}{3} m \, (C_{rms})^2 = \dfrac{3}{2} kT = \dfrac{3RT}{2N_A}$

Gravitational fields

force between two masses	$F = \dfrac{Gm_1m_2}{r^2}$
gravitational field strength	$g = \dfrac{F}{m}$
magnitude of gravitational field strength in a radial field	$g = \dfrac{GM}{r^2}$
work done	$\Delta W = m\Delta V$
gravitational potential	$V = -\dfrac{GM}{r}$ $g = -\dfrac{\Delta V}{\Delta r}$

Electric fields and capacitors

force between two point charges	$F = \dfrac{1}{4\pi\varepsilon_0} \dfrac{Q_1Q_2}{r^2}$
force on a charge	$F = EQ$
field strength for a uniform field	$E = \dfrac{V}{d}$
work done	$\Delta W = Q\Delta V$
field strength for a radial field	$E = \dfrac{1}{4\pi\varepsilon_0} \dfrac{Q}{r^2}$
electric potential	$V = \dfrac{1}{4\pi\varepsilon_0} \dfrac{Q}{r}$
field strength	$E = \dfrac{\Delta V}{\Delta r}$
capacitance	$C = \dfrac{Q}{V}$ $C = \dfrac{A\varepsilon_0\varepsilon_r}{d}$
capacitor energy stored	$E = \dfrac{1}{2} QV = \dfrac{1}{2} CV^2 = \dfrac{1}{2} \dfrac{Q^2}{C}$
capacitor charging	$Q = Q_0(1 - e^{-\frac{t}{RC}})$
decay of charge	$Q = Q_0 \, e^{-\frac{t}{RC}}$
time constant	RC

Magnetic fields

force on a current	$F = BIl$
force on a moving charge	$F = BQv$
magnetic flux	$\Phi = BA$
magnetic flux linkage	$N\Phi = BAN \cos \theta$
magnitude of induced emf	$\varepsilon = N \dfrac{\Delta \Phi}{\Delta t}$
	$N\Phi = BAN \cos \theta$
emf induced in a rotating coil	$\omega = BAN \omega \sin \omega t$
alternating current	$I_{rms} = \dfrac{I_0}{\sqrt{2}} \quad V_{rms} = \dfrac{V_0}{\sqrt{2}}$
transformer equations	$\dfrac{N_s}{N_p} = \dfrac{V_s}{V_p}$
	efficiency $= \dfrac{I_s V_s}{I_p V_p}$

Nuclear physics

inverse square law for γ radiation	$I = \dfrac{k}{x^2}$
radioactive decay	$\dfrac{\Delta N}{\Delta t} = -\lambda N, N = N_0 e^{-\lambda t}$
activity	$A = \lambda N$
half-life	$T_{\frac{1}{2}} = \dfrac{\ln 2}{\lambda}$
nuclear radius	$R = R_0 A^{1/3}$
energy-mass equation	$E = mc^2$

OPTIONS

Astrophysics

1 astronomical unit $= 1.50 \times 10^{11}$ m

1 light year $= 9.46 \times 10^{15}$ m

1 parsec $= 2.06 \times 10^5$ AU $= 3.08 \times 10^{16}$ m

$\qquad\qquad = 3.26$ ly

Hubble constant, $H = 65$ km s^{-1} Mpc^{-1}

$M = \dfrac{\text{angle subtended by image at eye}}{\text{angle subtended by object at unaided eye}}$

telescope in normal adjustment	$M = \dfrac{f_0}{f_e}$
Rayleigh criterion	$\theta \approx \dfrac{\lambda}{D}$
magnitude equation	$m - M = 5 \log \dfrac{d}{10}$
Wien's law	$\lambda_{max} T = 2.9 \times 10^{-3}$ m K
Stefan's law	$P = \sigma A T^4$
Schwarzschild radius	$R_s \approx \dfrac{2GM}{c^2}$
Doppler shift for $v \ll c$	$\dfrac{\Delta f}{f} = -\dfrac{\Delta \lambda}{\Delta \lambda} = \dfrac{v}{c}$
red shift	$z = -\dfrac{v}{c}$
Hubble's law	$v = Hd$

Medical physics

lens equations	$P = \dfrac{1}{f}$
	$m = \dfrac{v}{u}$
	$\dfrac{1}{f} = \dfrac{1}{u} + \dfrac{1}{v}$
threshold of hearing	$I_0 = 1.0 \times 10^{-12}$ W m^{-2}
intensity level	$10 \log \dfrac{I}{I_0}$
absorption	$I = I_0 e^{-\mu x}$
	$\mu_m = \dfrac{\mu}{\rho}$
ultrasound imaging	$Z = pc$
	$\dfrac{I_r}{I_i} = \left(\dfrac{z_2 - z_1}{z_2 + z_1} \right)^2$
half-lives	$\dfrac{1}{T_E} = \dfrac{1}{T_B} + \dfrac{1}{T_P}$

Engineering physics

moment of inertia $\qquad I = \Sigma mr^2$

angular kinetic energy $\qquad E_k = \frac{1}{2}I\omega^2$

equations of angular
motion

$$\omega_2 = \omega_1 + \alpha t$$

$$\omega_2^2 = \omega_1^2 + 2\alpha\theta$$

$$\theta = \omega_1 t + \frac{\alpha t^2}{2}$$

$$\theta = \frac{(\omega_1 + \omega_2)t}{2}$$

torque $\qquad T = I\alpha$

$$T = Fr$$

angular momentum $\qquad I\omega$

angular impulse $\qquad T\Delta t = \Delta(I\omega)$

work done $\qquad W = T\theta$

power $\qquad P = T\omega$

thermodynamics $\qquad Q = \Delta U + W$

$$W = p\Delta V$$

adiabatic change $\qquad pV^\gamma = \text{constant}$

isothermal change $\qquad pV = \text{constant}$

heat engines

$$\text{efficiency} = \frac{W}{Q_H} = \frac{(Q_H - Q_C)}{Q_H}$$

$$\text{maximum theoretical efficiency} = \frac{(T_H - T_C)}{T_H}$$

work done per cycle = area of loop

input power = calorific value × fuel flow rate

indicated power = (area of $p - V$ loop)
$\qquad$ × (number of cycles per second)
$\qquad$ × (number of cylinders)

output or brake power $\qquad P = T\omega$

friction power = indicated power – brake power

heat pumps and refrigerators

$$\text{refrigerator: } COP_{ref} = \frac{Q_C}{W} = \frac{Q_C}{Q_H - Q_C}$$

$$\text{heat pump: } COP_{hp} = \frac{Q_H}{W} = \frac{Q_H}{Q_H - Q_C}$$

Turning points in physics

electrons in fields $\qquad F = \frac{eV}{d}$

$$F = Bev$$

$$r = \frac{mv}{Be}$$

$$\tfrac{1}{2}mv^2 = eV$$

Millikan's experiment $\qquad \frac{QV}{d} = mg$

$$F = 6\pi\eta rv$$

Maxwell's formula $\qquad C = \frac{1}{\sqrt{\mu_0\varepsilon_0}}$

$$\lambda = \frac{h}{p} = \frac{h}{\sqrt{2meV}}$$

special relativity $\qquad t = \frac{t_0}{\sqrt{1 - \dfrac{v^2}{c^2}}}$

$$l = l_0\sqrt{1 - \frac{v^2}{c^2}}$$

$$E = mc^2 = \frac{m_0 c^2}{\sqrt{1 - \dfrac{v^2}{c^2}}}$$

Electronics

resonant frequency $\qquad f_0 = \frac{1}{2\pi\sqrt{LC}}$

Q-factor $\qquad Q = \frac{f_0}{f_B}$

operational amplifiers
open loop $\qquad V_{out} = A_{OL}(V_+ - V_-)$

inverting amplifier $\qquad \dfrac{V_{out}}{V_{in}} = -\dfrac{R_f}{R_{in}}$

non-inverting amplifier $\qquad \dfrac{V_{out}}{V_{in}} = 1 + \dfrac{R_f}{R_1}$

summing amplifier $\qquad V_{out} = -R_f\left(\dfrac{V_1}{R_1} = \dfrac{V_2}{R_2} + \dfrac{V_3}{R_3} + \cdots\right)$

difference amplifier $\qquad V_{out} = (V_+ - V_-)\dfrac{R_f}{R_1}$

Bandwidth requirement:

$\quad$ for AM $\qquad\qquad$ bandwidth $= 2f_M$

$\quad$ for FM $\qquad\qquad$ bandwidth $= 2(\Delta f + f_M)$

Index